# basic principles of electronics

# *basic principles of electronics*

Vester Robinson

Reston Publishing Company, Inc., Reston, Virginia

A Prentice-Hall Company

With love and affection to my lovely wife

**Florence Belle**


RESTON PUBLISHING COMPANY, INC.
*A Prentice-Hall Company*
Box 547
Reston, Virginia 22090



10 9 8 7 6 5 4 3 2 1

ISBN: 0-87909-063-4

Library of Congress Catalog Card Number: 72-97401
Printed in the United States of America.

# preface

The purpose of this book is to develop an understanding of the basic principles of electronics as simply as possible. Mathematics has not been avoided, but it has been used only as necessary to illustrate principles and show applications. Some history of electronics is included, as well as some speculation on the future trends.

Considerable selectivity had to be exercised in what material to leave out rather than what to include. As a result, certain assumptions had to be made as to the background of the user of this text. We assume that he has a working knowledge of *ac* and *dc* electricity and is reasonably proficient in high school algebra. Beyond that, the only requirement is a lively interest in the subject of electronics.

Since the invention of the transistor, the trend has been away from electron tubes. Most modern equipment uses solid-state devices exclusively. In recognition of this fact, this book covers the field of electronics with transistors and other solid-state components. At the same time, we recognize the fact that some equipment still uses electron tubes, and Chapter 1 is devoted to the electron tube.

Starting with Chapter 2, the remainder of the book is completely solid-state oriented. The theory of diodes and transistors is developed in considerable detail. Application is a point of strong emphasis and the devices are shown in many practical circuit configurations. Two chapters are used to develop amplifiers, and a third chapter analyzes a great many common circuits with transistor components.

Since space is always limited, no book can cover all aspects of solid-state devices nor all their applications in electronic circuits.

Chapter 9 is dedicated to a brief coverage of the many fine devices that would otherwise be crowded out. This includes cryosars, four layer transistors, lasers, and masers.

The final chapter (10) is a relatively nontechnical investigation of what we can expect in the near future. Some of this chapter reads like science fiction, but it is based on the current state of the electronic art. A few years ago a wrist watch radio was the product of a vivid imagination. Today, a wrist watch computer is a definite possibility.

If you find this book a bit difficult, or you want the same material in a simplified format, you may find *Electronic Concepts* to be very helpful. It is a programmed, self-teaching manual by the same author. You can obtain it from your book store or order directly from the Reston Publishing Company. A study of these texts will go a long way towards preparing a student for the Occupational Education Achievement Tests.*

No text book can be a one man effort, and this one is no exception. A great many people contributed to this book who must go unnamed for various reasons. However, a few must be given a special vote of thanks. Foremost is Mr. Matthew I. Fox, President of Reston Publishing Company. He was quick to recognize the possibilities of the book, and devoted a great deal of time procuring reviews, discussing problems, and making suggestions.

Two other people who made many valuable contributions are Dean Thomas Kubala of the Anne Arundel Community College of Arnold, Maryland and Herbert Jackson of the Ontario Ministry of Colleges and Universities of Toronto, Canada. These two gentlemen reviewed the outline when the book was in the planning stages and the manuscript before the final revision. Much of the content, and to some extent, the method of treatment is a result of many valuable suggestions from these individuals.

Many manufacturers of electronic equipment and research organizations were generous contributors of time, technical material, and illustrations. A special thanks goes to Bell Laboratories, Tektronic, International Business Machines, Sprague Electric Company, Texas Instruments, and Fairchild.

*Vester Robinson*

* A Directory of Achievement Tests for Occupational Education has been compiled by Educational Testing Service in cooperation with the United States Office of Education. It lists a number of achievement tests for Electricity and Electronics. Information on obtaining and using these tests can be acquired by contacting Educational Testing Service, Princeton, New Jersey.

# contents

# 1
# electron tubes

The invention of the electron tube has proved to be the greatest single step in the development of electronics since the discovery of electricity. The electron tube has been to electronics what the wheel was to transportation. It is true that new equipment uses only a few tubes, which are special purpose tubes. Indeed transistors have taken over many of the jobs once performed by tubes, but without the tube, the field of electronics would not have developed enough to need transistors. Our present world of radio, television, motion pictures, and computers never could have happened without the electron tube.

## ELECTRON EMISSION

Most electron tubes function on the principle that certain types of heated metals will emit electrons. There are some tubes that are based on other principles, but this has been the most popular.

Like so many other great discoveries, electron emission was discovered by accident.

### *the Edison effect*

From 1870 to 1890, several noted scientists were conducting experiments on electrical conductivity with gases, liquids, and solids. All of these experiments had some bearing on the development of electron tubes, but progress is slow when the goal lacks definition. The great breakthrough came quietly in 1883 in the laboratory of Thomas A. Edison, an American scientist (1847–1931).

A hot spot had developed in the filament of one of Edison's light bulbs. In an effort to eliminate this hot spot and increase the life of the bulb, he introduced a metal plate into the same glass envelope. He discovered that even though the plate and filament were physically isolated from each other, the meter which was connected to the plate registered a small amount of current (see Figure 1-1).

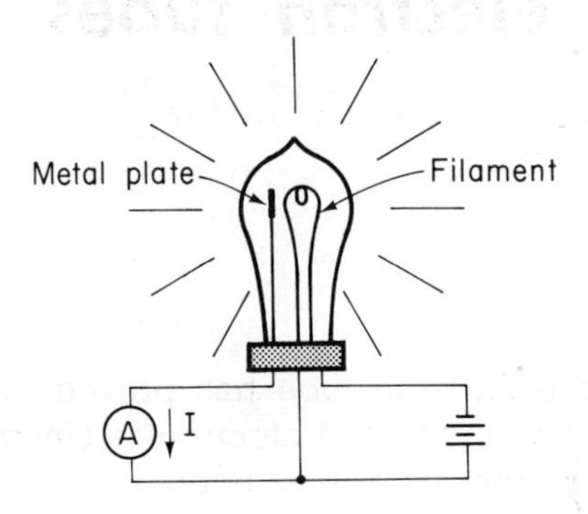


**Fig. 1-1.** The Edison Effect.

Little was known of electrons in those days, and even less was known about emission and conduction. Edison recorded the observation in his notebook along with many other interesting but unexplained facts. This particular observation went unexplained for another six years.

The electron theory of current was being developed at that time, and a British physicist J. J. Thomson used it to explain the phenomenon which had become known as the *Edison effect.* Thomson suggested that the heated filament not only gave off light but also emitted electrons. He stated that these negative particles were boiled off the filament and attracted to the positively-charged plate. The movement of these electrons back to the power supply constituted the meter current. He not only adequately explained a baffling fact, he also described the presently accepted idea of electron current.

Prior to that time, and in some modern texts, current was assumed to be some mysterious force with a direction from positive to negative in the external circuit. As the electron movement theory became accepted, it was necessary to distinguish between the two types of current. The old standby (positive to negative) became known as *conventional current.* The movement of electrons (negative to positive) became *electron current.*

Since the movement of electrons has been firmly established here, current in this book will mean "flow of electrons." The unit of measure is the ampere (A), and one ampere equals $6.24 \times 10^{18}$ (one coulomb of) electrons passing a point in one second.

### *Fleming valve*

John A. Fleming, a British scientist, followed up on the Edison effect in an effort to verify the electron explanation. He enclosed a plate and a filament in a glass envelope and connected external power sources. This experiment is illustrated in Fig. 1-2.

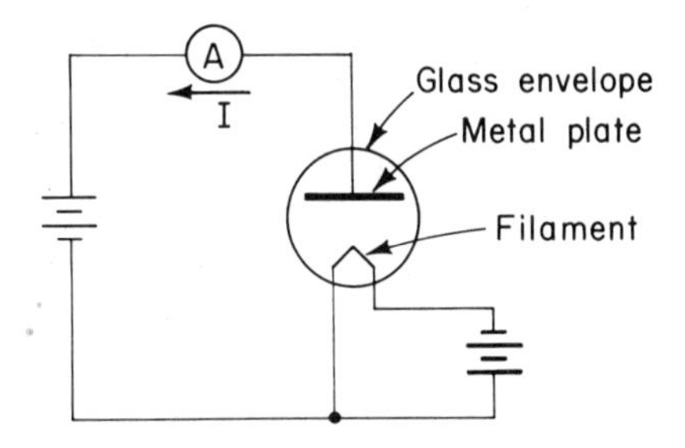


**Fig. 1-2.** Fleming's Experiment.

One battery was used to heat the filament, and a second battery was connected between the plate and filament. This arrangement provided a difference in potential between filament and plate, with the plate being the more positive. A definite current was registered as long as the filament was hot and the plate was more positive than the filament.

He then replaced the plate battery with an ac generator. This placed an ac sine wave between the plate and filament. The result was a positive potential on the plate during each positive alternation of the sine wave. He proved that current was present when the plate was positive, and that all current ceased when the plate was negative. In the process, he had invented the first diode vacuum tube. He explained that it functioned like a valve enabling a person to turn the current on or off at will. The diode became known as *Fleming's valve.* (In England, all electron tubes are still referred to as valves.)

Fleming proved two important points with his experiments. First, negative particles were emitted from a heated surface. Second, current was composed of these negative particles. Let's take a closer look at this process of electron emission.

### *surface barrier*

Electron emission is the process whereby an electron breaks away from the surface of the material which contained it. This electron actually escapes from the material into the space which surrounds the material. We frequently hear of free electrons, but this means that electrons are free within the metal. Here we have a different concept. An emitted electron is not only free from an atom, it is actually free from the metal.

A thin film forms on the surface of a solid when it comes in contact with either a gas or a liquid. This film is a potential barrier which we call a *surface barrier.* The surface barrier forms a restraining wall on the surface of the solid and prevents the free electrons from escaping from the metal. Theoretically this surface barrier would not exist in a perfect vacuum. Since we have no perfect vacuums, the theory cannot be adequately tested. The fact remains, however; a film does form to create a potential barrier.

### *work function*

In order for an electron to escape from the surface of a piece of metal, a certain amount of work is required to push it through the surface barrier. This quantity of work is different with each different type of metal, and it is known as the *work function* of that metal. The work function is defined as the amount of work required to force an electron through the surface barrier.

The work function is measured in units called *electron-volts.* One electron-volt (eV) is the energy acquired by an electron when it is accelerated through a potential of one volt (V). As previously stated, this work function is different for different materials. When a free electron in a given material acquires energy equivalent to the work function of that material, the electron breaks through the surface barrier and escapes from the metal.

The energy necessary to lift an electron through the surface barrier may be acquired in several different ways. Heat, light, collison, and electricity are some of the means of imparting energy to the electron. Let's examine some types of emission.

### *types of emission*

*Thermionic emission* is the most common type used in electron tubes, but not the only one. This type of emission has already been described to some extent in Edison's discovery as well as in Fleming's experiments. Heat supplied from any source may be sufficient to cause thermionic emission. Since we are dealing with

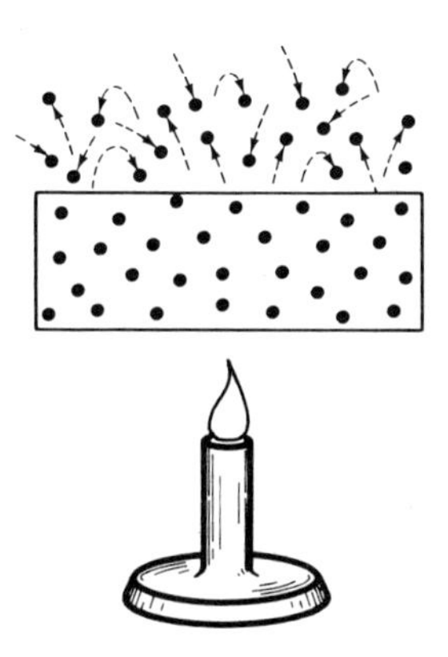

**Fig. 1-3.** Thermionic Emission.

electric apparatus, the most efficient means of heating is with an electric current. The current through the wire filament will cause energy dissipation which raises the temperature of the emitting element. This emitting element may, or may not, be the heater itself. When sufficient heat is present, electrons are boiled from the surface of the emitter. The higher the temperature, the greater the quantity of electrons released.

*Secondary emission* occurs when high-velocity electrons strike the surface of an emitting material. The force of the collision imparts enough energy to other electrons to drive them through the surface barrier. Some highly specialized tubes use this type of emission exclusively, and it is always present, to some extent, with thermionic emission. The term secondary emission comes from the fact that the electrons are released through bombardment by other electrons. The bombarding electrons are called primary electrons.

*Field emission* can be used to pull electrons through the surface barrier by electrostatic attraction. When a strong positive charge is placed near an emitting material, negative electrons will be attracted with enough force to rip them away from the surface. This type of emission is used with some electron tubes. This process is illustrated in Fig. 1-4.

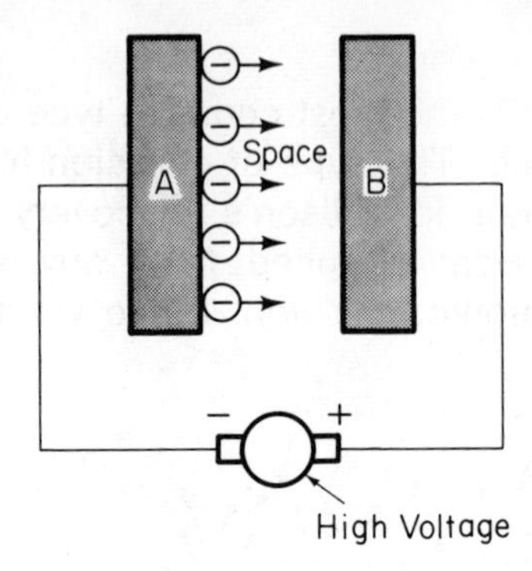


**Fig. 1-4.** Field Emission.

*Photoemission* occurs when light rays strike the surface of certain light sensitive materials. The photoelectric cell is an example of the use of photoemission. This is utilizing bombardment of light rays. The light energy is imparted to surface electrons enabling them to escape through the surface barrier. The intensity of the light determines the number of electrons emitted, but the amount of energy acquired by an electron is determined by the frequency of the light. The emitted electrons will break free of the material at a speed which is directly proportional to the light frequency. In order for a material to be capable of photoemission, it must have a very low work function. Certain television camera tubes utilize this type of emission.

Of the four types of emission discussed here, all are used in electron tubes, and some tubes use more than one type. One of the television camera image tubes uses thermionic, photo, and secondary emission. Regardless of the type of emission, the emitting material must be especially manufactured in order to give up a sufficient quantity of electrons with an optimum use of power.

### *emitter materials*

In this discussion, we will confine ourselves to materials used in thermionic emission. It is not a simple matter to select such materials. In addition to emitting a sufficient quantity of electrons at a reasonable temperature, they must meet several other rigid requirements. The material must be capable of withstanding high temperatures, intense electric fields, strong mechanical shocks, and severe vibrations. Not only must they survive these destructive factors, but they must not vaporize, break, or sag under such treatment.

One of the few materials that can meet these stringent requirements is tungsten. Unfortunately it has a high work function which requires a very high temperature. Pure tungsten has a work function of 4.53 eV. In order to liberate an effective quantity of electrons under these conditions, a temperature of 2227 °C is required. Even this unreasonable heat produces only seven milliamperes (mA) of current for each watt (W) of power.

A thorium coating on the emitting section of a piece of tungsten greatly improves its efficiency while retaining most of its durability. Such material is called thoriated tungsten. The work function is lowered to 2.86 eV which enables an operating temperature of 1700 °C. A thoriated tungsten emitter will produce more than 50 mA of current for each watt used in heat.

1700 °C is still a high temperature. In most tubes, a less rugged material is used to reduce the required heat to a more reasonable level. Nickel alloy has proved strong enough to survive most normal environments. When this material is coated with an oxide of barium and strontium, it produces a profusion of electrons at a fairly low temperature. The operating temperature goes down to about 750 °C with an efficiency of over 150 mA/W of power.

## DIODE TUBE

The diode is so named because it contains two active elements (electrodes). It has an electron emitting element and a collecting element. The emitter is called the *cathode* and the collector is called a *plate.* The collector is also known as an *anode.* Figure 1-5 shows two schematic symbols for thermionic diodes.

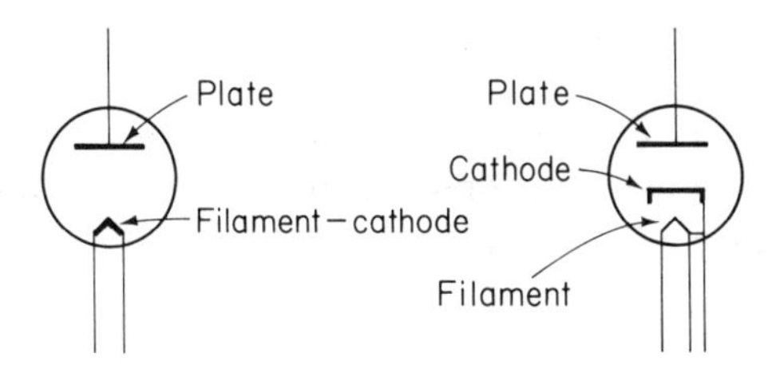


**Fig. 1-5.** Diode Symbols.

### *construction of cathodes*

Cathodes of thermionic diodes are heated in two ways: directly and indirectly. These two methods are shown symbolically in Fig. 1-5. The tube with the directly heated cathode has a combination cathode and heater filament. This element is generally constructed by using a thin strand or strip of emitting material. This wire is suspended by wire supports. Current through the filament heats it and causes the coating to give off electrons. Thus the filament serves as both heater and cathode.

The symbol in Fig. 1-5b represents a diode with an indirectly heated cathode. In this arrangement, a relatively small voltage can be used to heat the filament. The filament passes this heat on to the cathode which emits the electrons. The indirectly heated cathode provides a relatively steady flow of electrons even when the filament is heated with ac.

### *tube structure*

The indirectly heated cathode is formed with a cylinder of tungsten or nickel alloy. The heater is mounted in the center of the cylinder, and the outer surface is coated with emitting alloys.

The plate of the diode is a second cylinder of metal. This one is slightly larger and completely surrounds the cathode. Figure 1-6 illustrates the physical placement of heater, cathode, and plate.

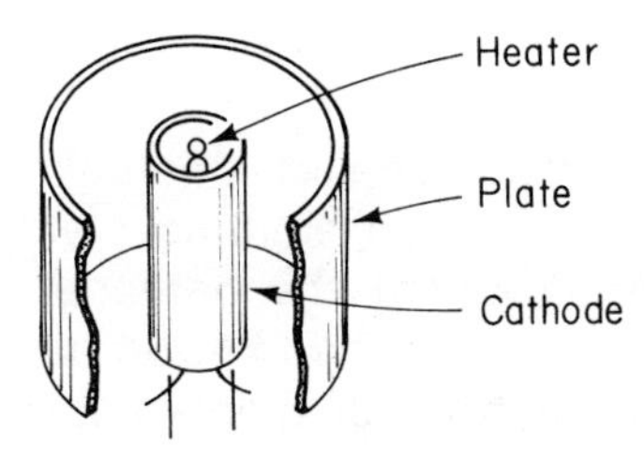


**Fig. 1-6.** Physical Placement of Elements.

The elements of the tube are then sealed inside an evacuated glass or metal envelope. Pins provide external connections for the elements. Figure 1-7 illustrates the general appearance of two types of diodes.

### *electrical characteristics*

When a diode has the recommended current through the filament, large quantities of electrons are emitted from the cathode. These electrons form a dense, negative cloud which hovers near the surface of the cathode. This cloud of electrons is called a *space*

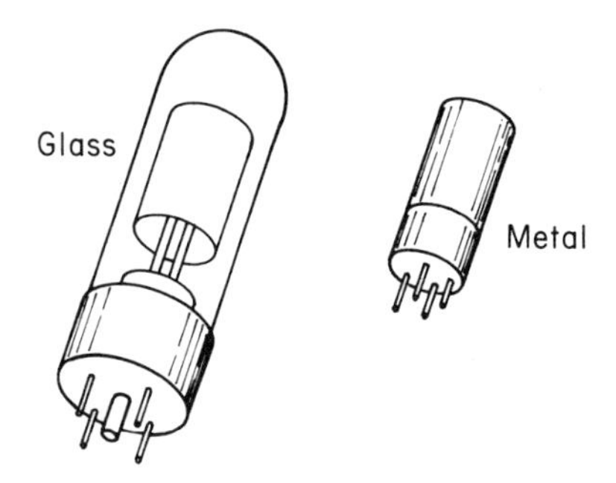


**Fig. 1-7.** Diode Tubes.

*charge,* and it exerts a repelling force on other electrons leaving the cathode.

In order to have current through the diode, two basic requirements must be met. There must be a complete external path for current from plate to cathode, and the plate must be positive with respect to the cathode. A proper circuit is illustrated in Fig. 1-8.

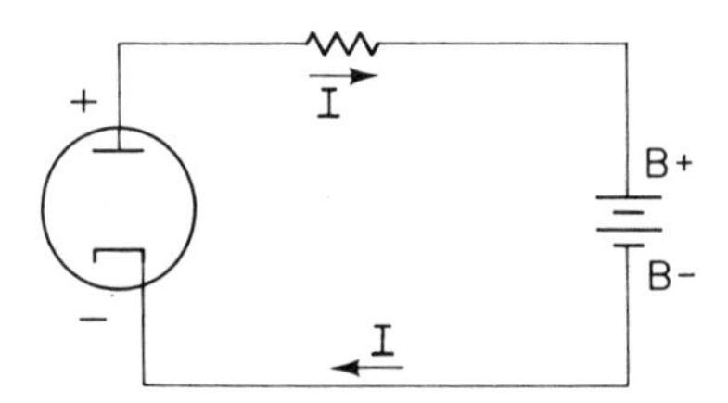


**Fig. 1-8.** Diode Current.

The positive potential on the plate attracts electrons from the space charge which serves as a reservoir of electrons. These electrons are collected by the plate and passed on to the positive terminal of the battery. Electrons leave the negative battery terminal and return to the cathode.

The quantity of electrons in the space charge is determined by the amount of heat. For any level of heat, the charge accumulates enough electrons to balance the emission from the cathode. The number of electrons leaving the space charge is determined by the potential on the plate. For each electron attracted to the plate, another electron leaves the cathode to replenish the space charge. The electrons attracted to the plate constitute current in the external circuit. This current is referred to as plate current ($I_p$), and it is the same throughout the circuit.

Notice in Fig. 1-8 that the heater filament has been omitted. This is common practice on schematics. It simplifies the drawing by eliminating some of the lines. As a result, the schematic is easier to read. The heaters are generally shown on a separate schematic as represented in Fig. 1-9.

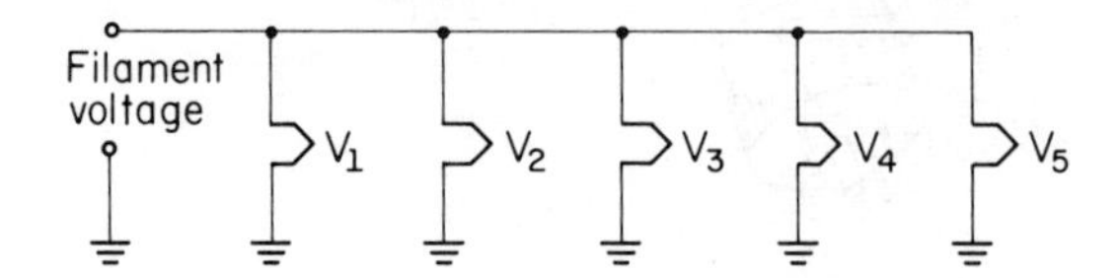


**Fig. 1-9.** Tube Heater Circuit.

These heaters are in parallel indicating that each tube uses the same voltage on the filaments. Tubes are rated for a specified level of heater current, but this item will take care of itself when the proper level of voltage is applied. There are many types of tubes and many different requirements for heater voltage. The designer plans his heater circuits so that each tube has the proper heater voltage.

Figure 1-8 also shows a B+ and a B— on the battery. In most schematics the battery will be left out, and the terminal points will be labeled as either B+ or B—. Electrically there is no difference between applying a B+ to the plate and a B— to the cathode. This is illustrated in Fig. 1-10.

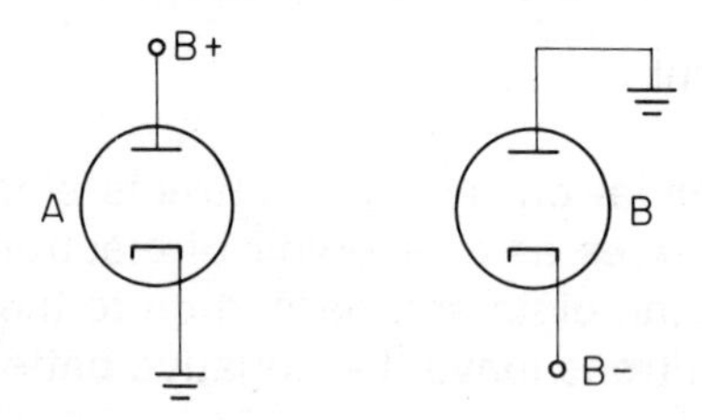


**Fig. 1-10.** Use of B+ and B—.

Both tubes in Fig. 1-10 are connected into complete circuits, and the two circuits are electrically identical. In fact, circuits *A* and *B* are both the same as the circuit shown in Fig. 1-8. This can be seen very readily by redrawing these circuits and replacing the ground with the battery.

It was stated that the magnitude of current was dependent

upon the level of the plate voltage. This could be stated more descriptively by saying that current is directly proportional to the difference of potential from plate to cathode. Let's examine this relationship a bit further.

### *plate current vs plate voltage*

The tube may be considered as a variable resistor. When the plate is negative with respect to the cathode, there is no plate current. This is equivalent to saying that the tube is an open circuit with infinite resistance. When the plate is made positive with respect to the cathode, even slightly positive, plate current will register. Making the plate more positive increases the plate current. This indicates that the internal resistance of the tube is now a finite value that grows smaller as the positive plate potential increases.

This plate voltage ($E_p$)–plate current ($I_p$) relationship can be observed with a circuit such as that in Fig. 1-11.

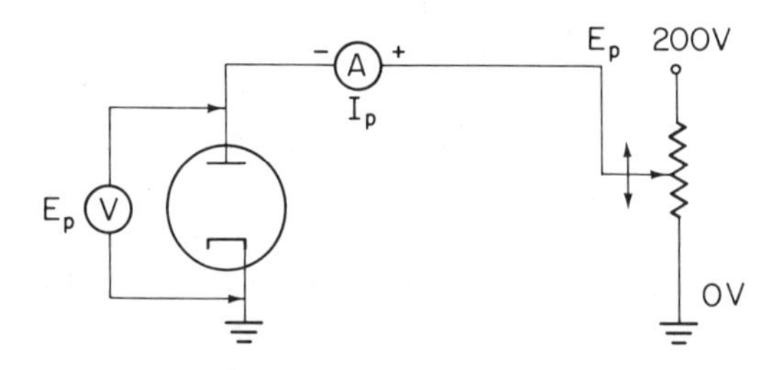


**Fig. 1-11.** Diode With Variable Plate Voltage.

Starting with the potentiometer set on zero, both plate and cathode have ground potential, and there is no current through the meter. As the arm of the potentiometer moves toward 200 V, the plate becomes positive with respect to the cathode and continues to become more and more positive. As soon as the plate becomes positive, the ammeter begins to register current. This is $I_p$ and it increases as $E_p$ increases. After the plate reaches a certain potential (exact potential determined by the type of diode) the current levels off. A further increase in $E_p$ has practically no effect on $I_p$. Plotting the values of $E_p$ and $I_p$ will produce a graph similar to that in Fig. 1-12.

As $E_p$ increases from 0, $I_p$ starts a slow increase. From point *A* to point *B,* there is a linear increase of $I_p$ with an increase in $E_p$.

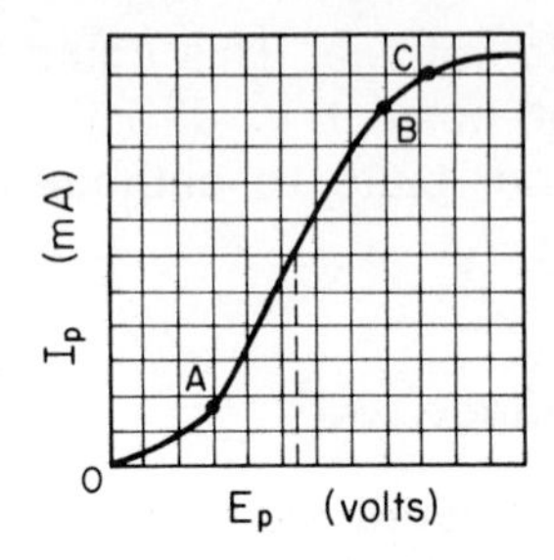


**Fig. 1-12.** $E_p$–$I_p$ Characteristic Curve.

The center of this linear area would be the best operating point for the tube in most situations. When $E_p$ increases beyond point *B*, $I_p$ begins to level off. After point *C*, there is no appreciable increase in $I_p$ even though $E_p$ continues to increase. *C* marks the point where the entire space charge is being attracted to the plate. This means that the plate is attracting every electron emitted by the cathode. Therefore, a further increase in $E_p$ cannot increase the current because the supply of electrons has been exhausted.

Point *C* on the graph is called the *plate saturation point.* This point could be changed by changing the level of heat on the filaments, but this is not recommended. Saturation does not mean that the plate is incapable of accepting more electrons. It means that the plate is attracting all the electrons that are being emitted.

We have already described the diode in terms of a variable resistor. Actually two distinct types of resistance are involved. This characteristic needs to be analyzed more carefully.

### *plate resistance*

The two types of plate resistances are dc and ac. As the terms imply, dc plate resistance is the tube opposition to direct current. and ac plate resistance is the opposition it offers to a changing current. These are important distinctions because most tubes are powered from a dc source but are used to process ac signals.

In Fig. 1-11, the voltmeter is indicating the plate voltage ($E_p$). Notice that the meter is connected between the plate and the cathode of the diode. In that illustration, any level of voltage selected by the potentiometer is the plate voltage. This is not always the case (see Fig. 1-13).

The voltage across the diode (plate to cathode) is still $E_p$, but it is not the same as the applied voltage. The battery voltage (B+) is 50 V which is divided between the tube and the plate load resistor ($R_L$). Since 45 V are dropped across $R_L$, the remaining 5 V

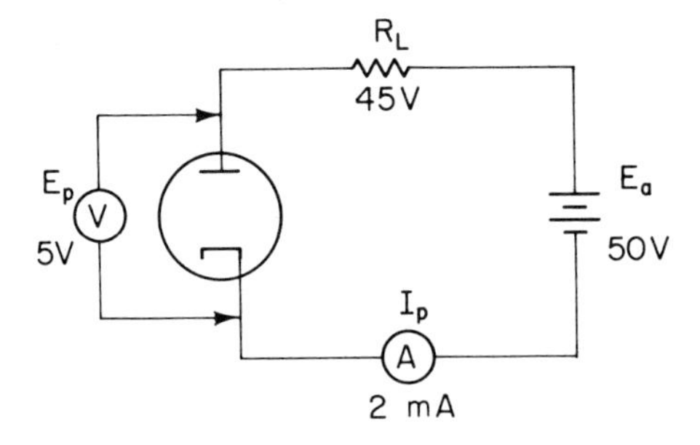


**Fig. 1-13.** Voltage Distribution in a Diode Circuit.

appear across the tube as $E_p$. These figures, along with the 2 mA of $I_p$ which is indicated by the ammeter, provide all the ingredients needed to calculate the dc plate resistance. The process is simply an application of Ohm's law.

$$R_p = E_p/I_p$$

where $R_p$ is the dc plate resistance in ohms ($\Omega$), $E_p$ is the plate voltage in V, and $I_p$ is the plate current in A. In Fig. 1-11, $R_p =$ 5 V/2 mA $=$ 2500 $\Omega$. Of course, the same formula can be used just as readily to calculate values of $E_p$ and $I_p$.

It should not be assumed that the dc resistance is a fixed constant. Refer back to Fig. 1-12. For all values of $E_p$ and $I_p$ which fall along the linear portion of the current curve, dc plate resistance holds relatively constant. Dividing values of $E_p$ by $I_p$, below point *A* and above point *B* on the graph, will show drastic changes in the value of $R_p$.

Ohm's law is also used to calculate the ac plate resistance. But now we are interested in increments or decrements in $E_p$ and $I_p$ rather than in specific values. The formula is:

$$r_p = \frac{\Delta e_p}{\Delta i_p}$$

where $r_p$ is the ac plate resistance in ohms, $\Delta$ is a small change, $e_p$ is instantaneous plate voltage in V, and $i_p$ is instantaneous plate current in A. For instance, if a 2-V change in plate voltage produces a 4-mA change in plate current, the ac plate resistance is 500 $\Omega$. This is determined as follows:

$$r_p = \frac{\Delta e_p}{\Delta i_p}$$

$$r_p = \frac{2\ \text{V}}{4\ \text{mA}} = 500\ \Omega$$

The ac plate resistance holds reasonably constant as long as the tube is operating along the linear portion of its characteristic curve (see Fig. 1-12).

With the plate load resistor ($R_L$) added to the circuit, the diode functions much the same as it will in normal operation. This is a dynamic condition as opposed to the static condition with no load resistor. In a dynamic condition, the $E_p$–$I_p$ curve will be somewhat different than it is in a static situation. This results in different values of plate resistance.

Static tube characteristics ($E_p$–$I_p$) are published by tube manufacturers. These indicate the capabilities of the tube. The dynamic characteristics cannot be published because they are different for very different value of load resistance. To know what the tube will do in a given circuit, the $E_p$–$I_p$ curve will need to be measured and plotted while using the actual plate load resistor.

### *uses*

The diode tube's principle use is in *rectification.* Rectification is the process of converting ac to a pulsating dc. The diode is ideally suited for this task because it permits current in only one direction. Not only that; it permits current only when the plate is more positive than the cathode. This means that a diode, which is powered from an ac source, will conduct only during the positive alternation of the input voltage. One manner of accomplishing rectification is illustrated in Fig. 1-14.

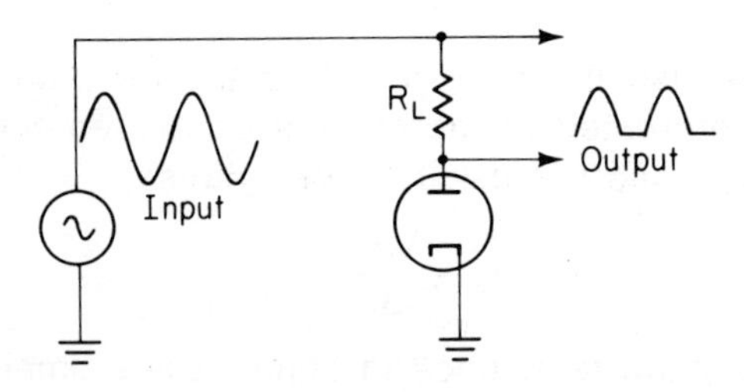


**Fig. 1-14.** Diode Rectifier.

The output is slightly lower in amplitude than the positive alternation of the input because some voltage will be dropped across the internal resistance of the tube. During the positive alternations of the input, the diode plate is more positive than the cathode. This causes plate current through $R_L$ which develops a positive pulse of voltage across $R_L$. During the negative alternation of the input, the plate is more negative than the cathode. This stops the plate current, and no voltage is developed across $R_L$. Thus, each negative alternation of the input is eliminated from the output. The output is a pulsating dc.

Diodes are rated in three ways; two ratings are for current and one for voltage. There is a maximum current rating and a maximum average current rating. Neither of these rated values should be exceeded. The third rating applies to the diode when it is cut off. This is a peak inverse voltage rating. It refers to the amount of negative voltage that can be safely applied to the plate of a cutoff diode. If the peak inverse voltage rating is exceeded, it could cause an arc of current from plate to cathode. Such an arc would very likely destroy the tube.

The invention of the diode opened a whole new field in electronics, and it was closely followed by the triode, the tetrode, the pentode, and many varieties of special purpose tubes.

## TRIODE TUBE

The triode electron tube was invented in 1907 by Lee DeForest, a United States inventor (1873–1961). Like most great inventions, this was a very simple one. He added a wire mesh between the cathode and the plate of a diode. A small potential on this mesh exerted a strong control over plate current. A few volts change on the wire mesh could drive the tube into cutoff or all the way to saturation. This controlling action was largely independent of the plate potential. The mesh logically became known as a grid, and because of its positive control on current, it was called a *control grid.*

### *structure*

Figure 1-15 shows the physical placement of cathode, grid, and plate in a triode as well as a schematic symbol.

The cutaway shows a helix wire grid placed around the cathode and supported by pieces of insulated wire. The addition of the grid is the only internal difference between the diode and the triode. Outwardly the only difference may be the number of pins. The diode has four pins; the triode has five.

The heater filament is shown in this illustration (Fig. 1-15) as a reminder that the triode (most triodes) uses thermionic emission. It will not appear in most schematic symbols.

Notice that the electrons which are emitted by the cathode must pass through the open space between the grid windings in order to reach the plate. A small negative potential on the grid with respect to the cathode exerts a powerful repelling force on these electrons. A small positive potential on the grid accelerates the electrons with such force that the plate dissipation is quickly exceeded, and the plate becomes red hot. The position of the grid enables it to have absolute control of plate current with reatively small variations in grid potential.

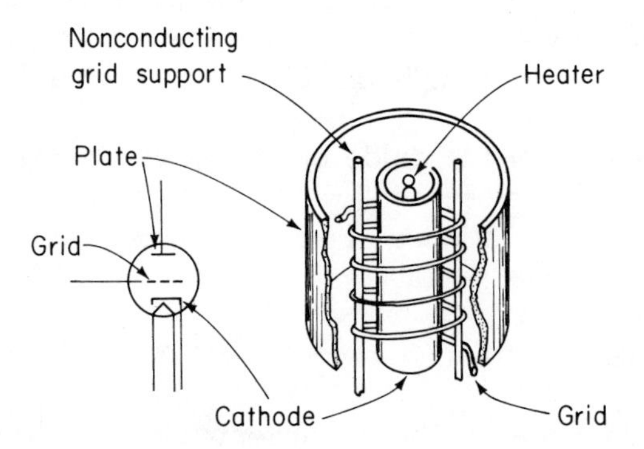


**Fig. 1-15.** Elements of a Triode.

It must be emphasized that the grid is *not* an emitting element. In the first place, it is designed of materials which have a very high work function. Secondly, the grid is not heated. If the grid is made more positive than the cathode, it will draw some current from the tube, but the few electrons that a grid may emit can be safety ignored.

### *zero grid voltage*

Earlier, we stated that the plate voltage of the diode was measured from the plate to the cathode. The same is true with the triode. In fact, all tube potentials are in respect to the cathode unless otherwise indicated. When the cathode is grounded, this frame of reference is not important, but when a resistor appears in the cathode circuit, the distinction is vital.

With zero potential on the grid, the triode acts exactly like a diode. The grid has no influence on the electron movement, and plate current is determined solely by the plate potential. This is illustrated in Fig. 1-16.

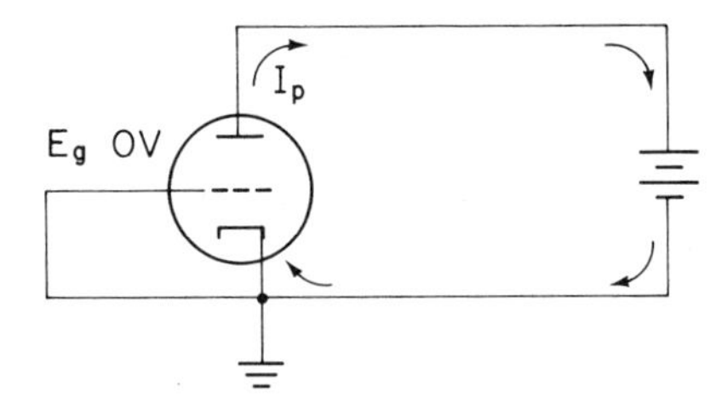


**Fig. 1-16.** Triode Functioning as a Diode.

The grid is a completely neutral element which in no way influences the electrons. Electrons move freely from cathode to plate, and $I_p$ is directly proportional to $E_p$.

### *positive grid voltage*

This has already been explained briefly, but it needs a bit more emphasis. When the grid potential ($E_g$) is positive, electrons will be attracted to it. In comparison to the number of electrons reaching the plate, the grid actually captures only a very small portion of the electrons. This is caused by two factors: the grid area is small, and the grid potential is small. Electrons that are captured by the grid constitute grid current ($I_g$.) Since grid current is extracted from the total electrons emitted by the cathode, it introduces a third factor of current to be considered in some cases. This is cathode current ($I_k$). $I_k = I_p + I_g$. It can be said that the possible plate current is reduced by an amount equal to the grid current. This is illustrated in Fig. 1-17.

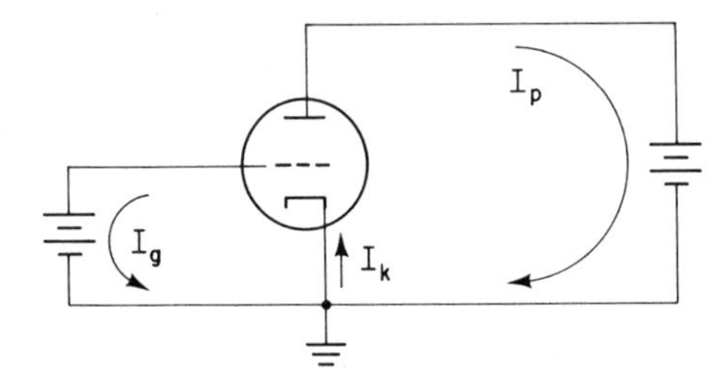


**Fig. 1-17.** Triode With a Positive Grid.

There is no space charge in this tube. All the emitted electrons are being collected by either the plate or the grid. Therefore, we can say that a triode with any positive level on the grid is operating at saturation. Keep in mind that grid potential is in respect to the cathode. In order to have a positive grid potential, it must be more positive than the cathode. In this circuit, there is no problem. The cathode is ground (0 V), therefore any potential above zero is a positive grid potential. However, the cathode potential is not always zero. Suppose that the cathode was 150 V positive. The same 150 V on the grid would constitute a zero grid potential.

### *negative grid voltage*

In most cases triodes are operated with a negative grid potential. Small variations of this negative potential can exert maximum control over plate current. At zero grid potential, the tube has a high plate current. Moving from zero to a few volts negative changes the plate current from a high value to complete cut off. When the grid is negative, it repels emitted electrons back toward the cathode. These electrons form a space charge between the grid and cathode. Making the grid less negative allows more electrons to pass which increases plate current. Making the grid more negative allows fewer electrons to pass which reduces plate current.

### *grid curves*

A grid curve is a characteristic curve showing the value of plate current for various grid potentials. A different curve can be plotted for each different value of plate voltage. A graph of several of these curves is known as a family of grid characteristic curves. A grid curve is constructed by applying a fixed quantity of plate voltage, changing grid potential in steps, measuring the plate current, plotting these points on a graph, and connecting the points with a smooth curve. The plate voltage is then changed to a new level, and the process is repeated in order to plot another curve. A family of these curves reveals considerable information about the capabilities and limitations of a particular type of triode. Figure 1-18 illustrates a partial family for one type of triode.

These lines are for illustration only, but actual curves for specific tubes can be obtained wherever tubes are sold. Reproductions of actual curves for most tubes are contained in standard tube manuals.

What does the graph show? Well, for one thing, it shows the exact amount of negative grid potential that is required in order to cut the tube off (stop all plate current) for representative levels of plate voltage. It also reveals the amount of plate current that can

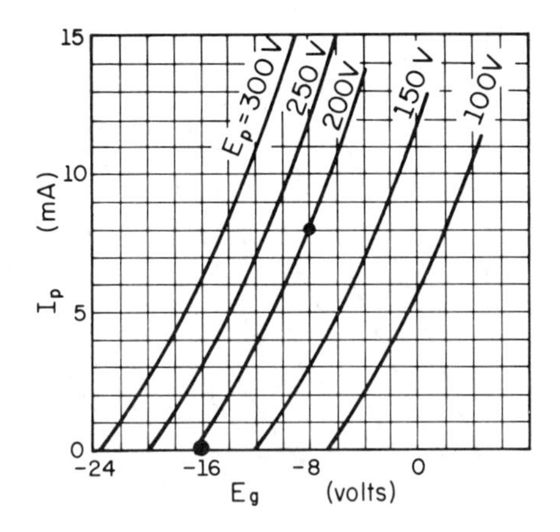


**Fig. 1-18.** Grid Characteristics.

be expected with specified levels of grid and plate potentials. For instance, when $E_p$ is 200 V and $E_g$ is —8 V, there will be 8 mA of $I_p$. A close examination of the grid family will answer many questions that arise when selecting a triode for a particular circuit. Another source of similar information is the family of plate characteristic curves.

### *plate curves*

These characteristic curves plot the values of plate voltage against plate current for various quantities of grid voltage. Sound like the same thing? Not quite. Examine Fig. 1-19 and compare it with Fig. 1-18.

Each grid voltage curve represents a fixed value of $E_g$. It covers a wide range of plate potentials and shows the plate current over the entire range. For instance, the —12-V curve shows that with an $E_g$ of —12 V, plate current starts when $E_p$ reaches 200 V. This line gives the value of $I_p$ for all plate potentials from 200 to 325 V. Therefore, it is easy to determine what value of grid potential is required to maintain operating current over a wide fluctuation of plate voltage.

### *tube constants*

The design factors of a tube take into account the geometric organization of the tube elements, the dimensions, the spacing, and

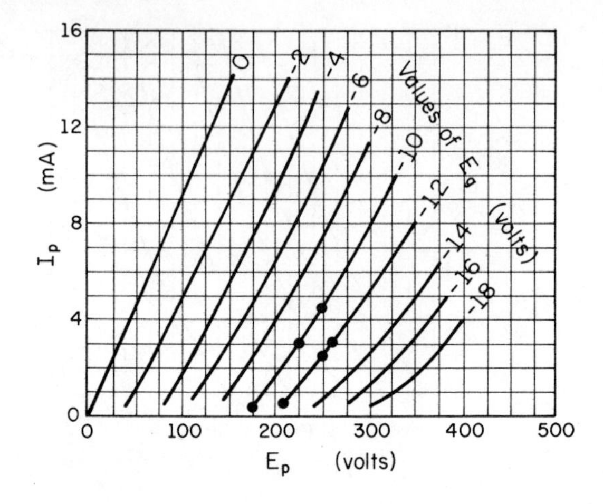


**Fig. 1-19.** Plate Characteristics.

many other factors. These design characteristics determine the potentials that can be applied to each element, operating parameters, and conditions for cut off. We call these design factors tube constants. Three of these factors may be calculated from the characteristic curves. These are ac plate resistance, amplification factor, and transconductance.

We calculated ac plate resistance in connection with the diode. It is no different here, $r_p = \Delta E_p/\Delta I_p$. Pick a grid voltage line in the plate characteristic family of curves. Let's say the —10-V line. Now, we make a small change in $E_p$ and divide it by the change it causes in $I_p$ (on the $E_g$ line). Starting where $E_g$ (10) intersects $E_p$ (225), raise the plate voltage to 250 V. $\Delta E_p$ is 25 V. This caused $I_p$ to change from 3 mA to 4.3 mA. $\Delta I_p$ is 1.3 mA. Substituting into the formula, we have

$$r_p = \frac{\Delta E_p}{\Delta I_p}$$
$$= \frac{25\text{ V}}{1.3\text{ mA}}$$
$$= 19\text{ k}\Omega$$

The amplification factor is a number which indicates the ratio of voltage change in the tube. That is, it shows the effectiveness of the grid in controlling the plate current. It is symbolized by the Greek letter mu ($\mu$) and is a ratio between a small change in plate voltage and a small change in grid voltage. The formula is $\mu = \Delta E_p/\Delta E_g$. It is understood here that $I_p$ is constant. We may use either grid or plate characteristic curves to obtain values for calculating the amplification factor. Using Fig. 1-19, start with $E_g$ of —10 V, $E_p$ of 225 V, and $I_p$ of 3 mA. Move to the right on the 3-mA line to the $E_g$ —12-V line. $\Delta E_p$ is 35 V, and $\Delta E_g$ is 2 V.

$$\begin{aligned}\mu &= \frac{\Delta E_p}{\Delta E_g}\\ &= \frac{35\text{ V}}{2\text{ V}}\\ &= 17.5\end{aligned}$$

These figures indicate that a 1-V change on the control grid causes a 17.5-V change at the plate.

Although we obtained our figures for this calculation along a constant $I_p$ line, what we have actually calculated is this: 17.5 V of $E_p$ change produces the same change in $I_p$ that a change of 1 V of $E_g$ will produce.

The transconductance of a tube tells how much plate current is changed by a small change in grid voltage. It is symbolized by $g_m$, and the formula is $g_m = \Delta I_p/\Delta E_g$.

Using Fig. 1-19, start on the $E_g$ 12-V line where $E_p$ is 250 V and $I_p$ is 2.5 mA. Move up the 250-V line to the $E_g$ —10 V-line. $\Delta I_p$ is 1.8 mA and $\Delta E_g$ is 2 V. Substituting into the formula, we have

$$\begin{aligned}g_m &= \frac{\Delta I_p}{\Delta E_g}\\ &= \frac{1.8\text{ mA}}{2\text{ V}}\\ &= 900\ \mu\mho\text{ (micromhos)}\end{aligned}$$

This is simply a ratio between $I_p$ change and $E_g$ change, but it is measured in mhos ($\mho$) as is any other conductance.

There is a mathematical relationship among these tube constants. We express it as $\mu = g_m \times r_p$. Using our sample figures $\mu = 17.5$, $g_m = 900\ \mu\mho$ and $r_p = 19\text{ k}\Omega$. Does it check?

$$\begin{aligned}17.5 &= (900\ \mu\mho)(19\text{ k}\Omega)\\ &\cong 17.1\end{aligned}$$

That is pretty close, especially when you consider that our graphs are only rough estimates of the actual characteristics.

The potential at the grid of a tube (always in respect to the cathode) we call bias. Bias then is our control voltage. Under operating conditions, maintaining the proper bias is a vital consideration.

### *types of bias*

We will be concerned with three types of bias. These are fixed bias, grid leak bias, and cathode self bias. The first, fixed bias, has already been shown. It is simply a dc voltage applied directly to the control grid. Moving our illustration a bit closer to a functional circuit, we have Fig. 1-20.

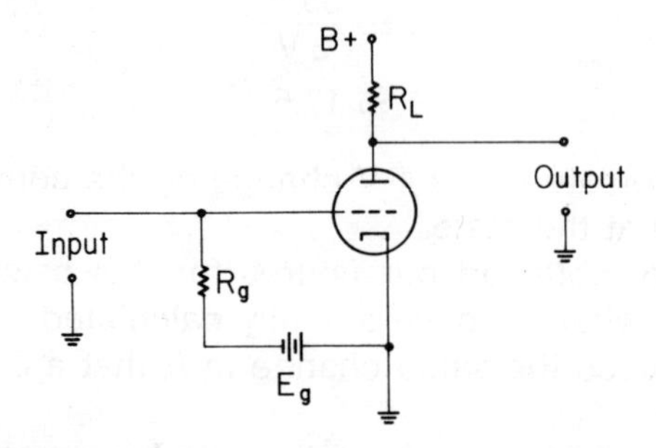


**Fig. 1-20.** Fixed Bias.

Assuming that the battery ($E_g$) has a 5-V potential, this is the difference in potential between the grid and cathode. We, therefore, have a bias of —5 V. Normally bias is a negative potential. Therefore the word bias, without further qualification, means the negative potential between grid and cathode.

The primary function for a circuit such as that in Fig. 1-20 is to increase the amplitude of a small ac signal. Assume that the input is an ac sine wave with a three volt peak to peak amplitude. When it arrives at the control grid, it will vary the grid potential by plus and minus 1.5 V as illustrated in Fig. 1-21.

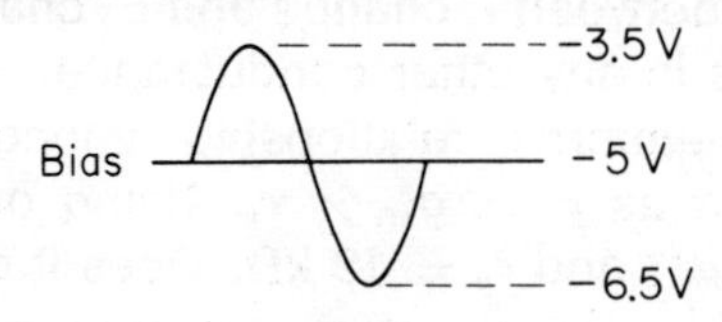


**Fig. 1-21.** Signal Riding the Bias.

The bias level becomes the reference voltage for the input signal. When it swings positive, the grid is less negative and plate current increases. When it swings negative, the grid is more negative and plate current decreases. The bias then determines the operating level, and the input signal causes variations about this level. The small voltage variations on the grid cause much greater variations in $E_p$. Plate and grid variations are of the same frequency, but they have a 180° phase difference. This means that the plate swings positive when the grid swings negative and vice versa.

A triode may be connected as shown in Fig. 1-22, and it will develop its own bias.

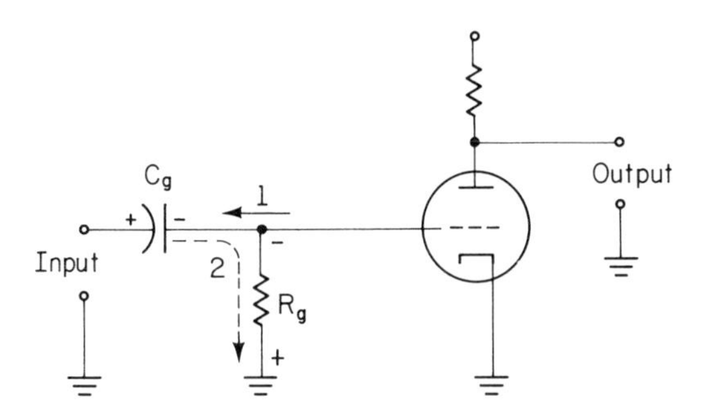


**Fig. 1-22.** Grid Leak Bias.

With this arrangement, the grid potential is zero until an input signal arrives. The positive alternation of the input signal causes the grid to go positive and to draw some current. The grid current charges the grid capacitor with negative on the right and positive on the left.

When the input swings negative, $C_g$ tries to discharge through the grid resistor ($R_g$). $C_g$ and $R_g$ are of values which form a long time constant for the input frequency. The negative alternation is too brief for the capacitor to lose any appreciable charge. $C_g$ remains charged and holds a negative potential on the control grid. Path 1 shows the capacitor charging action. Path 2 shows the discharging action which develops a negative potential at the top of $R_g$. The voltage across $R_g$ is the grid potential.

Each subsequent input increases and decreases the grid potential in the same fashion that it did with fixed bias. Since the tube is generating its own bias, grid leak bias is a type of self-bias. This type of bias can cause the tube to operate over a range from very low $I_p$ to saturation, but it *cannot* cut the tube off.

Another type of self-bias is cathode bias. We accomplish this by connecting a *RC* network into the cathode circuit as shown in Fig. 1-23.

Remember that we have two types of action in a tube: a dc and an ac. The dc through the tube goes through the cathode resistor

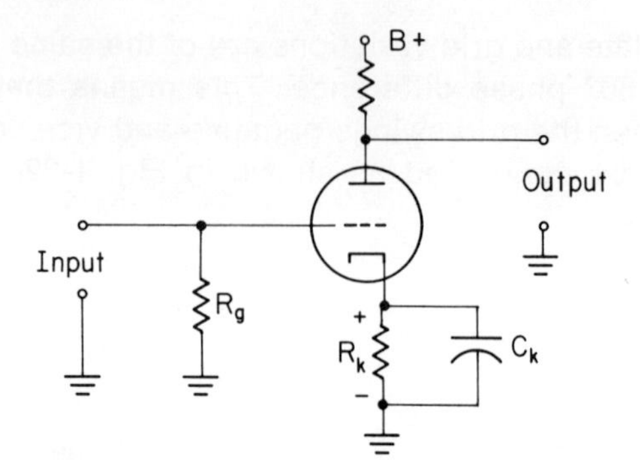


**Fig. 1-23.** Cathode Self Bias.

($R_k$), and develops a positive potential on the cathode. The ac change (signal input) finds the cathode capacitor ($C_k$) to be a low opposition path and uses it to bypass the cathode resistor. The tube would function without $C_k$, but the ac signal would be badly attenuated by $R_k$. This would be signal degeneration, and it is easily avoided by using the bypass capacitor in the cathode circuit.

The positive potential, which this circuit develops on the cathode, has the same effect as placing the same amount of negative voltage on the grid. The grid potential in respect to ground remains at zero, but in respect to the cathode, the grid is negative.

## APPLICATIONS OF THE TRIODE

The triode tube is a very versatile electronics device. Electronic switch, variable resistor, and voltage amplifier are a few of its many uses.

### *variable resistor*

The dc plate resistance of a triode is inversely proportional to plate current. Since plate current is under complete control of bias, we can set the $R_p$ to any desired value and vary it at will by adjusting the value of bias.

Assume that we have a triode with 150 V on the plate, the plate resistance will vary with the bias level. We can determine the plate resistance by measuring $I_p$ and using Ohm's law. For instance, when bias is —8 V, we may measure 0.8 mA of $I_p$. At this time

$$R_p = \frac{E_p}{I_p} = \frac{150\ \text{V}}{0.8\ \text{mA}} = 187.5\ \text{k}\Omega$$

We may use the $E_p$–$I_p$ curves for our particular triode and determine the $I_p$ without using a live circuit. Figure 1-24 is a sample graph for a 6C5 triode.

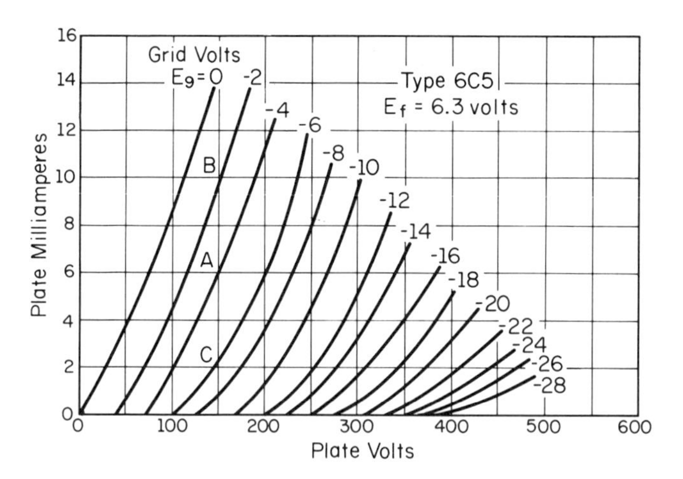


**Fig. 1-24.** $E_p$–$I_p$ Curve for a 6C5 triode.

Using the graph as a guide, let's calculate the dc plate resistance for values of —2 V, —4 V, and —6 V bias. We will hold $E_p$ constant at 150 V. Point *A* on the graph represents an $E_g$ of —4 V. The $I_p$ at this point is 6 mA.

$$R_p = \frac{E_p}{I_p} = \frac{150\text{ V}}{6\text{ mA}} = 25\text{ k}\Omega$$

Point *B* on the graph represents the conditions for —2 V bias. The $I_p$ is now 10 mA.

$$R_p = \frac{150\text{ V}}{10\text{ mA}} = 15\text{ k}\Omega$$

Point *C* on the graph represents the conditions for —6 V bias. The $I_p$ is 2.6 mA.

$$R_p = \frac{150\text{ V}}{2.6\text{ mA}} = 57.5\text{ k}\Omega$$

We should conclude from this exercise that $R_p$ is a variable factor, and for a given value of $E_p$, $R_p$ varies directly with $E_g$.

### amplifier

We defined mu ($\mu$) as the amplification factor of a tube and $\mu$ is directly proportional to gain when we use the tube as an amplifier.

The voltage gain is $E_{out}/E_{in}$, and this value is always less than the amplification factor when the tube uses a resistive plate load. $E_{out}$ is the ac component of the voltage across $R_L$, or if you prefer, it is the change in voltage drop across $R_L$. Figure 1-25 is a triode used as an amplifier.

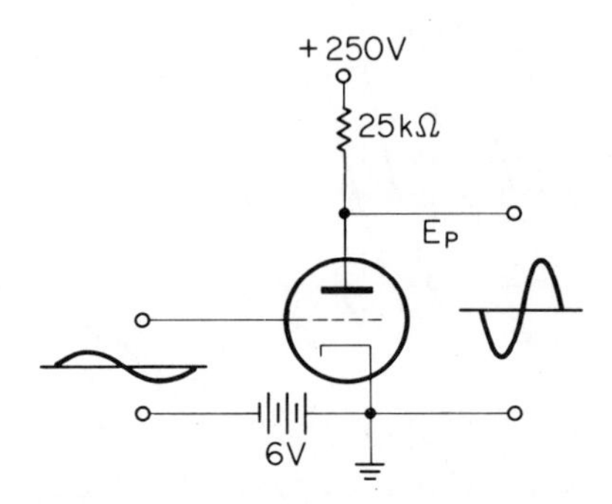


**Fig. 1-25.** Amplifying a Signal.

With the —6 V bias as indicated, suppose that we measure $I_p$ of 3 mA when in a quiescent condition (no signal). What is the value of $E_p$ and $E_{R_L}$ at this time?

$$E_{R_L} = 3 \text{ mA} \times 25 \text{ k}\Omega = 75 \text{ V}$$

$$E_p = E_a - E_{R_L} = 250 \text{ V} - 75 \text{ V} = 175 \text{ V}$$

The applied signal is 4 V peak to peak. When the input is maximum positive, $I_p$ is 4 mA. When it is maximum negative, $I_p$ is 2 mA. Calculate minimum and maximum $E_p$, $\Delta E_{R_L}$, and voltage gain.

$$E_{R_L}(\text{max}) = 4 \text{ mA} \times 25 \text{ k}\Omega = 100 \text{ V}$$

$$E_p(\text{min}) = 250 \text{ V} - 100 \text{ V} = 150 \text{ V}$$

$$E_{R_L}(\text{min}) = 2 \text{ mA} \times 25\text{k}\Omega = 50 \text{ V}$$

$$E_p(\text{max}) = 250 \text{ V} - 50 \text{ V} = 200 \text{ V}$$

$$\Delta E_{R_L} = 100 \text{ V} - 50 \text{ V} = 50 \text{ V} = E_{out}$$

$$\text{Voltage gain} = \frac{E_{out}}{E_{in}} = \frac{50 \text{ V}}{4 \text{ V}} = 12.5$$

### *dc load line*

The dc load line is a convenient method of predicting exactly how a triode will react in a given circuit. It is a straight line drawn diagonally across the $E_p$–$I_p$ curves to connect all possible points of

$E_p$ and $I_p$. Suppose that we wish to use a 6C5 triode in the circuit of Fig. 1-26.

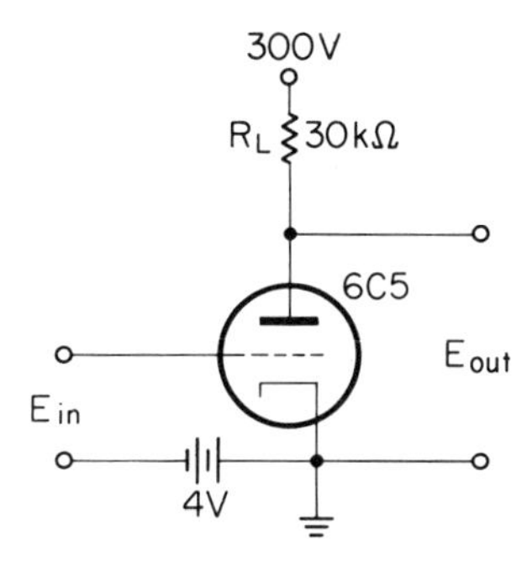


**Fig. 1-26.** Typical Triode Amplifier.

The extreme conditions are:

$$E_p = 0 \quad \text{and} \quad I_p = 10\,\text{mA}$$
$$E_p = 300\text{ V} \quad \text{and} \quad I_p = 0$$

When we plot these two points on the characteristic curves and connect them with a straight line, we have the graph in Fig. 1-27.

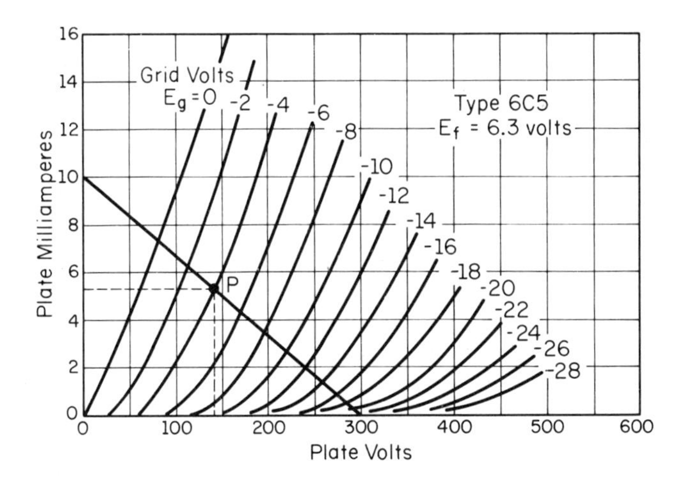


**Fig. 1-27.** Load Line for Circuit in Fig. 1-26.

The quiescent condition is set by the bias. The bias is —4 V and the point on the graph where the load line intersects the —4 V line (point *P*) is our operating point. The incoming signal will cause

$E_p$ and $I_p$ to swing both ways from the operating point. The graph now tells us that at the operating point we have:

$$E_p = 145 \text{ V}$$
$$I_p = 5.2 \text{ mA}$$
$$E_g = -4 \text{ V}$$

Now we may apply any reasonable amplitude signal and obtain a detailed picture of the circuit action. Assume that we have a 4-V peak to peak input. Maximum values of $I_p$ and $E_{R_L}$ and minimum value of $E_p$ occur when signal is maximum positive. Follow the description on the graph in Fig. 1-27.

$E_g$ swings from —4 to —2 V

$I_p$ swings from 5.2 to 6.2 mA

$E_p$ swings from 145 to 112 V

Minimum values of $I_p$ and $E_{R_L}$ and maximum $E_p$ occur when the signal is maximum negative.

$E_g$ swings back from —2 to —6 V

$I_p$ swings from 6.2 to 4.3 mA

$E_p$ swings from 112 to 170 V

$$\Delta E_g = 4 \text{ V}$$
$$\Delta I_p = 2.9 \text{ mA}$$
$$\Delta E_p = 58 \text{ V}$$
$$\text{Gain} = \frac{E_{\text{out}}}{E_{\text{in}}} = \frac{58 \text{ V}}{4 \text{ V}} = 14.5$$

In circuit design, we may also work in reverse, that is, construct a load line to determine what value of plate load resistor we need. Suppose that we have a 300-V source and wish to construct a circuit that will draw 4.8 mA of $I_p$ with a bias of —6 V. When we take a 6C5 $E_p$–$I_p$ graph and connect these two points, we have plotted the load line from $E_p$(max) to the operating point. Extending this straight line to the left margin of the graph, we find the point of $I_p$(max). This completes the load line as shown in Fig. 1-28.

What does the graph tell us?

The extremes are:

$$E_p(\text{max}) = 300 \text{ V} \quad \text{and} \quad I_p = 0$$
$$I_p(\text{max}) = 12 \text{ mA} \quad \text{and} \quad E_p = 0$$

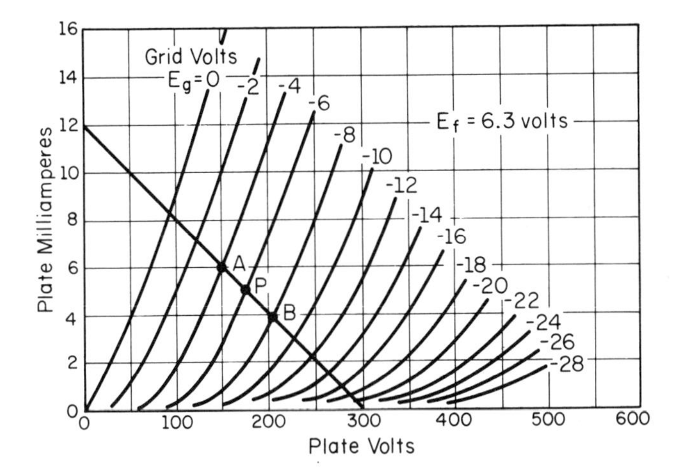


**Fig. 1-28.** Load Line for Determining Value of $R_L$.

The operating point is:

$$E_g = -6 \text{ V}$$
$$I_p = 4.8 \text{ mA}$$
$$E_p = 180 \text{ V}$$
$$E_{R_L} = 300 \text{ V} - 180 \text{ V} = 120 \text{ V}$$

We can now use the values at either the operating point or the maximum points to calculate $R_L$.

$$R_L = \frac{E_p(\text{max})}{I_p(\text{max})} = \frac{300 \text{ V}}{12 \text{ mA}} = 25 \text{ k}\Omega$$

or

$$R_L = \frac{120 \text{ V}}{4.8 \text{ mA}} = 25 \text{ k}\Omega \qquad \text{(values at operating point)}$$

If we use these values to construct a schematic, we will have the circuit of Fig. 1-29.

When a 4-V peak to peak signal is applied to this circuit, it causes a swing along the load line from point *P* to *A,* back to *P,* from *P* to *B,* and back to *P.* From the graph, determine $\Delta E_g$, $\Delta I_p$, $\Delta E_p$, and voltage gain.

$\Delta E_g$ is 4 V; this was already given

$\Delta I_p$ is 6 mA − 3.8 mA = 2.2 mA

$\Delta E_p$ is 205 V − 150 V = 55 V

$$\text{Voltage gain} = \frac{55 \text{ V}}{4 \text{ V}} = 13.75$$

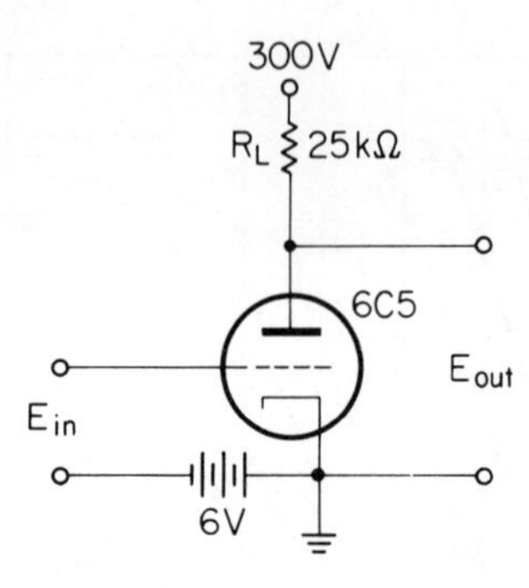


**Fig. 1-29.** Circuit from Graph in Fig. 1-28.

## CATHODE RAY TUBE

This is the picture tube of the electronics industry. It is used for computer displays; presentation of radar data; observation of electric waveshapes; various measurements of time, amplitude, and frequency; and the television screen. For all these uses and many others, you might expect to find a great variety of cathode ray tubes. This is not the case. The cathode ray tube (CRT) comes in two standard types; electrostatic and electromagnetic. The name refers to the type of deflection system used with the scope.

CRTs are built with special features, but this is the exception rather than the rule. Circuits can be designed to cause a standard CRT to perform a great variety of function. All CRTs have more common features than they have differences. First we will examine the common characteristics; then we will compare the two types.

### *electron gun*

The CRT operates on thermionic emission, and the electrons are boiled from the cathode in much the same manner as in other tubes. But the stream of electrons doesn't go directly to a plate. This stream of electrons is molded into a sharply focused beam and fired toward a screen. Pictures are drawn on the screen by moving the beam about in much the same way that a pencil moves on a piece of paper. The electron gun is composed of a cathode, a control grid, an accelerating anode, and a focusing anode. Figure 1-30 illustrates the electron gun.

The cathode is a closed cylinder with the heating element on the inside. Electrons are emitted only from the end of the cathode. The grid is a cap which fits over the cathode. A small hole in the

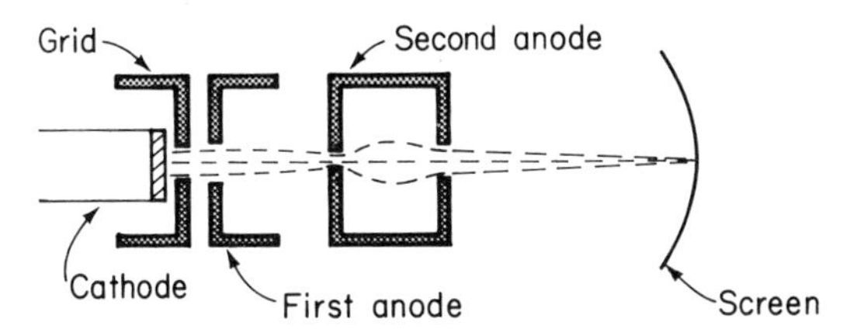


**Fig. 1-30.** Electron Gun.

center of the grid allows electrons to pass through. The potential on the grid can be varied to control the number of electrons that are allowed to pass. This control is complete. It can shut off the electron stream completely, or it can allow huge quantities of electrons to pass.

The first anode is shaped similar to the grid, but it is reversed. The flat end with a small opening faces the on-coming electrons. This is the accelerating anode. It has a high positive potential which attracts the electrons with an impelling force. Most of the electrons strike the metal and are returned to the power source. The small percentage of electrons which pass through the opening are fired down the tube with terrific force. It has been estimated that these electrons reach a velocity in excess of 10,000 miles/s.

The second anode must shape this rapidly moving stream into a fine beam. Further, it must bring the electrons to a sharp focal point exactly where they touch the screen. This is accomplished by using a highly negative potential. This focusing anode is a hollow cylinder with a small opening at either end. The negative potential on this cylinder exerts a compression on the negative electrons. The area of least pressure is a line exactly down the center.

CRTs are not limited to a single electron gun. Dual gun and multiple gun CRTs are now in use.

### *envelope and screen*

The evacuated glass envelope and phosphorescent screen are illustrated in Fig. 1-31.

In comparison to other tubes, the CRT is a huge affair. It may be any size from several inches to a few feet in length. The screen size is also varied to suit the intended use of the CRT.

The screen is formed by depositing a phosphor coating on the inside front of the tube. The high-velocity electrons cause the phosphor to give off light where they strike. The viewer in front of the screen will see this phosphorescent light.

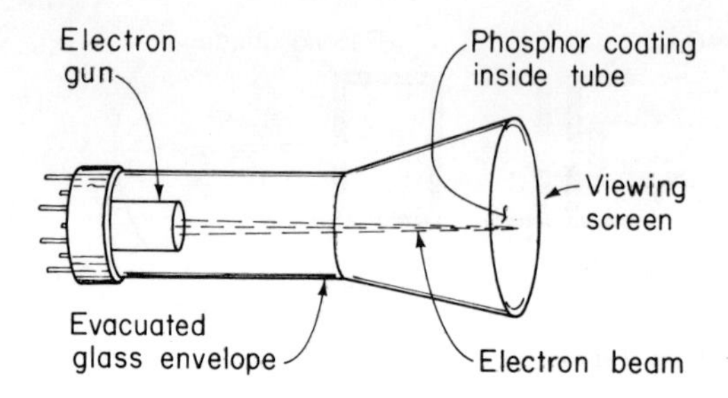


**Fig. 1-31.** Envelope and Screen.

### *aquadag*

The electrons must be supplied with a return path. Otherwise, they would form a cloud between the screen and the electron gun. This is usually accomplished by depositing a coat of metal on the inside of the glass. The coating is called aquadag. This coating is illustrated by the heavy lines in Fig. 1-32.

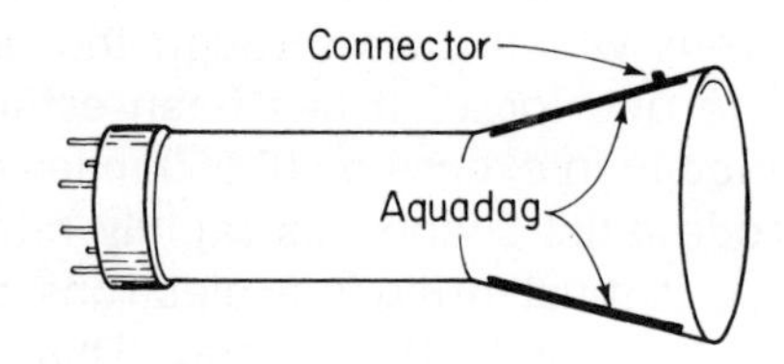


**Fig. 1-32.** Aquadag Coating.

This coating goes all the way around in one continuous layer. An external connector is brought through the side of the glass envelope. An extremely high positive potential is connected to the aquadag coating. As the electrons bounce from the screen, the aquadag collects them and returns them to the power supply.

### *electrostatic CRT*

This CRT uses electrostatic fields to position the presentation on the screen. These fields are produced by potentials on deflection plates inside the glass envelope. The tip of the beam can be moved about on the screen by varying the voltages on the deflection plates. The moving beam can be moved to trace patterns, draw pictures, and write words. These plates are shown in Fig. 1-33.

The potential difference between the plates forms an electric field. The intensity of this field controls the deflection of the beam. Varying the potential on the vertical plates move the beam up and

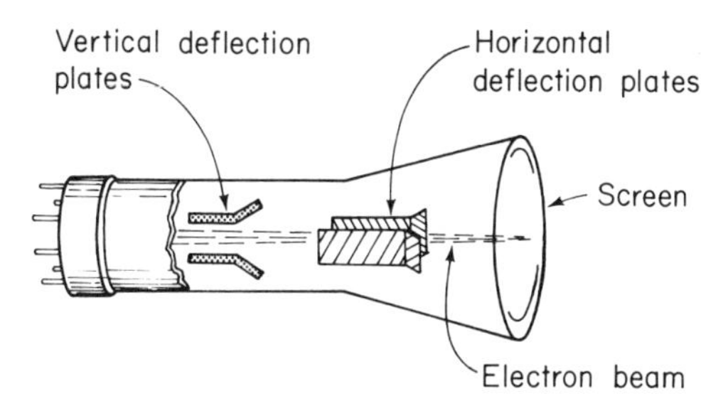


**Fig. 1-33.** Electrostatic Deflection System.

down. Right and left movements are controlled by varying the potential on the horizontal plates. These potentials can be varied manually by control knobs as well as automatically by ac signals. Figure 1-34 is a sample presentation. This picture was produced by a sine wave on the vertical plates and a sawtooth voltage on the horizontal plates.

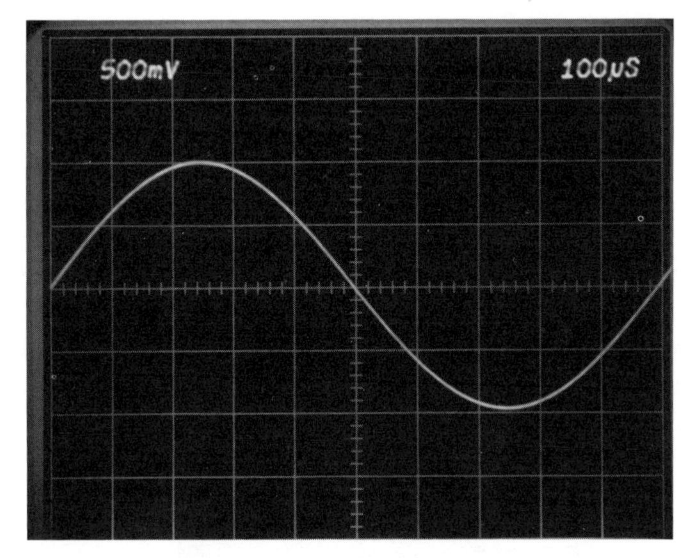


**Fig. 1-34.** CRT Presentation. (Courtesy of Tektronix, Inc.).

### *electromagnetic CRT*

The elcteromagnetic CRT finds its greatest use in fixed installations where a limited number of patterns are presented. Deflection coils replace the deflection plates; otherwise the CRT is the same as the electrostatic. The coils, being a bit on the bulky side, are mounted external to the glass envelope. There are two horizontal coils and two vertical coils. These coils are represented in Fig. 1-35.

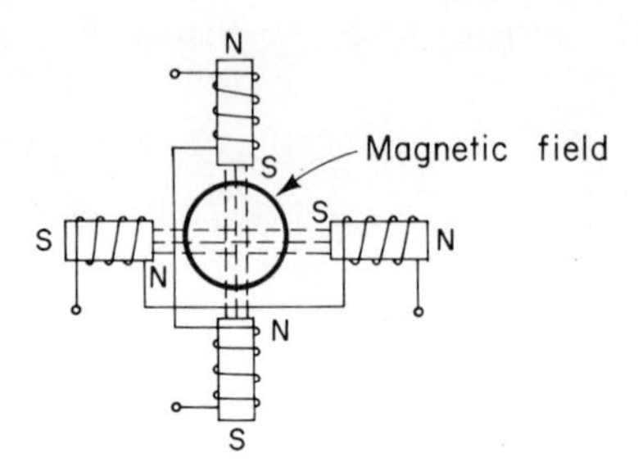


**Fig. 1-35.** Electromagnetic Deflection.

This drawing is merely a schematic representation. The coils are wound on yokes which fit snugly against the neck of the CRT. There is sufficient physical displacement between horizontal and vertical coils to cause the two fields to act independentiy. Movement of the electron beam is controlled by varying the current through two sets of coils.

## APPLICATIONS OF THE OSCILLOSCOPE

The oscilloscope is the most valuable tool the electronics technician has at his disposal. He uses it to measure dc level, wavelength, time difference, and signal amplitude. The oscilloscope faithfully reproduces waveshapes for comparison of amplitude,

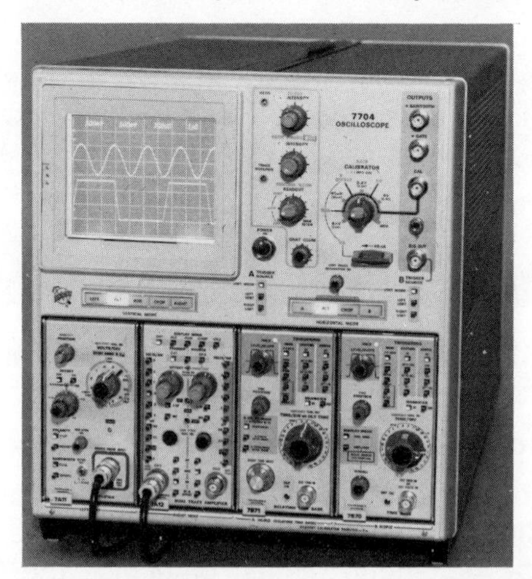


**Fig. 1-36.** Tektronix 7704 Oscilloscope. (Photographs of Oscilloscope and Oscilloscope Presentations Courtesy of Tektronix, Inc.).

phase, and symmetry. Figure 1-36 is a photograph of a modern oscilloscope.

This oscilloscope uses a dual beam CRT. It provides up to four pictures simultaneously with digital read out as to amplitude and time. The oscilloscope face is inscribed with both horizontal and vertical graticules. The graticules are lighted, and the oscilloscope may be adjusted to cause the distance between lines to represent precise increments of time or amplitude. These lines are clearly defined in Fig. 1-37.

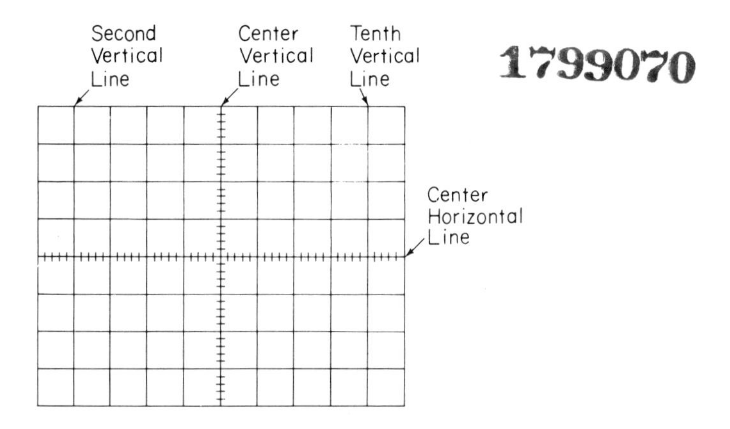




**Fig. 1-37.** Measurement Lines on the 7704 Oscilloscope. (Courtesy of Tektronix, Inc.).

### *measuring phase and frequency*

One method of measuring phase and frequency is to use a single trace and place one ac signal on the horizontal channel and another on the vertical channel. The oscilloscope will present the resultant of these two signals in a lissajous pattern. The shape of the pattern reveals the phase relation of the two signals. Figure 1-38 shows several lissajous patterns.
The two signals are of the same frequency, and the horizontal signal is 2.5 times the amplitude of the vertical signal.

Figure 1-39 is a pattern resulting from two signals of different amplitude and frequency.
The vertical signal is twice the frequency and $\frac{3}{4}$ the amplitude of the horizontal signal.

Another way that we can measure phase and frequency is to use two or more traces. A reference signal can be placed on one

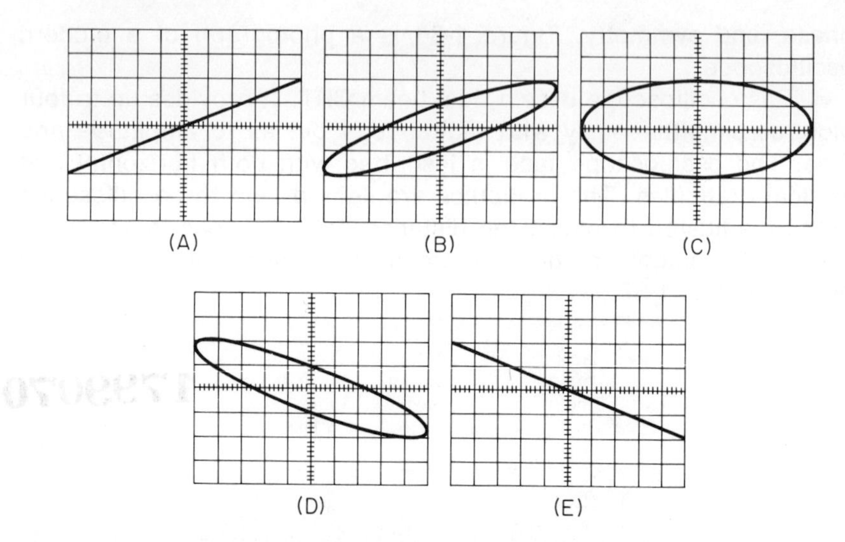


**Fig. 1-38.** Patterns Produced by Two Signals of the Same Frequency. (Courtesy of Tektronix, Inc.).

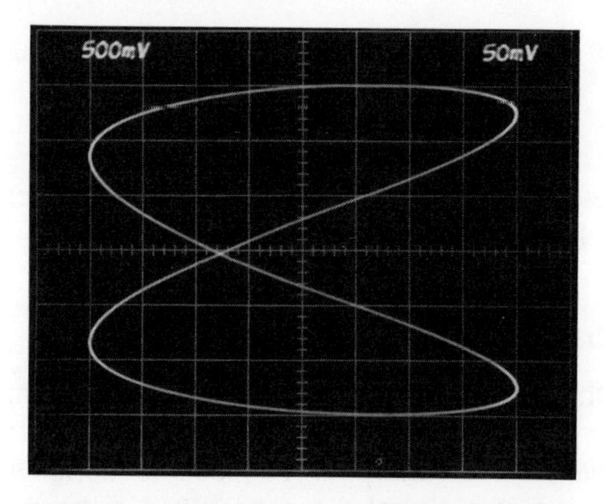


**Fig. 1-39.** Two Signals of Different Frequencies.

trace and compared to a signal on another trace. Figure 1-40 illustrates this method of measuring phase.
These two signals are of the same frequency and amplitude. The channel 2 signal is lagging the reference signal by 45°.

### *time measurement*

The horizontal displacement represents a specific period of time. We can adjust the sweep for a specified number of microseconds between lines. We can then measure the time duration of

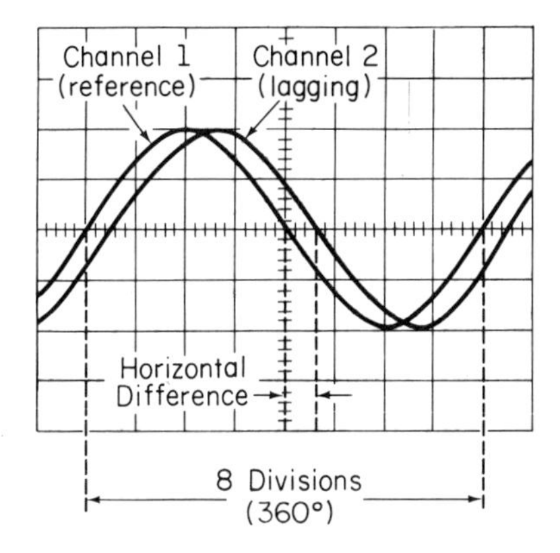


**Fig. 1-40.** Measuring Phase Difference.

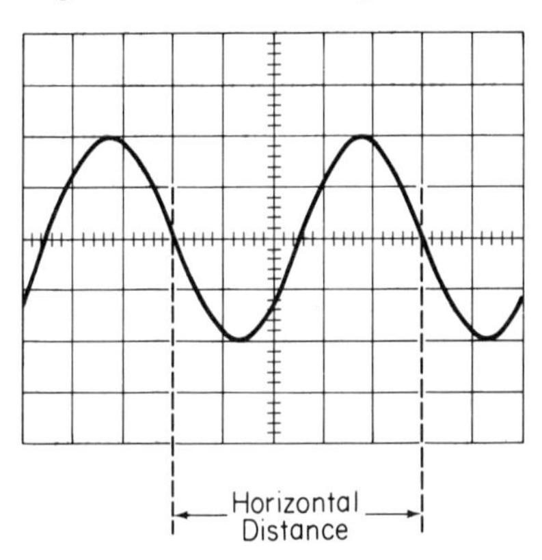


**Fig. 1-41.** Measuring Time Duration Between Points on a Signal.

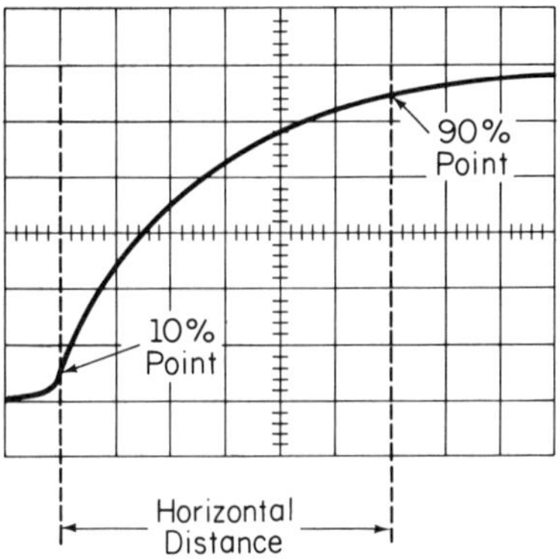

**Fig. 1-42.** Measuring Rise Time on a Waveform.

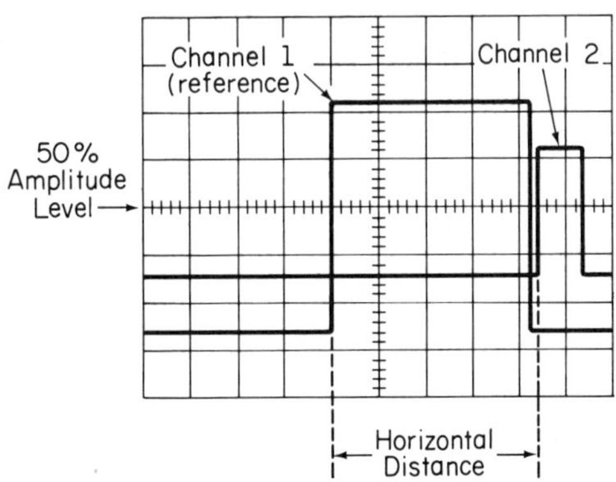


**Fig. 1-43.** Measuring Time Difference Between Two Pulses.

a waveshape and quickly convert this to either wavelength or frequency. Some time measurements are illustrated in Figs. 1-41, 1-42, and 1-43.

When our sweep is adjusted for 1 $\mu$s between vertical lines, the signal in Fig. 1-41 is 5 $\mu$s in duration. We then calculate frequency like this:

$$f = \frac{1}{t}$$

$$= \frac{1}{5} \times 10^{-6} = 200 \text{ kHz}$$

We convert frequency to wavelength by this process:

$$\lambda = \frac{300{,}000{,}000 \text{ m}}{f \text{ in hertz}}$$

$$\lambda = \frac{300{,}000{,}000}{200{,}000}$$

$$\lambda = \frac{300}{0.2} = 1500 \text{ m}$$

***measuring amplitude***

We can calibrate the vertical deflection on the oscilloscope to cause each square to represent a specified quantity of voltage. We can then measure the level of a dc voltage by counting the number of squares the sweep moves up or down when we apply the dc potential.

When measuring ac amplitudes, we can take a choice of values: peak to peak or peak. A peak to peak measurement is illustrated in Fig. 1-44.

Assuming that each square represents 1 V, the signal in Fig. 1-44 is 4.7 V in amplitude.

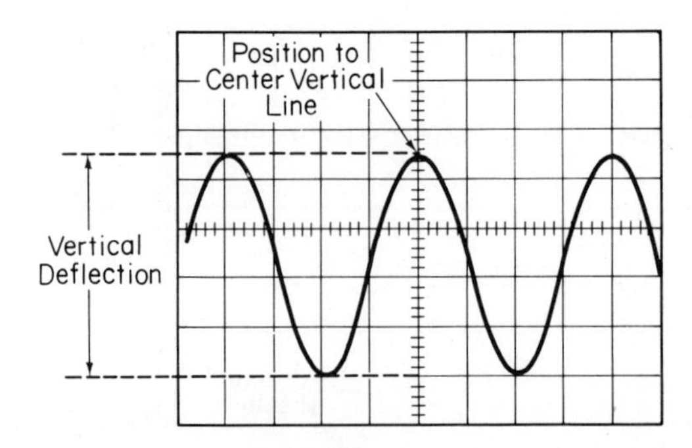


**Fig. 1-44.** Measuring Amplitude.

When we desire a value other than peak to peak or peak, it is easier to take one of these readings and convert it. Suppose that we want the effective value of the reading just taken.

$$E_{\text{eff}} = 0.707 \times \frac{E_{p\text{-}p}}{2}$$
$$= 0.707 \times \frac{4.7\text{ V}}{2}$$
$$= 1.66\text{ V}$$

## CHAPTER 1 REVIEW EXERCISES

1. (a) Name four types of electron emission.
   (b) Which of these are used in electron tubes?
2. Describe the process of thermionic emission.
3. What is the name of the restraining potential which opposes emission?
4. What is meant by the work function of an element?
5. Draw a schematic symbol for a diode and label the parts.
6. Under what conditions will electrons flow through a diode?
7. Draw a half-wave diode rectifier circuit which will produce a negative pulsating dc. Show input and output waveshapes.
8. What is the ac plate resistance of a diode when a 3-V change in $E_p$ produces a 6-mA change in $I_p$?
9. What is the dc plate resistance of a diode when an $E_p$ of 10 V produces an $I_p$ of 1 mA?
10. Draw a full-wave diode rectifier circuit which will produce a positive pulsating dc. Show input and output waveshapes.
11. Draw a schematic symbol of a triode and label the parts.
12. What is meant by bias?
13. What are the three tube constants that can be calculated from static characteristic curves? Write the formula for each.
14. Write the formula which shows the relationship among the three tube constants.
15. Name the three types of bias. Draw a circuit to illustrate each type.
16. Describe the characteristic of a triode which enables its use as an amplifier.
17. What is meant by voltage amplification?

18. What is the gain of an amplifier which has a 6-V input signal and an output of 114 V?
19. What is a load line?

Items 20 through 27 refer to Fig. 1-45.

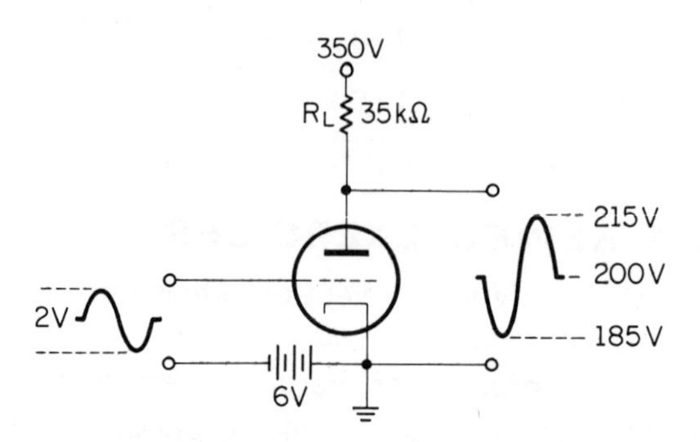


**Fig. 1-45.** Amplifier Circuit.

20. What are the values of $I_p$ and $E_{R_L}$ during the quiescent state?
21. What are the values of $I_p$, $E_p$, and $E_{R_L}$ when the input is maximum positive?
22. What are the values of $I_p$, $E_p$, and $E_{R_L}$ when the input is maximum negative?
23. The input signal causes the bias to vary between what two values?
24. $\Delta E_p$ is how many volts for each volt of $\Delta E_g$?
25. What is the gain of this amplifier?
26. If a load line were constructed for this amplifier, what two extreme points would it connect on the $E_p$–$I_p$ graph?
27. What are the values of $E_p$, $I_p$, and $E_g$ at the operating point?

Items 28 through 30 refer to Fig. 1-46.

28. Draw the circuit represented by this load line. Label the values of B+, $R_L$, and $E_g$.
29. What are the values of $E_p$, $I_p$, and $E_{R_L}$ at the operating point?
30. The input signal causes bias to swing between points $R$ and $Q$.
    (a) What is the peak to peak amplitude of the input signal?
    (b) What is the minimum value of $I_p$?

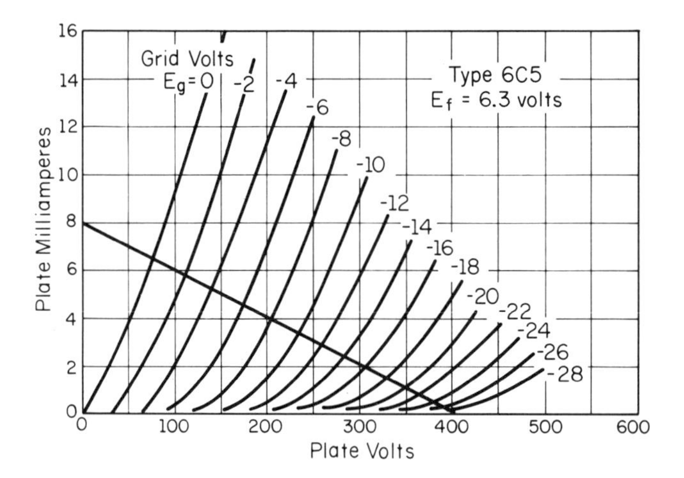


**Fig. 1-46.** Load Line.

(c) What is the maximum value of $I_p$?
(d) What is the minimum value of $E_p$?
(e) What is the maximum value of $E_p$?
(f) What is the gain of this amplifier?

31. An oscilloscope has been calibrated so that each vertical square is equivalent to 2 V and each horizontal square is equivalent to 2 $\mu$s. A displayed sine wave covers 10 squares horizontally and 6 squares vertically. What is the:
(a) Peak to peak amplitude?
(b) Peak amplitude?
(c) Effective value?
(d) Time duration?
(e) Frequency?
(f) Wavelength?

32. A sine wave is displayed on each of two sweeps of an oscilloscope which has the same calibrations as described in item 31. Each waveshape covers eight horizontal squares, and the signal being measured reaches a peak two squares ahead of the reference signal. What is the:
(a) Frequency of the two signals?
(b) Phase difference?

# 2

# solid-state devices

In modern equipment the vacuum tube has, for the most part, been replaced by a small blob of solid material known as a solid-state device. Yet knowledge of some solid-state properties is much older than the vacuum tube. Michael Faraday (British physicist, 1791–1867) discovered that silver sulfide has a negative temperature coefficient. This was a highly significant fact, but its importance was not recognized. This happened in 1833, and in 1835, a much more important discovery was made. This was the rectifying property of solid-state devices. It received practically no notice until it was rediscovered in 1874.

The first solid-state rectifier was patented in the same year as the famous Fleming valve, 1904. The interest in developing solid-state devices was greatly curtailed when Dr. Lee DeForest designed his triode electron tube.

In the early 1940s the electron tube ran into trouble and caused a serious study of solid states. The electronic equipment was becoming more sophisticated and the higher frequencies were reaching beyond the capabilities of the electron tube. The breakthrough came just before Christmas in 1947. At this time William Shockley, John Bardeen, and Walter H. Brattain produced the first solid-state electronic amplifier. This achievement by these three scientists of the Bell Telephone Laboratory was acclaimed as one of the greatest inventions of all times. The technological and social significance was so great that they were awarded the 1956 Nobel Prize for Physics.

Figure 2-1 is a photograph of these scientists at work.

**Fig. 2-1.** Inventors of the Transistor. (Photograph and description Courtesy Bell Laboratories.)

This picture shows (left to right) Dr. John Bardeen, Dr. William Shockley, and Dr. Walter H. Brattain along with some of the apparatus they used in developing the first transistor. Figure 2-2 shows the product of this labor.

This may appear as a crude device if you compare it with today's standards, but it is one of the greatest inventions that mankind has ever enjoyed. This is a point contact transistor. The two gold

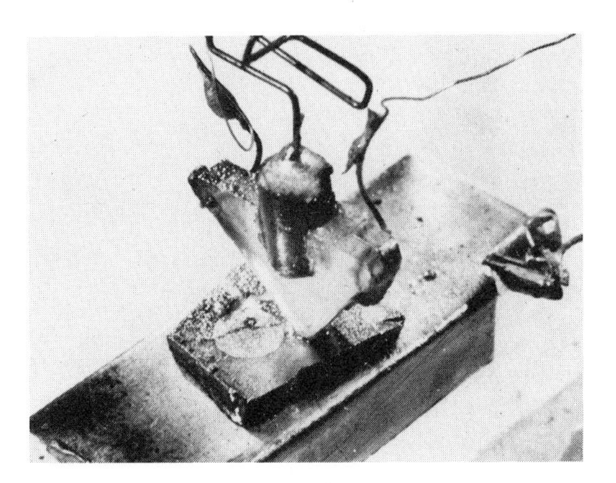

**Fig. 2-2.** The First Transistor. (Photograph and description Courtesy Bell Laboratories.)

contacts are supported by the wedge shaped piece of insulating material. The contacts are pressed into a piece of germanium semiconductor material and are separated by a few thousandths of an inch. The semiconductor is supported by a metal base. This device amplified electrical signals by passing them through a solid-state device. Basically this is the same operation that modern transistors perform.

In this chapter we will explore the basic structure of the transistor and how it functions. Along the way, we will look into the science of semiconductors in general and solid-state diodes in particular.

We will start with an in depth examination of semiconductors.

## SEMICONDUCTORS

Some materials carry current readily and are called *conductors.* Others offer a very high opposition to current and are called *insulators.* There is a class of materials falling some place between these extremes that is called semiconducting material. These semiconductors have been used to accomplish an outstanding advancement in electronics. Devices formed of certain combinations of this material can perform almost any function that an electron tube can perform and many other functions beyond the capabilities of electron tubes. Furthermore, they do it with less power and greater reliability. As a group, they are known as solid-state devices. Among them we have solid-state diodes and transistors. In order to visualize the function of these devices, it is essential to begin at the atomic level.

### *review of atomic structure*

In the basic structure of matter, it has been established that all things are composed of atoms. Theory states that electrons rotate about the nucleus of an atom in much the same manner as planets rotate about the sun. These electron orbits represent bands (or shells) of energy. The neutral atom is electrically stable and contains the same number of electrons in the shells as it has protons in the nucleus. Some complex atoms are known to have as many as seven energy shells while simple, one or two electron, atoms have only one energy shell.

There is a law that limits the number of electrons that can occupy each shell. When this limit is reached, the band is full. The shells are filled from the nucleus outward. The outer shell is called the valence shell and it may contain a maximum of eight electrons. The word valence means ability to combine. The electrons in the valence shell combine with the valence electrons of other atoms to create a bond between the atoms of a material. Figure 2-3 illustrates a balanced atom with four full shells.

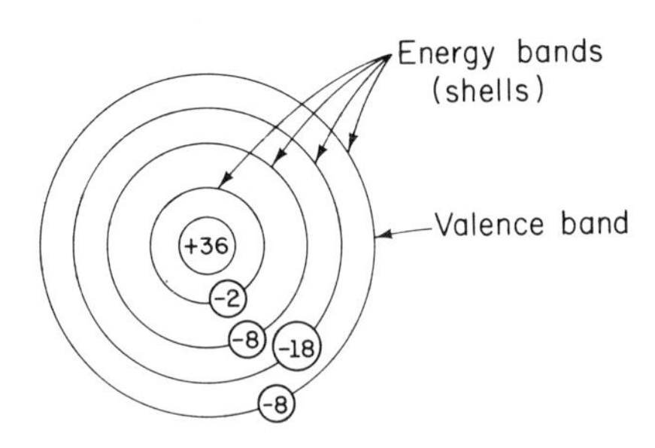


**Fig. 2-3.** A Balanced Atom.

Figure 2-4 is a partial list of the elements in the order of their atomic number.

The table shows the maximum number of electrons in each energy shell. Notice that the atomic number is the same as the maximum number of electrons the atom can contain. For instance, selenium has an atomic number of 34. It has four energy levels which contain a total of 34 electrons; they contain 2, 8, 18, and 6 electrons respectively. The valence shell of selenium is the fourth band which normally contains 6 electrons. This band is *not* full because it contains *less than 8 electrons.* In fact very few atoms in this group have a full valence shell. Among those that do, we have neon (number 10) and radon (number 86).

The valence structure of an atom is a reference to the number of electrons it normally contains in the valence shell. For instance, an atom with a trivalent structure contains three electrons in the valence shell. A pentavalent structured atom contains five valence electrons.

An atom is electrically neutral when it has the same number of electrons as protons. It is chemically stable when the valence shell is full (contains 8 electrons). An atom with less than eight electrons in the valence shell is chemically unstable, and will tend to either gain more electrons or give up some of its own valence electrons. When the valence shell is less than half full (less than four electrons), the atom gives up its few remaining valence electrons very readily. When the valence shell is more than half full (more than

| ATOMIC NUMBER | ATOM | SHELL | | | | | ATOMIC NUMBER | ATOM | SHELL | | | | | | |
|---|---|---|---|---|---|---|---|---|---|---|---|---|---|---|---|
| | | 1 | 2 | 3 | 4 | 5 | | | 1 | 2 | 3 | 4 | 5 | 6 | 7 |
| 1 | Hydrogen H | 1 | | | | | 50 | Tin Sn | 2 | 8 | 18 | 18 | 4 | 0 | 0 |
| 2 | Helium He | 2 | | | | | 51 | Antimony Sb | 2 | 8 | 18 | 18 | 5 | | |
| 3 | Lithium Li | 2 | 1 | | | | 52 | Tellurium Te | 2 | 8 | 18 | 18 | 6 | | |
| 4 | Beryllium Be | 2 | 2 | | | | 53 | Iodine I | 2 | 8 | 18 | 18 | 7 | | |
| 5 | Boron B | 2 | 3 | | | | 54 | Xenon Xe | 2 | 8 | 18 | 18 | 8 | | |
| 6 | Carbon C | 2 | 4 | | | | 55 | Cesium Cs | 2 | 8 | 18 | 18 | 8 | 1 | |
| 7 | Nitrogen N | 2 | 5 | | | | 56 | Barium Ba | 2 | 8 | 18 | 18 | 8 | 2 | |
| 8 | Oxygen O | 2 | 6 | | | | 57 | Lanthunum La | 2 | 8 | 18 | 18 | 9 | 2 | |
| 9 | Fluorine F | 2 | 7 | | | | 58 | Cerium Ce | 2 | 8 | 18 | 19 | 9 | 2 | |
| 10 | Neon Ne | 2 | 8 | | | | 59 | Praseodymium Pr | 2 | 8 | 18 | 20 | 9 | 2 | |
| 11 | Sodium Na | 2 | 8 | 1 | | | 60 | Neodymium Nd | 2 | 8 | 18 | 21 | 9 | 2 | |
| 12 | Magnesium Mg | 2 | 8 | 2 | | | 61 | Promethium Pm | 2 | 8 | 18 | 22 | 9 | 2 | |
| 13 | Aluminum Al | 2 | 8 | 3 | | | 62 | Samarium Sm | 2 | 8 | 18 | 23 | 9 | 2 | |
| 14 | Silicon Si | 2 | 8 | 4 | | | 63 | Europium Eu | 2 | 8 | 18 | 24 | 9 | 2 | |
| 15 | Phosphorous P | 2 | 8 | 5 | | | 64 | Gadolinium Gd | 2 | 8 | 18 | 25 | 9 | 2 | |
| 16 | Sulphur S | 2 | 8 | 6 | | | 65 | Terbium Tb | 2 | 8 | 18 | 26 | 9 | 2 | |
| 17 | Chlorine Cl | 2 | 8 | 7 | | | 66 | Dysprosium Dy | 2 | 8 | 18 | 27 | 9 | 2 | |
| 18 | Argon A | 2 | 8 | 8 | | | 67 | Holmium Ho | 2 | 8 | 18 | 28 | 9 | 2 | |
| 19 | Potassium K | 2 | 8 | 8 | 1 | | 68 | Erbium Er | 2 | 8 | 18 | 29 | 9 | 2 | |
| 20 | Calcium Ca | 2 | 8 | 8 | 2 | | 69 | Thulium Tm | 2 | 8 | 18 | 30 | 9 | 2 | |
| 21 | Scandium Sc | 2 | 8 | 9 | 2 | | 70 | Ytterbium Yb | 2 | 8 | 18 | 31 | 9 | 2 | |
| 22 | Titanium Ti | 2 | 8 | 10 | 2 | | 71 | Lutetium Lu | 2 | 8 | 18 | 32 | 9 | 2 | |
| 23 | Vanadium V | 2 | 8 | 11 | 2 | | 72 | Hafnium Hf | 2 | 8 | 18 | 32 | 10 | 2 | |
| 24 | Chromium Cr | 2 | 8 | 13 | 1 | | 73 | Tantalum Ta | 2 | 8 | 18 | 32 | 11 | 2 | |
| 25 | Manganese Mn | 2 | 8 | 13 | 2 | | 74 | Tungsten W | 2 | 8 | 18 | 32 | 12 | 2 | |
| 26 | Iron Fe | 2 | 8 | 14 | 2 | | 75 | Rhenium Re | 2 | 8 | 18 | 32 | 13 | 2 | |
| 27 | Cobalt Co | 2 | 8 | 15 | 2 | | 76 | Osmium Os | 2 | 8 | 18 | 32 | 14 | 2 | |
| 28 | Nickel Ni | 2 | 8 | 16 | 2 | | 77 | Iridium Ir | 2 | 8 | 18 | 32 | 15 | 2 | |
| 29 | Copper Cu | 2 | 8 | 18 | 1 | | 78 | Platinum Pt | 2 | 8 | 18 | 32 | 16 | 2 | |
| 30 | Zinc Zn | 2 | 8 | 18 | 2 | | 79 | Gold Au | 2 | 8 | 18 | 32 | 18 | 1 | |
| 31 | Gallium Ga | 2 | 8 | 18 | 3 | | 80 | Mercury Hg | 2 | 8 | 18 | 32 | 18 | 2 | |
| 32 | Germanium Ge | 2 | 8 | 18 | 4 | | 81 | Thallium Tl | 2 | 8 | 18 | 32 | 18 | 3 | |
| 33 | Arsenic As | 2 | 8 | 18 | 5 | | 82 | Lead Pb | 2 | 8 | 18 | 32 | 18 | 4 | |
| 34 | Selenium Se | 2 | 8 | 18 | 6 | | 83 | Bismuth Bi | 2 | 8 | 18 | 32 | 18 | 5 | |
| 35 | Bromine Br | 2 | 8 | 18 | 7 | | 84 | Polonium Po | 2 | 8 | 18 | 32 | 18 | 6 | |
| 36 | Krypton Kr | 2 | 8 | 18 | 8 | | 85 | Astitine At | 2 | 8 | 18 | 32 | 18 | 7 | |
| 37 | Rubidium Rb | 2 | 8 | 18 | 8 | 1 | 86 | Radon Rn | 2 | 8 | 18 | 32 | 18 | 8 | |
| 38 | Strontium Sr | 2 | 8 | 18 | 8 | 2 | 87 | Francium Fr | 2 | 8 | 18 | 32 | 18 | 8 | 1 |
| 39 | Yttrium Y | 2 | 8 | 18 | 9 | 2 | 88 | Radium Ra | 2 | 8 | 18 | 32 | 18 | 8 | 2 |
| 40 | Zirconium Zr | 2 | 8 | 18 | 10 | 2 | 89 | Actinium Ac | 2 | 8 | 18 | 32 | 18 | 9 | 2 |
| 41 | Niobium Nb | 2 | 8 | 18 | 12 | 1 | 90 | Thorium Th | 2 | 8 | 18 | 32 | 19 | 9 | 2 |
| 42 | Molybdenum Mo | 2 | 8 | 18 | 13 | 1 | 91 | Protactinium Pa | 2 | 8 | 18 | 32 | 20 | 9 | 2 |
| 43 | Techetium Te | 2 | 8 | 18 | 14 | 1 | 92 | Uranium U | 2 | 8 | 18 | 32 | 21 | 9 | 2 |
| 44 | Ruthenium Ru | 2 | 8 | 18 | 15 | 1 | 93 | Neptunium Np | 2 | 8 | 18 | 32 | 22 | 9 | 2 |
| 45 | Rhodium Rh | 2 | 8 | 18 | 16 | 1 | 94 | Plutonium Pu | 2 | 8 | 18 | 32 | 23 | 9 | 2 |
| 46 | Palladium Pd | 2 | 8 | 18 | 18 | 0 | 95 | Americium Am | 2 | 8 | 18 | 32 | 24 | 9 | 2 |
| 47 | Silver Ag | 2 | 8 | 18 | 18 | 1 | 96 | Curium Cm | 2 | 8 | 18 | 32 | 25 | 9 | 2 |
| 48 | Cadmium Cd | 2 | 8 | 18 | 18 | 2 | 97 | Berkelium Bk | 2 | 8 | 18 | 32 | 26 | 9 | 2 |
| 49 | Indium In | 2 | 8 | 18 | 18 | 3 | 98 | Californium Cf | 2 | 8 | 18 | 32 | 27 | 9 | 2 |

**Fig. 2-4.** Atomic Energy Levels.

four electrons), the atom tends to attract other electrons in an effort to completely fill the valence shell. Atoms with one, two, or three valence electrons are good conductor materials because their valence electrons are easily dislodged. Materials that are the best insulators have eight electrons in the valence shell. The semiconductor group of materials have atoms with their valence shells about half full. Thus, they are neither good conductors nor good insulators.

Going back to the chart of Fig. 2-4, let's select some sample materials for good insulators. Among the very best, there are neon, argon, and a few others with full valence shells. How about good conductors? They are plentiful. To name a few, we have hydrogen, aluminum, copper, silver, cadmium, gold, and mercury. All of these and many others have three or less electrons in their valence shells. Some of the best semiconductors are carbon, nitrogen, silicon, germanium, tin and lead. Notice that these have four electrons in their valence shells.

Only the electrons in the valence shells are available for current. In the electron tube, we discussed a surface barrier that had to be overcome in order to free an electron from the material. Here our problem is similar; we have a forbidden shell surrounding the valence shell. In order for an electron to take part in current, it must absorb enough energy to lift it through the forbidden shell. This is illustrated in Fig. 2-5.

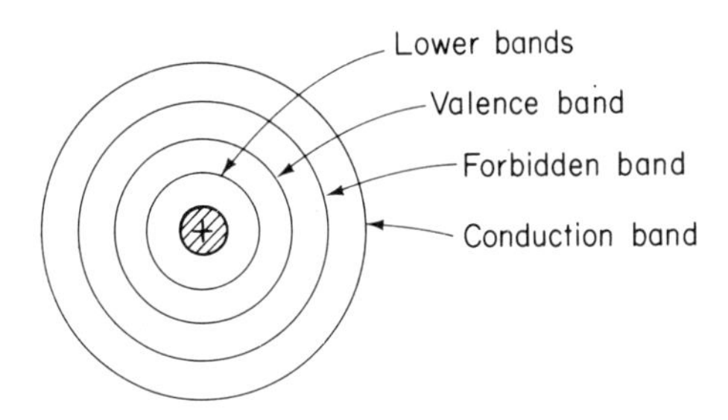


**Fig. 2-5.** Energy Shells.

The width of the forbidden shell is expressed in the number of electron-volts required to move an electron from the valence shell to the conduction shell. When the electron reaches the conduction shell, it is a free electron and may become a part of current. The atoms of insulators have a wide forbidden shell. The conductor atoms have a narrow forbidden shell. The forbidden shell for semiconductor atoms falls between these two extremes.

When an electron absorbs a specific amount of energy, it moves up to a higher shell. When it loses that same amount of energy, it drops down to a lower shell. Electrons cannot exist for any appreciable time in the area between two energy shells. In the valence shell, the electron has reached its maximum energy level for that atom. Any further absorbtion of energy will tend to lift the electron through the forbidden shell. When this occurs, the electron is free from attachment to any atom. It can now be directed as part of a useful current, and is said to be in the conduction shell.

### *crystal lattice structure*

A microscopic analysis of a piece of semiconductor material would reveal a crystal lattice structure. The structure is formed by covalent bonding between the atoms of the material. The lattice is usually composed of cubic or diamond cells; each corner of a cell is secured by the nucleus of an atom. These atoms are bound together because they share the valence electrons with adjacent atoms. This is illustrated in Fig. 2-6.

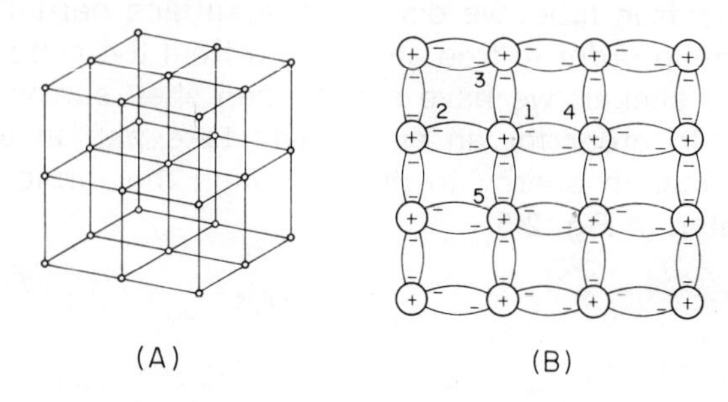


**Fig. 2-6.** Covalent Bonding.

Part A of this drawing shows eight cubes with an atom nucleus at each corner of each cube. Notice that all eight cubes are joined at the center by a single nucleus. The edges of these cubes represent the covalent bonding brought about by the adjacent atoms sharing the same valence electrons. The same idea is represented in a different fashion in Part B. Examine the area where the atoms are labeled 1, 2, 3, 4, and 5. Atom 1 has a full valence shell, but it retains this condition by sharing valence electrons. It shares two electrons each with atoms 2, 3, 4, and 5. This covalent bonding effectively gives each atom a full valence shell. The fact that each atom uses its own four valence electrons, and borrows four more (on a time sharing basis) from its neighbors makes no difference. The valence shells overlap, but they are all full. The material is chemically stable.

### *movement of electrons*

Even though a lattice structure is stable, thermal agitation can break the covalent bond and move electrons into the conduction shell. When this happens to a single electron, a hole is left in the valence shell of two atoms. This electron will drift at random within the material until it finds a similar vacancy created by another electron. Therefore, each freed electron leaves a hole, and each hole is in turn filled by another free electron. This action is illustrated in Fig. 2-7. Within a crystal lattice the holes and electrons drift about in a random fashion.

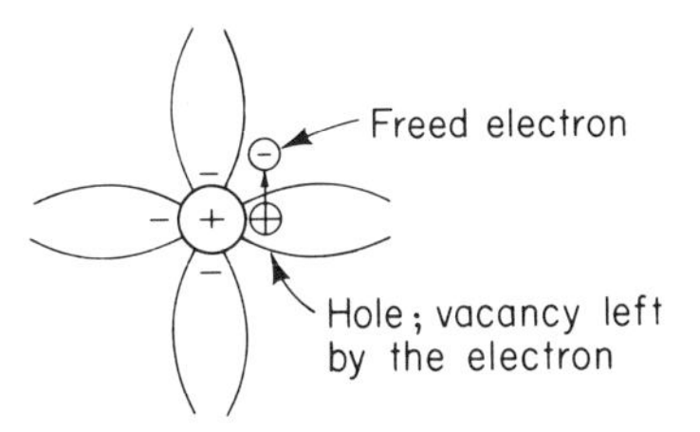


**Fig. 2-7.** Electron Freed from Bond.

The vacancy in the valence shell, which was created when the electron broke the bond, is a small, localized positive charge. It is commonly referred to as a hole. It exerts an attraction to any drifting electron. When an electron settles into this hole, the covalent bond is again complete, and the hole has been filled. These electrons, and holes, are referred to as current carriers.

### *improving the current carriers*

Two materials which are commonly used in solid-state devices are germanium and silicon. Each atom of these materials contains four valence electrons. The crystal lattice of silicon or germanium is formed by covalent bonding as previously described. With the covalent bond, each atom has eight electrons in its valence shell, and the material is stable.

Doping is a process of growing crystal lattice structures with other materials added in order to create either an excess of valence electrons or a shortage of valence electrons. Do not misunderstand, the overall charge of the material is still neutral. But the material with the overflowing valence shells will tend to give up electrons while the material with holes in the valence shell will tend to collect free electrons.

Atoms of indium have trivalent structures. That is, each atom contains three electrons in the valence shell. When indium is added to germanium or silicon during the growth process, the indium atoms become part of the crystal structure. But now the covalent bonds will have holes. The material will now tend to adopt free electrons from any source in an effort to fill these holes. For this reason, crystals formed in this fashion are called acceptor materials.

Arsenic atoms have pentavalent structures; they have five electrons in the valence shell. When arsenic is added to germanium or silicon during the growth process, the arsenic atoms become part of the crystalline structure. But now, the valence shells are overflowing. Every arsenic covalent bond has an extra electron. This type of material tends to give up electrons, and is called donor material.

Relatively small amounts of arsenic or indium can create the effects described here. When they are used in this fashion, they are called impurities; not that either material is impure. They are impurities only from the standpoint of the crystal structure which would otherwise be either pure germanium or pure silicon. Other impurities which produce donor materials are phosphorus and antimony. In addition to indium, galium and boron also produce acceptor materials.

The acceptor materials are said to contain positive carriers. This refers to the holes which readly accept electrons. This material is commonly known as *p*-type material. The *p* signifies that it uses positive carriers; the material is still neutral. Although *p*-type material has very few free electrons, the excess of holes creates an unstable condition. The doping with the impurity has narrowed the forbidden shell. Electrons break the covalent bond and drift about from one bond to another.

. The donor material is said to use negative carriers, and it is commonly called *n*-type material. The *n* signifies that negative carriers are used. This material has many free electrons, and this causes it to be unstable. The doping with the impurity narrowed the forbidden shell and makes freeing of electrons a very easy matter. These free electrons drift about in a random fashion.

### *directing the carriers*

The random drifting of holes and electrons can be turned to a definite direction by applying a voltage across a semiconductor material. Let's examine the *p*-type material under the conditions shown in Fig. 2-8.

This material is designed to have very few free electrons and a great many holes. As a result, the valence electrons have little tendency to move. The circles represent the stable portion of the atoms. The plus signs (+) signify the positive holes in the covalent bands. Applying a voltage in this fashion forces current through the material. Electrons from the battery pass through the material from left to right. In passage, they move from one hole to the next. As a result, the electrons move in one direction, and the holes (positive charges) appear to move in the opposite direction. Since there are always more holes than electrons, it appears that the holes are actually flowing toward the negative terminal of the battery. Thus, the name of positive carrier or *p*-type material. The holes carry the current.

Application of a voltage to the *n*-type material is illustrated in Fig. 2-9.

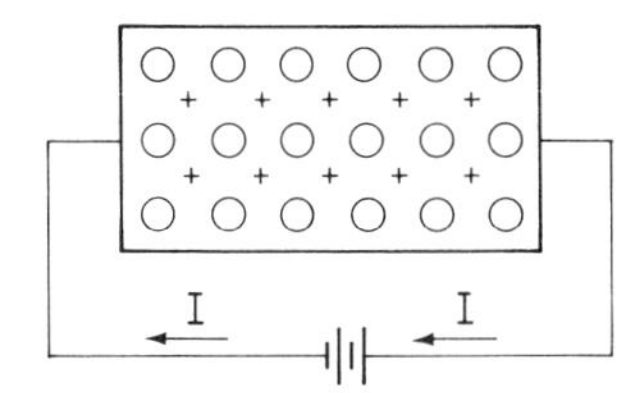


**Fig. 2-8.** Voltage Across *p*-Type Material.

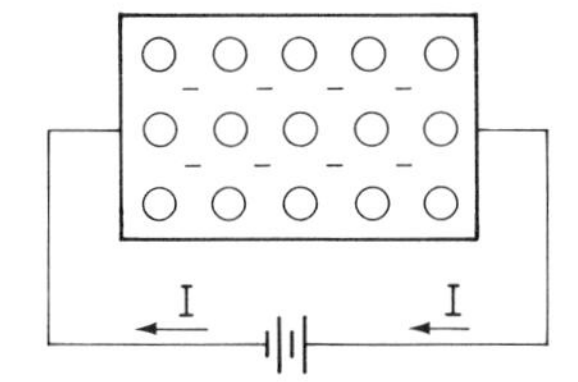


**Fig. 2.9.** Voltage Across *n*-Type Material.

In this case, the circles indicate the stable portion of the atom while free electrons are indicated by minus signs (—). This material is designed to have no holes and many free electrons. The positive terminal of the battery attracts these free electrons from the material. Other electrons from the negative terminal of the battery move into the material to replace the electrons that are moving out. Thus, the name of negative carrier or *n*-type material. The electrons carry the current. Let's see what happens when the two types are joined.

## SOLID-STATE DIODES

The solid-state diode does all the work formerly accomplished by an electron tube diode. Yet it is simply a very small chunk of apparently solid material. It uses very little power, has a long life expectancy, can withstand strong physical shocks, and uses very little space. Let's see how this amazing device is made.

### *construction*

The solid-state diode is created by joining *p*- and *n*-type materials. This is not a mechanical joining but a chemical one. It is actually a growing or heat fusion process in which the two types of materials join together in a single crystal structure. In the joining process, electrons from the *n* section move across the junction and fill in holes in the *p* section. The result is a chemically stable area near the junction which has an absence of both holes and free electrons. This area is called a *depletion region* and is illustrated in Fig. 2-10.

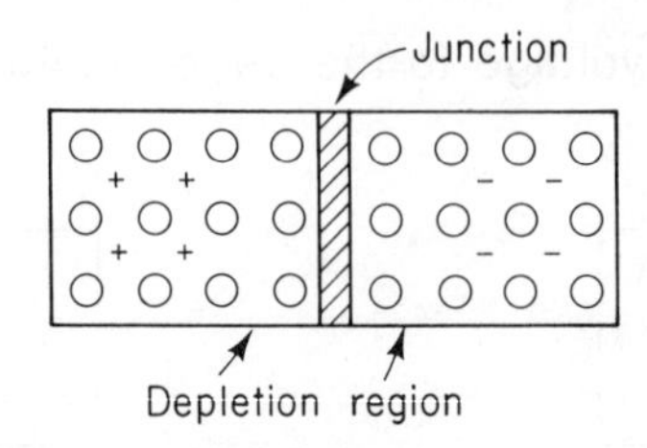


**Fig. 2-10.** Junction of *p* and *n* Material.

From the area near the junction, free electrons from the *n* material have crossed the junction and filled in the nearest holes of the *p* material. This has resulted in an area of stable material (full valence shells) which extends to either side of the junction. Remember that good insulating material is composed of atoms with full valence shells. This depletion region has become good insulating material.

***forming the junction***

There are three basic techniques for joining *n* and *p* materials. These are diffusion, growth, and alloy. All utilize heat to control the process. First a rod or ingot of a basic material (*n* or *p*) is grown from a pot of properly doped, molten material. A seed of the same type is then lowered into the melt and slowly withdrawn. A solid crystal of the material then grows on the seed as it moves upward. This is illustrated in Fig. 2-11.

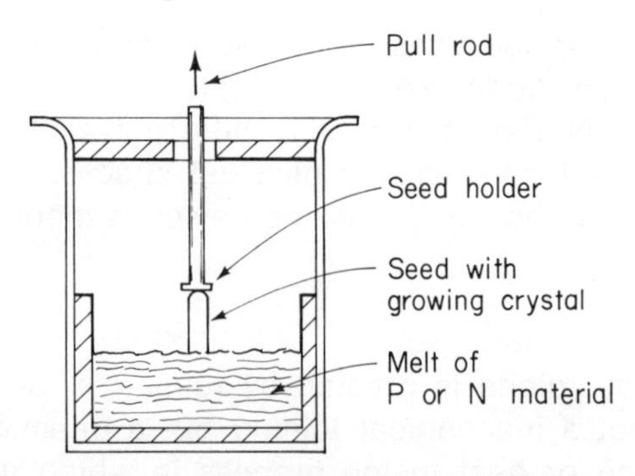


**Fig. 2-11.** Growing the Crystal.

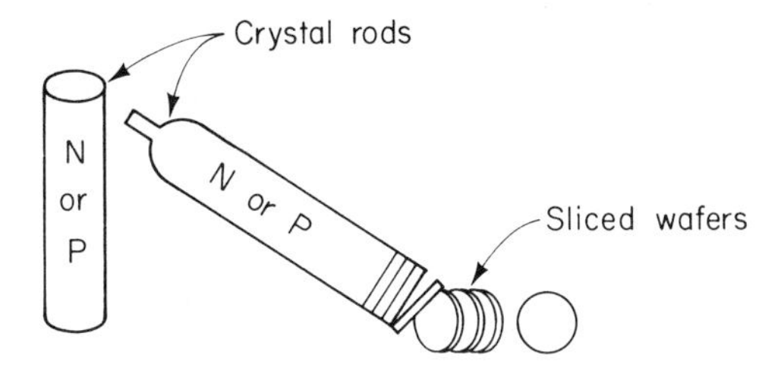


**Fig. 2-12.** Slicing a Basic Material.

Long, slim rods of the basic crystal can be grown in this fashion. The rod is then sliced into thin wafers as illustrated in Fig. 2-12. Each of these wafers may now have a part, or all, of its surface joined with a material of the opposite type.

***diffusion***

Wafers can be enclosed in an envelope which is filled with a gas of the opposite type material. The envelope is then placed into a carefully controlled fusion oven. The temperature approaches, but does not quite reach the melting temperature of the wafers. This temperature is held for several hours. During this time, the entire surface of the wafers becomes diffused with the gas, as illustrated in Fig. 2-13.

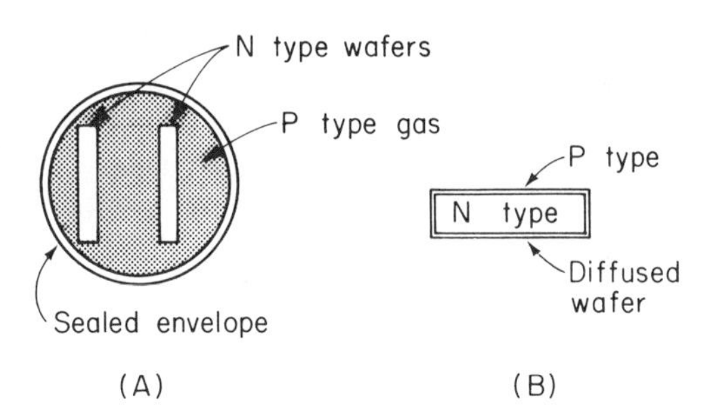


**Fig. 2-13.** Diffusion Process.

During the fusion process the *n*-type crystal wafers are in a container of *p*-type gas as shown, or it can be *p*-type crystal wafers in a container of *n*-type gas. When the wafer is removed from the furnace a thin layer of the entire surface has been diffused with the gas. The wafer is now *n*-type joined with *p*-type. This can be sliced again to produce many diodes with *n–p* junctions.

In the growth process, a common melt may contain both *p* and *n* impurities. The seed is slowly withdrawn from the melt in steps synchronized with exact heat variations. At one temperature, a layer of *p*-type material grows; at another temperature, a layer of *n*-type material grows. In this fashion alternate layers of *p*-type and *n*-type can be grown into a single rod of crystal. This process is illustrated in Fig. 2-14.

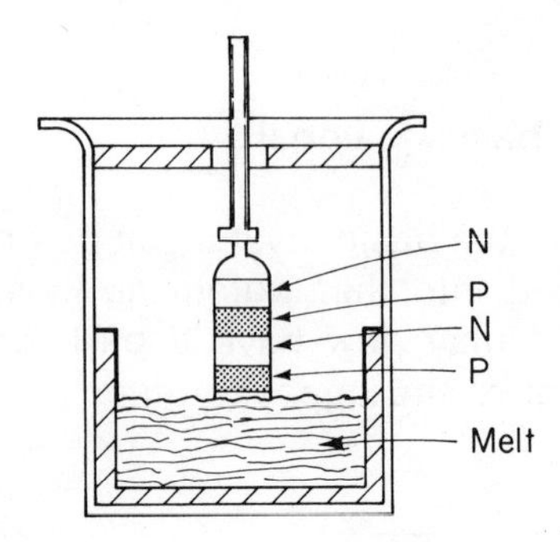


**Fig. 2-14.** Growing Junctions.

The crystals of the two types form at different temperatures. The exact amount of heat for each is determined by the material and the doping impurities. This rod can be sliced into small cubes and made to produce thousands of *p–n* junction diodes.

In the alloy process, a small bit of impurity material is placed on a wafer. Heat is then applied to melt the impurity but not the wafer, as illustrated in Fig. 2-15.

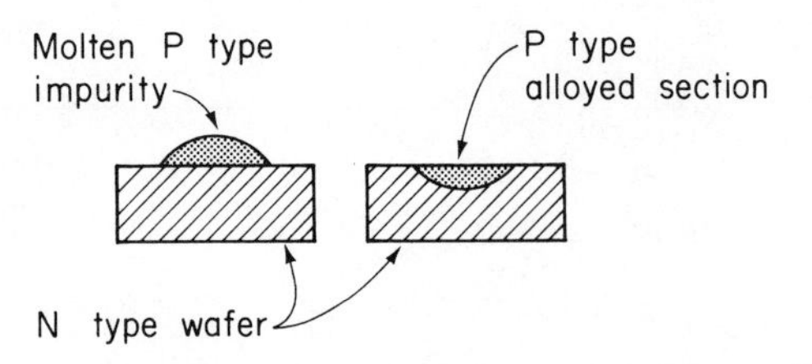


**Fig. 2-15.** Alloy Junction.

The molten material penetrates a section of the wafer and combines with that material to form an alloy. The alloyed section is one type material and the wafer is the other.

### *diode bias*

Connecting a battery across a junction diode will cause electrons to cross the depletion region and set up a current. The current from *n* to *p* can be obtained easily, but in the opposite direction, there can be very little current. The ratio here is more than 10 to 1, and in some cases more than 1000 to 1. It is referred to as majority current and minority current. The majority current is from *n* to *p,* and the minority current is from *p* to *n* as illustrated in Fig. 2-16.

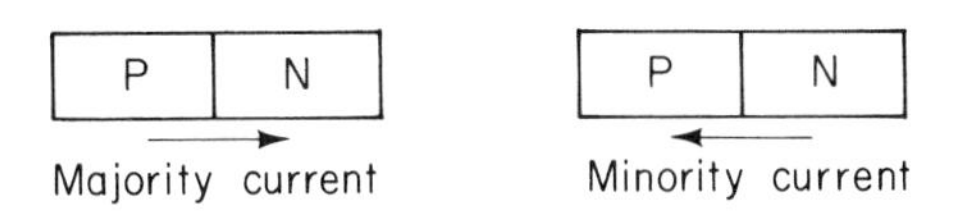


**Fig. 2-16.** Direction of Minority and Majority Currents.

Applying a potential in either direction across a diode will regulate the current. Therefore, such a potential is called bias. When the bias is aiding minority current, it is called *reverse bias.* When it aids majority current, it is called *forward bias.* Both types of bias are illustrated in Fig. 2-17.

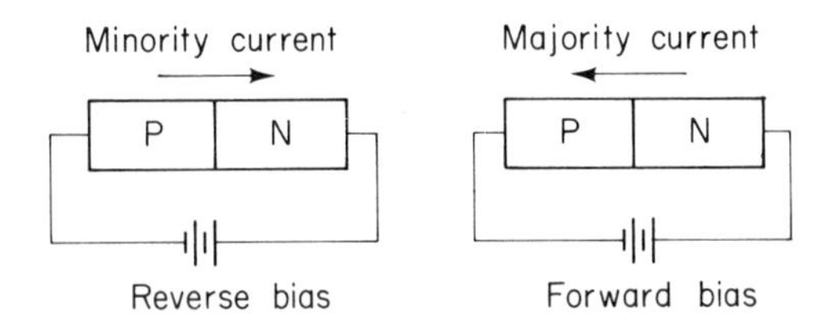


**Fig. 2-17.** Application of Forward and Reverse Bias.

The application of reverse bias produces only minimal current. The minority current is so low that, in most cases, it can be ignored. This is equivalent to trying to force current through an electron tube from plate to cathode. The difference is the slight current in the solid-state diode. A reverse biased diode is considered to be in a cutoff condition.

When forward bias is applied, it aids the majority current, and results in a heavy current. This is equivalent to the action of an electron tube diode when its plate is positive. A forward biased diode is in a conducting condition. The current through the diode is directly proportional to the forward bias potential. Care should be exercised not to exceed the rated current. Otherwise the diode may be damaged.

A schematic symbol for a diode is shown in Fig. 2-18. The arrow represents the *p*-type material. When the diode is forward biased (*p* positive and *n* negative), current direction is against the arrow. With reverse bias, the minute current is with the arrow or not at all.

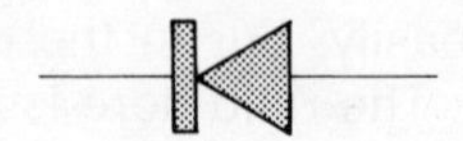

**Fig. 2-18.** Symbol of Solid-State Diode.

***characteristics***

Most of the characteristics have been mentioned. The solid-state diode functions almost like an electron tube diode. But the solid-state diode does not need to be heated. One other point bears a more careful examination. This is the current characteristic. Figure 2-19 illustrates current in both directions.

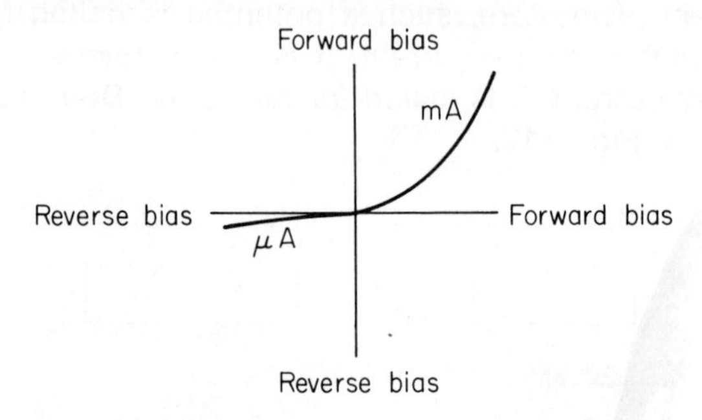


**Fig. 2-19.** Diode Current.

Over the normal operating range of forward and reverse bias, forward current is very prominent. The reverse current is so small that it can be ignored. As we mentioned earlier, the forward current is directly proportional to the amplitude of forward bias. Care must be exercised not to exceed the rated current value.

On the other hand, if reverse bias is increased, there will be a point where the reverse current takes a sharp increase. This is illustrated in Fig. 2-20.

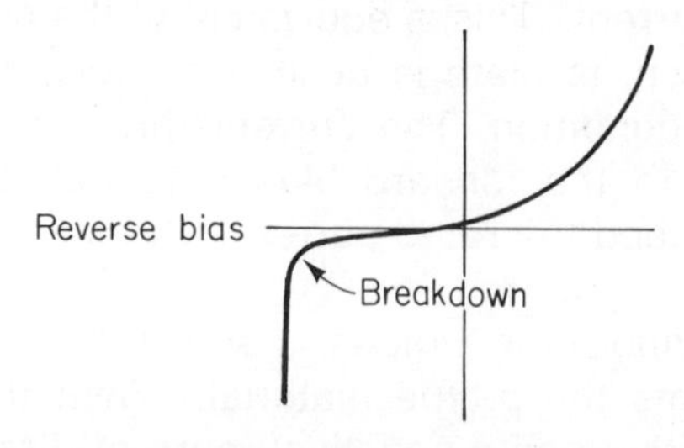


**Fig. 2-20.** Sharp Increase in Reverse Current.

The knee on the current curve represents the point where the crystal breaks down. The voltage at this point is called the *breakdown voltage.* The knee on the curve is called the *avalanche breakdown.* An ordinary diode is subject to be greatly damaged when breakdown occurs. Some special diodes are designed to operate in the breakdown region. One of these is called a Zener diode. We will return to it in a little while.

### *diode rectifiers*

Since current passes readily in one direction and hardly at all in the other, the solid-state diode is a very effective rectifier. Figure 2-21 illustrates the rectifier action. Notice that the input ac sine wave is changed to a pulsating, positive dc at the output.

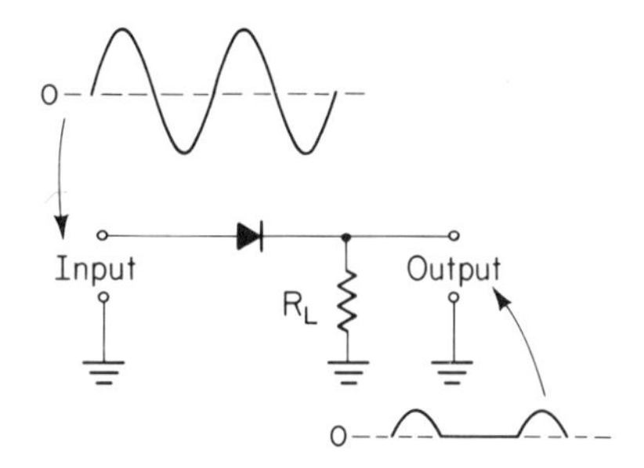


**Fig. 2-21** Rectifier Circuit.

The positive alternations of the input signal place forward bias on the diode. Forward current from ground passes through $R_L$ and through the diode against the arrow. The amplitude of this current is directly proportional to the amplitude of the positive portion of the input. The output is the voltage wave shape across $R_L$ with respect to ground. A positive alternation of the input produces a positive pulse out.

The negative alternation of the input places reverse bias on the diode. Current ceases, and there is no voltage across $R_L$. In effect, the positive alternations are passing through, and the negative alternations are being eliminated.

Reversing the diode connections would produce the opposite effect. Then, the negative alternations would pass through. This is illustrated in Fig. 2-22.

In this case, the positive alternation cuts the diode off with reverse bias. The negative alternation applies forward bias. During

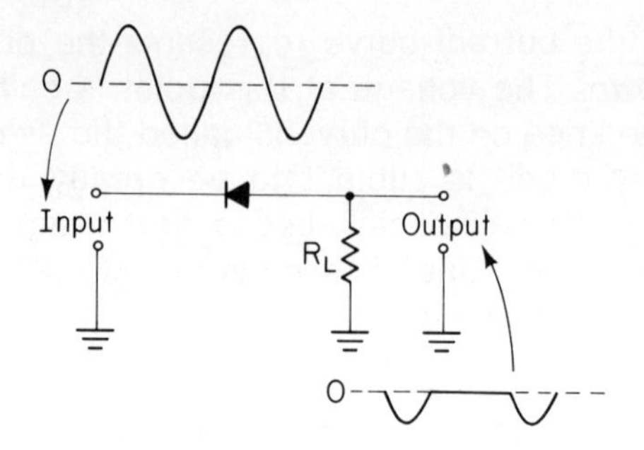


**Fig. 2-22.** Reversed Diode.

the negative alternation, current is through the diode (against the arrow) and through $R_L$ to ground. This develops a negative pulse of voltage across $R_L$ (with respect to ground) for each negative alternation of the input.

Average voltage in one alternation of a sine wave is 0.636 × peak value. A rectifier like those in Fig. 2-21 and 2-22 produce one pulse out for each sine wave in. This means that the pulsating (or ripple) frequency out is equivalent to the input frequency. Since only half of the alternations appear in the output, the output voltage is equivalent to half the value of the average voltage for one alternation. Peak output equals peak input minus drop across the diode.

$$f_{\text{in}} = \text{pulsating frequency out}$$

$$\text{Peak } E_{\text{out}} = \text{Peak } E_{\text{in}} - \text{peak } E_{\text{diode}}$$

$$\text{Average dc} = \frac{0.636 \times \text{peak } E_{\text{out}}}{2}$$

**Problem:** What is the polarity, pulse frequency, peak amplitude, and average dc level of the output from the rectifier in Fig. 2-23.

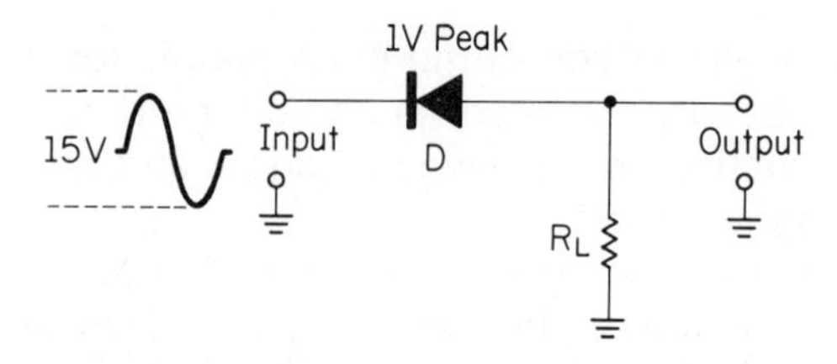


**Fig. 2-23.** Determine Nature of the Output.

Solution:

The diode passes negative alternations, so the output is negative pulses. Pulse frequency $= f_{in} = 400$ pulses/s. Peak input $= 7.5$ V. Peak output $= 7.5\text{ V} - 1\text{ V} = 6.5$ V.

$$\text{Average dc out} = \frac{0.636 \times \text{peak } E_{out}}{2}$$
$$= \frac{0.636 \times 6.5}{2}$$
$$= 2.067\text{ V}$$

## THE ZENER DIODE

In 1934 a scientist by the name of C. Zener advanced a theory of electrical breakdown in solids. He pointed out that, under a certain field intensity, the carriers can cross the junction through a process of quantum mechanics. It appeared as if the carriers had tunneled through the barrier. Once breakdown occurred, it rapidly developed into an avalanche of current. Of course, he was speaking of current on the order of a few microamperes, but this was reverse current. The Zener diode is designed to operate on this premise.

This diode is operated on reverse bias. The proper operating bias produces current in the center of the breakdown region. Since the Zener diode was the first device that operated under these conditions, a part of the reverse current curve is known as the Zener region. Figure 2-24 illustrates the zener region and the operating bias. A symbol for the Zener diode is shown in the circuit of Fig. 2-25. In this circuit, the positive dc input places reverse bias on the Zener diode. Avalanche breakdown occurs with current with the arrow to the level shown in Fig. 2-24.

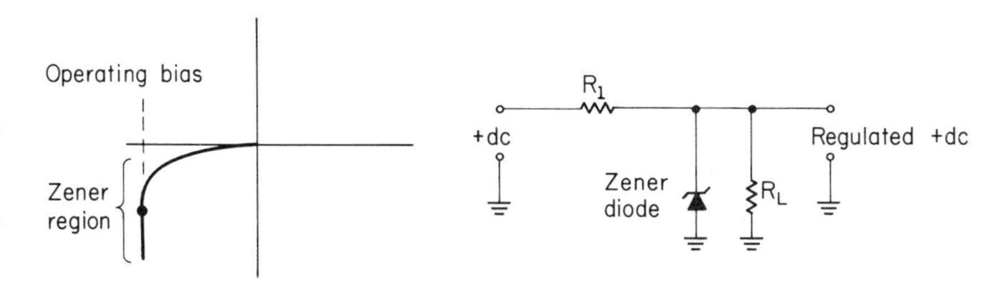


**Fig. 2-24.** Zener Region. **Fig. 2-25.** Zener Diode Regulator.

Current through $R_L$ develops a positive voltage at the output. Any variation of the amplitude of the +dc input causes a change in the Zener bias. For instance, when the input increases, the Zener diode conducts more. This causes the increase in voltage to be dropped across $R_1$. When the input decreases, the reverse action occurs. The Zener diode conducts less, and this causes less voltage drop across $R_1$. The result is a constant current through $R_L$ and a steady (regulated) output. Regulating voltage in this fashion is one of the primary uses of a Zener diode.

## TRANSISTORS

When crystal is grown or fused in a manner which places a thin section of donor material between two pieces of acceptor material, a transistor is born. Placing the materials in reverse order (acceptor between two donors) creates the same effect. This is illustrated in Fig. 2-26.

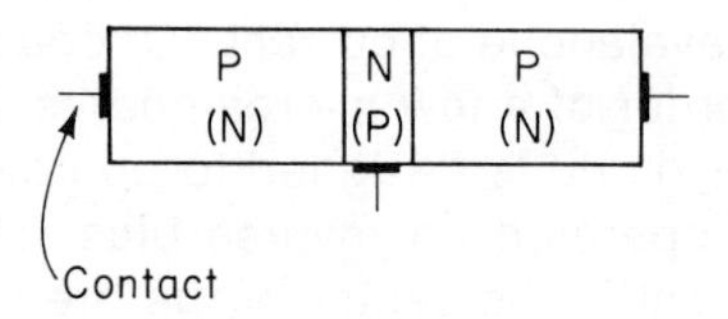


**Fig. 2-26.** Three Element Transistor.

This device, which ranges in size from very small to microscopic, performs all the functions of a triode electron tube. Figure 2-27 compares size and shape of several tubes and transistors. This is a random selection of both. They are shown approximately

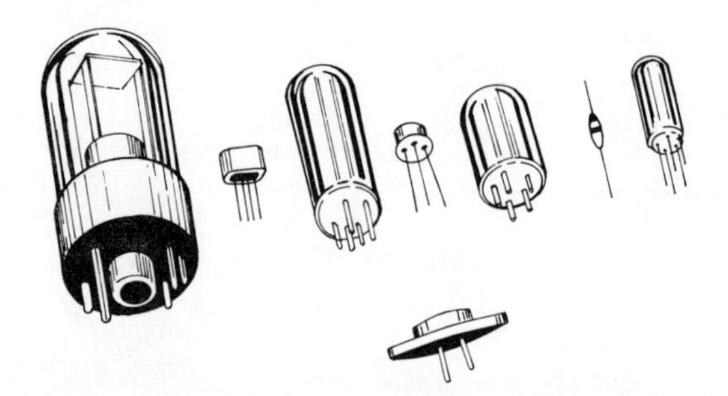

**Fig. 2-27.** Comparison of Tubes and Transistors.

actual size. Therefore it would be impossible to include either the largest tube or the smallest transistor. However, Fig. 2-28 does move us in that direction.

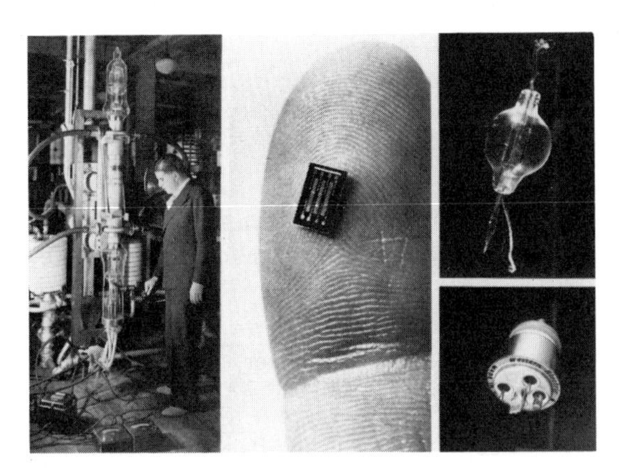

**Fig. 2-28.** The Large, Small, Old, and New. (Photograph and description Courtesy Bell Laboratories.)

The left side of this picture shows a 250-kW radio tube. It is one of the largest ever built. The center displays a semiconductor wafer which can contain many transistors. Upper right is a picture of the DeForest triode; the first triode. Lower right is a modern microwave amplifier tube.

### *types of triode transistors*

As previously illustrated in Fig. 2-26, there are two types of triode junction transistors. In either case, a thin layer of one type of material separates two pieces of the other type of material. The type is determined by the arrangement of the materials. The *npn* has the *p*-type material in the center with *n*-type material to either side. The *pnp* has the *n*-type material in the center and *p*-type material to either side.

The thin layer of material in the center is called the base. One end emits current carriers into the base and is logically called the emitter. The other end collects current carriers from the base, and it is called the collector. These three parts are standard. Figure 2-29 illustrates the parts for both types of triodes.

There are other types of transistors, such as tetrodes and field effect, but they are encountered with far less frequency than the triode.

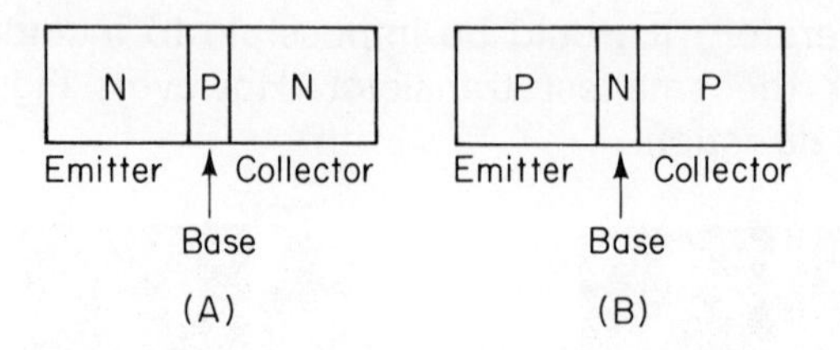


**Fig. 2-29.** Parts of a Triode.

Each type of transistor has its own schematic symbol. The symbols for both types of triodes are shown in Fig. 2-30.

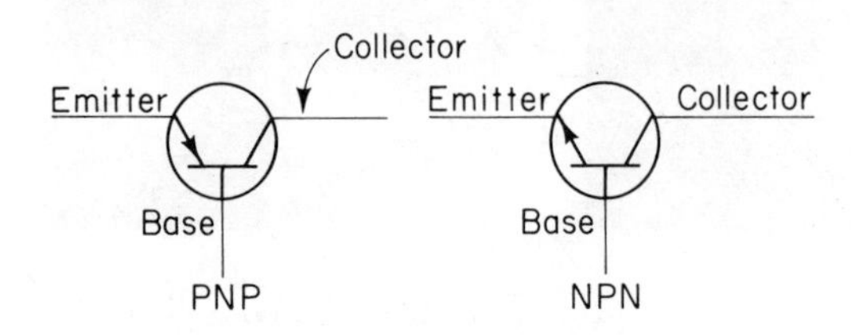


**Fig. 2-30.** Triode Schematic Symbols.

The element with the arrow is the emitter in either case. The flat surface is the base, and the remaining element is the collector. The only difference between the two symbols is the direction of the emitter arrow. This arrow always points toward *n*-type material. If you will remember this, you will have no trouble keeping the symbols straight. If the arrow points to the base, the base is *n*-type material; the transistor is *pnp*. When the arrow points away from the base, the base is *p*-type material; the transistor is *npn.*

### *characteristics*

The triode transistor is nothing more than two solid-state diodes. It has one diode section consisting of the emitter-base junction. The other diode section is the base-collector junction. The versatility of the triode comes from the fact that it can be biased in a manner which will allow the input signal to one diode section to control the current through the other diode section. Actually this is similar to the action in the triode electron tube, when signal between grid and cathode controls the current between cathode and plate.

Transistor junctions are formed in two ways. They can be drawn from a crystalline substance during the growing process.

They can also be formed by placing a spot of one type material on a piece of another type of material. The spot would then be diffused into the larger section by a heat process.

Whether grown or diffused, the emitter and collector sections are *not* interchangeable. In the growing process, the collector has less conductivity than the emitter. In the diffusion process, the collector has a larger area than the emitter.

What constitutes proper bias? That depends upon the type of triode. The input section is the emitter-base diode. The output section is the base-collector diode. The input section (emitter-base) is forward biased while the output section (base-collector) is reverse biased. These bias arrangements are different for the two types of triodes.

## *operation of the npn*

The diagram in Fig. 2-31 shows the proper bias arrangements for an *npn* transistor.

The emitter is similar to the cathode in a tube. It carries the total circuit current. Internally, this current is divided between the collector and the base with the collector getting the larger share. The division is approximately 98 percent to the collector with the remaining 2 percent to the base. This is illustrated in Fig. 2-32.

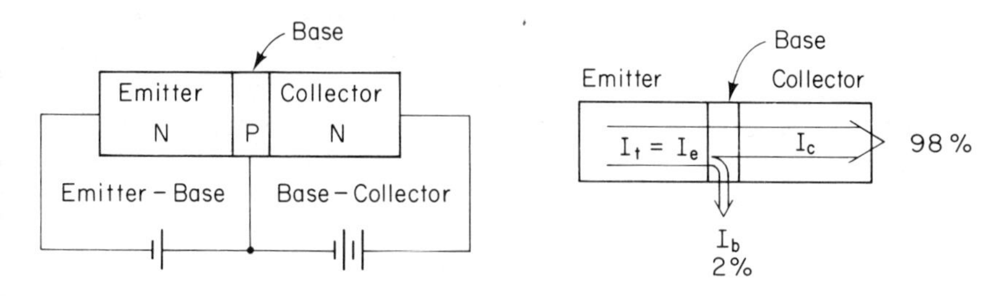


**Fig. 2-31.** Properly Biased *npn*. **Fig. 2-32.** Current Division in the *npn*.

Since the emitter current is also total circuit current, the emitter-base bias is a direct control over all currents. Since it is forward bias, the emitter current ($I_e$) is directly proportional to the amount of voltage applied between the base and the emitter. The collector current ($I_c$) accounts for approximately 98 percent of the emitter current. The other 2 percent is diverted through the base contact as base current ($I_b$). $I_e = I_c + I_b$; $I_e = I_t$. This same division is relatively constant for all levels of operating bias.

Converting our diagram to a schematic and adding provisions for input and output, produces Fig. 2-33.

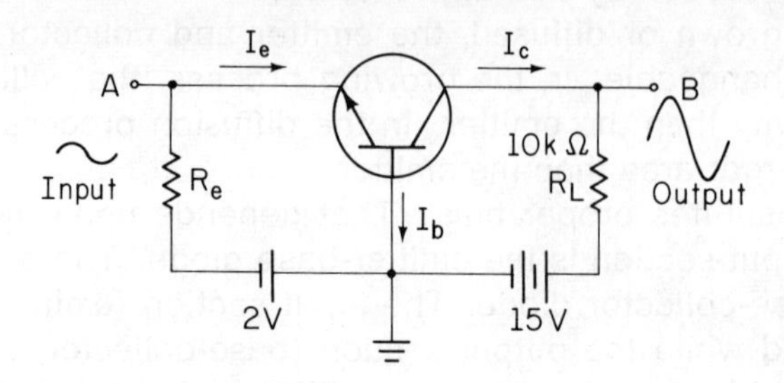


**Fig. 2-33.** *npn* Schematic.

The input signal increases the forward bias between base and emitter when it swings negative. It decreases this bias when it swings positive. A small change in emitter-base bias causes a relatively large change in emitter current. About 98 percent of the change in emitter current is reflected in the collector current. Since $R_L$ is a relatively large resistor, a small change in $I_c$ produces a large change in the voltage across $R_L$. The output signal will be an exact duplicate of the input except that it will have a greater amplitude.

Let's examine that action in a little more detail. Suppose that the values of bias, signified in Fig. 2-33, causes the transistor to conduct so that $I_c$ is 1 mA. $R_L$ is 10 kΩ, and 10 of the 15 V are dropped across $R_L$. The output from point *B* includes both the battery voltage and the voltage drop across $R_L$. To be more exact, it is the difference between these two voltages. With the 10-V drop across $R_L$, the voltage from *B* to ground is the battery voltage minus the drop across $R_L$. In this instance, $E_{out}$ is 5 V.

If a small sine wave is applied to point *A* (between *A* and ground), it will vary the forward bias on the input (emitter-base) junction. When the input goes positive, it decreases this bias. Since the emitter to base potential is negative, reducing the bias is moving the potential in a less negative direction. As the emitter to base potential becomes less negative, the emitter current decreases. If the positive portion of the input signal is 2 V in amplitude, it will completely cancel the 2 V of forward bias and stop the emitter current completely.

As the emitter current decreases, collector current decreases. The decrease in collector current causes less voltage drop across $R_L$. The voltage across $R_L$ cancels less of the battery voltage, and the potential at point *B* swings in a more positive direction. If $I_c$

drops to 0.5 mA, voltage across $R_L$ will be 5 V. $E_{out} = 15\text{ V} - 5\text{ V} = 10\text{ V}$. So, with a decrease in collector current from 1 to 0.5 mA, the voltage at point *B* rises from 5 to 10 V.

When emitter current stops completely, collector current is also zero. With no current through $R_L$, the potential at point *B* will be the same as the battery potential (15 V).

As the input signal drops back to zero, the bias and conduction of the transistor return to their original conditions. On the negative alternation of the input, the forward bias is increased; that is, the emitter to base potential becomes more negative. This causes an increase in emitter current. About 98 percent of this increase is reflected in the increase of the collector current. The increase in collector current causes more voltage to drop across $R_L$, and the output decreases.

When the input becomes sufficiently negative, the transistor will saturate. At saturation, the voltage across $R_L$ will be almost as large as the collector bias voltage. This will reduce the output to almost zero. Thus, a 4-V (peak to peak) signal at the input will produce a 15-V (peak to peak) signal at the output. Input and output signals will be of the same phase and frequency.

The example used here, which carried the transistor all the way from cut off to saturation, would not give a faithful reproduction of the input signal. In order to amplify a sine wave without drastic distortion, the transistor would need to stay in the center of its operation range. Specific amplifier functions will be covered in a later chapter. The intent here is to visualize the current–voltage swings in response to a changing input.

One of the principle uses of the transistor is its function as an electronic switch. When so used, it is operated in two states: saturation and cutoff. The input section is held cut off by a very small reverse bias. The activating input signal is a negative pulse. During the input pulse time, the transistor is driven to saturation. When the pulse passes, the transistor reverts to its static cutoff condition. The output then drops from 15 V at cut off to nearly 0 V during saturation. The output goes back to 15 V as soon as the input pulse has terminated.

### *operation of the pnp*

The *pnp* functions much the same as the *npn,* but the arrangement of the materials are reversed. This makes it necessary to reverse both bias batteries in order to accomplish conduction. Figure 2-34 illustrates the proper biasing for a *pnp* transistor.

The emitter still carries the total circuit current, but now all currents are reversed. Electrons flow out of the emitter and into both base and collector. Again collector current is about 98 percent of emitter current which leaves 2 percent for the base current. Current division is illustrated in Fig. 2-35.

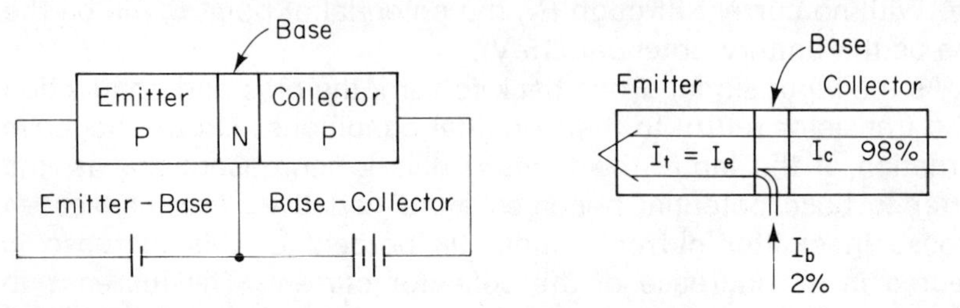


**Fig. 2-34.** Properly Biased *pnp.* **Fig. 2-35.** Current Division in the *pnp.*

The emitter-base junction is still the input section with the output section being the base-collector junction. The input section is forward biased while the output section is reverse biased. This is the same as the *npn* in principle, but bias polarities and current direction are reversed as shown in Figs. 2-34 and 2-35. Going to a schematic diagram, we have the circuit in Fig. 2-36.

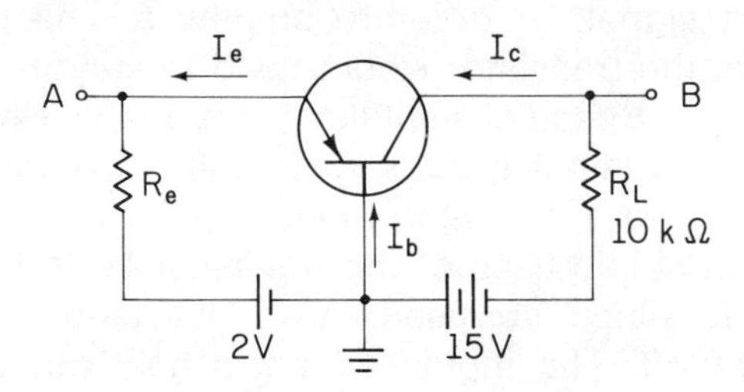


**Fig. 2-36.** *pnp* Schematic.

The input signal varies the bias in exactly the same fashion as described for the *npn* transistor except for polarity. In this case, a positive going input increases the forward bias between emitter and base. A negative going input decreases the emitter-base bias. Changing this bias controls the emitter current, and emitter current is the same as total circuit current.

Suppose that, under the bias conditions shown in Fig. 2-36, $I_c$ is 1 mA. The voltage drop across $R_L$ is 10 V with negative at the top. This 10 V subtracts from the 15 V of the battery and leaves a —5-V output at point *B* (with respect to ground).

Assuming a 4-V peak to peak input, the transistor can be driven to cut off and back to saturation. The positive swing increases the bias and drives the transistor to saturation. The voltage drop across $R_L$ is almost equal to the battery voltage. The output at point *B* is nearly zero potential. So, on the positive swing, $I_c$ increases, and the negative output decreases. That is, the output swings from a —5 V to nearly zero.

When the input swings negative, the forward bias on the emitter-base junction is reduced to zero. This causes the emitter current to cease and stops the collector current as well. With no current through $R_L$, the output at point *B* is the same as the battery potential. In respect to ground, this is —15 V. So, a 4-V (peak to peak) signal at the input cause a 15-V swing at the output. In this instance, the entire output is negative with respect to ground.

If the *pnp* circuit of Fig. 2-36 is used as an electronic switch, we would reverse the emitter bias. It will produce a —15-V output as long as the transistor is held cut off. The output will rise to some less negative potential during conduction.

## JUNCTION TEMPERATURE

When we pass current through a solid-state device, it becomes warm to the touch. This is caused from the carriers recombining at the junction and releasing energy in the process. When the energy is relased faster than the surrounding (ambient) air can dissipate the heat, the junction temperature rises. We must not exceed the maximum operating junction temperature if we expect sustained service from our semiconductor.

### *power limitation*

Maximum operating junction temperature is approximately 100 °C for germanium and 175 °C for silicon junctions. This temperature limits the junction power. Since the ambient air tends to dissipate the heat, the junction temperature is in respect to the ambient air temperature. For example; we have a transistor which is rated at 250 mW at 25 °C. This indicates that it can handle 250 mW of power as long as the ambient air temperature does not exceed 25 °C. The same transistor has a power derating value in mW/deg. If this rating is 2 mW/°C, and we operate in an ambient temperature of 50 °C, the safe junction power will be 200 mW. If this power ($E \times I$ at the junction) is exceeded, our transistor becomes too hot.

and damage will probably result. If our ambient temperature is 150 °C, any power at all will probably cause damage.

$$150° - 25° = 125° \text{ above the specified temperature}$$

$$\text{Safe power} = \frac{250\text{ mW} - 2\text{ mW}}{°\text{C}} = 250\text{ mW} - 250\text{ mW} = 0$$

### *thermal resistance*

Junctions are covered with metal cases or some other type of protective package. These packages are small and offer very little cooling surface to radiate the heat. This means that we have opposition to the transfer of heat from the junction. We refer to this opposition as *thermal resistance.*

Total thermal resistance between junction and ambient air is composed of two individual resistances: *junction to case* and *case to ambient air.* Thermal resistance is measured in degrees Celsius per watt of power at the junction (°C/W). We can assign symbols for each thermal resistance and construct a formula similar to Ohm's law. We will use the Greek letter theta ($\theta$) for thermal resistance and identify the resistances by subscript initials. Thus:

$\theta_{ja}$ = total thermal resistance (junction to ambient).

$\theta_{jc}$ = thermal resistance (junction to case).

$\theta_{ca}$ = thermal resistance (case to ambient).

Therefore, $\theta_{ja} = \theta_{jc} + \theta_{ca}$.

### *electric analogy*

The term thermal resistance came about from association of thermal and electric characteristics. We may carry the analogy a bit further to include temperature and junction power as illustrated in Fig. 2-37.

Figure 2-37 illustrates the following analogies:

Current : junction power
Voltage : temperature
Resistance : thermal resistance
(In this instance : has been used to mean "is analogous to".)

The product of thermal resistance (°C/W) and junction power will reveal the junction temperatures. Just as $E = I \times R$, so $T_j = \theta_{ja} \times P_j$. This is actually the rise in temperature starting at the ambient temperature. The true temperature is $T_j = T_a$, and this is our operating temperature.

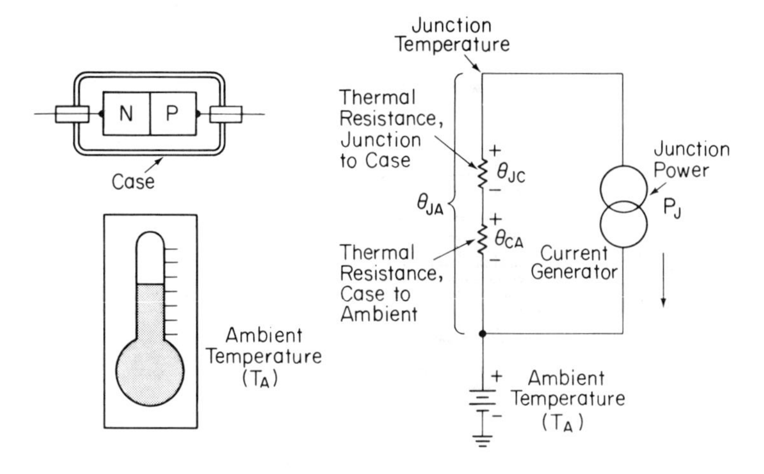


**Fig. 2-37.** Electrical Analogy of Heat Transfer.

### *heat sink*

A heat sink is a device used to aid the transfer of heat. We usually find it in the form of a larger piece of material which will conduct heat easily. The case of the semiconductor is connected directly to the heat sink. In effect, this enlarges the cooling surface and reduces the thermal resistance. This removes the $\theta_{ca}$ (case to ambient) from our thermal resistance formula and replaces it with two other thermal resistances: case to sink ($\theta_{cs}$) + sink to ambient ($\theta_{sa}$). The sum of $\theta_{cs}$ and $\theta_{sa}$ is always less than $\theta_{ca}$.

$$\theta_{cs} + \theta_{sa} < \theta_{ca}$$

This means that the addition of a heat sink enables more power per degree of junction temperature. Figure 2-38 illustrates one method of fabricating a heat sink, and shows the circuit analogy of our new values.

Our total thermal resistance is now:

$$\theta_{ja} = \theta_{jc} + \theta_{cs} + \theta_{sa}$$

The maximum allowable operating temperature [$T_j$(max)] is still a limiting factor, but now our semiconductor can handle more power without reaching this critical temperature.

### *maximum power*

The maximum power is easily calculated when these other values are known by using this formula:

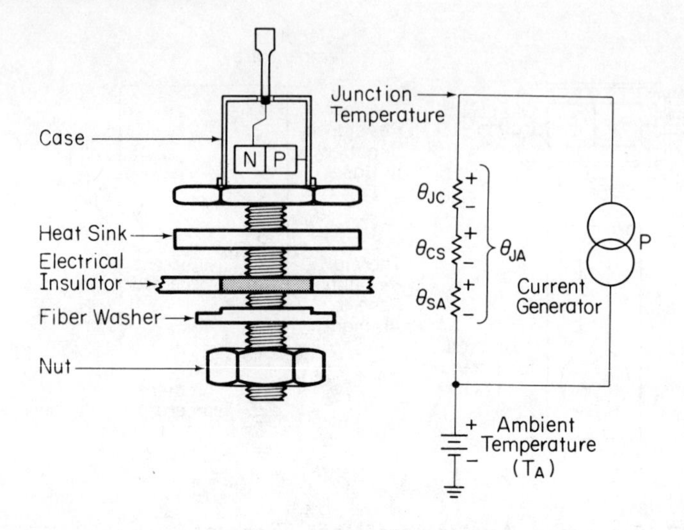


**Fig. 2-38.** Thermal Resistance with Heat Sink.

$$P_j(\text{max}) = T_j(\text{max}) - \frac{T_a}{\theta_{ja}}$$

We can, of course, manipulate the formula to solve for other values when the power is a known factor. For instance:

$$\theta_{ja} = T_j(\text{max}) - \frac{T_a}{P_j}$$

We may also rewrite the formula in terms of the individual thermal resistances.
Since:

$$\theta_{ja} = \theta_{jc} + \theta_{cs} + \theta_{sa}$$

Then:

$$\theta_{jc} + \theta_{cs} + \theta_{sa} = T_j(\text{max}) - \frac{T_a}{P_j}$$

**Problem:** What is the value of $\theta_{sa}$ in a diode which has $\theta_{jc} = 0.8$ °C/W, $T_j(\text{max}) = 150$ °C, $\theta_{cs} = 0.6$ °C/W, and a maximum power rating of 40 W? (These values were determined in an ambient temperature of 50 °C.)

Solution:

$$\theta_{sa} = T_j(\text{max}) - \frac{T_a}{P_j} - \theta_{jc} - \theta_{cs}$$

$$= 150 - \frac{50}{40} - 0.8 - 0.6$$

$$= 2.5 - 0.8 - 0.6$$
$$= 2.5 - 1.4 = 1.1\ °C/W$$

**Problem:** We are using a diode with the following specifications:

$$\theta_{ja} = 6\ °C/W$$
$$T_j(\max) = 150\ °C$$

When ambient temperature is 40°C, what is the maximum safe power?

Solution:

$$P_j(\max) = T_j(\max) - \frac{T_a}{\theta_{ja}}$$

$$P_j(\max) = 150 - \frac{40}{6} = 18.33\ W$$

If we lower the ambient temperature to 25°C, how will this affect the maximum safe power?

$$P_j(\max) = 150 - \frac{25}{6} = 20.83\ W$$

Lowering the ambient temperature by 15 °C increased the maximum safe power by 2.5 W; this is about 0.167 W/°C.

## CHAPTER 2 REVIEW EXERCISES

1. Figure 2-39 shows a diode with a potential applied across the junction. Indicate on the diagram:
   (a) Direction of electrons in the diode.
   (b) Direction of holes in the diode.
   (c) Whether bias is forward or reverse.
2. What constitutes an electrically stable atom?
3. When is an atom chemically stable?
4. Describe the valence shell of atoms in each of these materials:
   (a) Conductors.
   (b) Insulators.
   (c) Semiconductors.
5. (a) What is the meaning of valence?
   (b) Why does it apply to atomic structure?
6. What is meant by trivalent and petavalent structures?

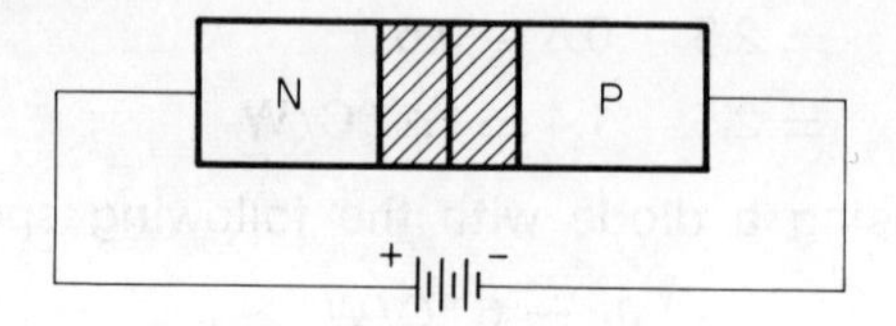


**Fig. 2-39.** Potential Across a Junction.

7. Arrange these elements according to the width of their forbidden shell from the narrowest to the widest:
   (a) Neon.
   (b) Lithium.
   (c) Bismuth.
   (d) Silicon.
   (e) Sulphur.
8. What unit is used to express the width of the forbidden shell?
9. How can pure silicon be change to:
   (a) A donor material
   (b) An acceptor material?
10. Why are donor materials labeled *n* type and acceptor materials labeled *p* type?
11. List three methods of forming *p–n* junctions.
12. What portion of a diode is called a depletion region? Why?
13. Draw a diagram of a *p–n* junction and indicate the direction of minority and majority currents.
14. What is the relationship of these currents:
    (a) Majority?
    (b) Minority?
    (c) Forward?
    (d) Reverse?
15. Draw two *p–n* junctions with forward bias on one and reverse bias on the other.
16. Draw a schematic symbol for a semiconductor diode, label the parts, and indicate the direction of forward current.

Items 17 through 19 refer to the circuit in Fig. 2-40.

17. The output ripple frequency is 500 pulses/s. What is the input frequency?
18. The average dc out is 20 V. Assuming a 1-V peak drop across the diode, what is the peak to peak amplitude of the input?

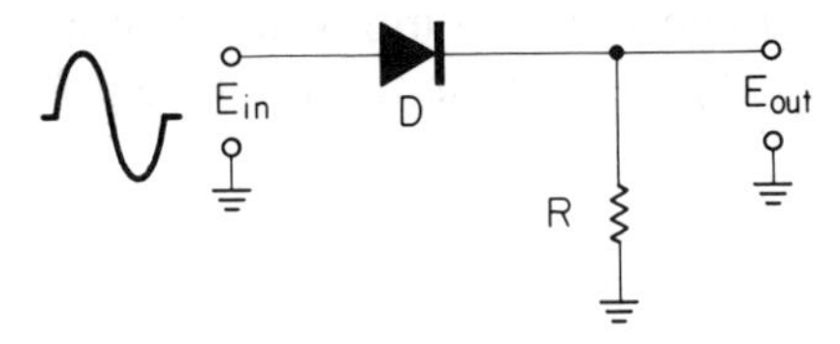


**Fig. 2-40.** A Diode Rectifier.

19. Draw the output showing polarity and peak value.
20. Draw a schematic symbol for a Zener diode, label the parts, and indicate the direction of current during normal operation.
21. Figure 2-41 represents the two types of junction triode transistors. Label the sections as to type of material and draw battery circuits to properly bias each junction.

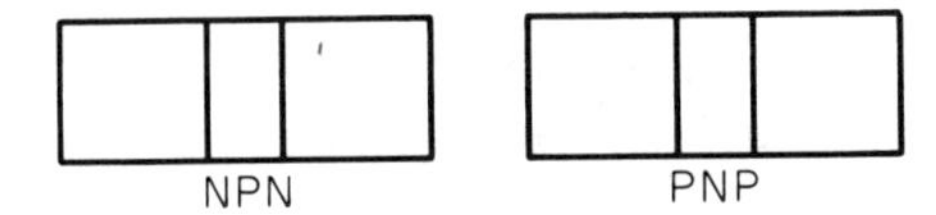


**Fig. 2-41.** Types of Triode Junction Transistors.

22. Draw a schematic symbol for these junction triodes, label the parts, and indicate current direction.
    (a) *npn*
    (b) *pnp*
23. What percentage of total circuit current is represented by:
    (a) $I_c$?
    (b) $I_e$?
    (c) $I_b$?

Items 24 through 25 refer to Fig. 2-42.

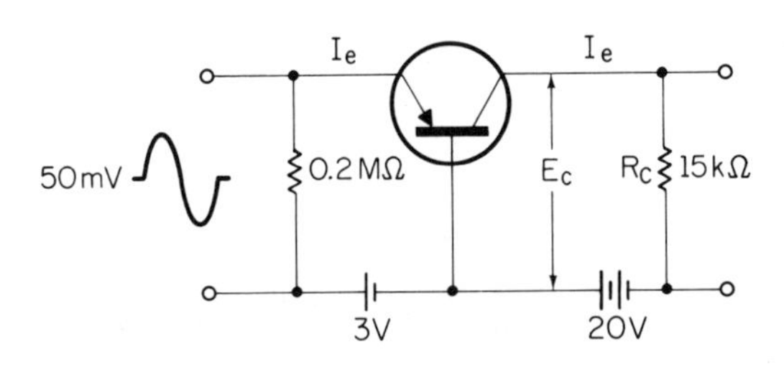


**Fig. 2-42.** Circuit Values.

24. Suppose that the input signal causes $I_e$ to swing by a total of 500 $\mu$A. What is the peak to peak change of:
    (a) $I_c$?
    (b) $E_{R_c}$?
25. When the input signal swings positive, what happens to:
    (a) $I_e$?
    (b) $I_c$?
    (c) $E_c$?
26. A transistor is rated for 200 mW at 25 °C with a power derating of 2 mW/°C. What is the maximum safe power in an ambient temperature of 100 °C?
27. The transistor in item 26 has $\theta_{jc} = 350$ °C/W, $\theta_{cs} = 275$ °C/W, and $\theta_{sa} = 375$ °C/W. What is the total thermal resistance?
28. For the transistor described in items 26 and 27, what is the maximum:
    (a) Junction temperature?
    (b) Operating temperature?

# 3

# *amplification principles*

Amplification is the process which increases the magnitude of a signal. The magnitude may be applied to voltage, current, or power depending upon which of the signal components is being amplified. A transistor has the ability to increase the size of a signal. Therefore, it is an amplifying device. Very few practical electronic circuits can be constructed without creating the need for an amplifier. In fact, about 50 percent of all electronics circuits produce some form of amplification.

## GENERAL

Simply stated, the basic principle of amplification is putting a small signal in and taking a larger signal out. This is oversimplification, of course, and it tells us nothing about the process of enlarging the signal. Since our vehicle for amplification is the triode transistor, let's examine some characteristics and conditions which cause it to amplify.

### *biasing*

It has been previously stated that the emitter-base junction must be forward biased if the transistor is to conduct. Also, the collector-base junction must be reverse biased. Since we have both the *npn* and the *pnp,* it is sometimes difficult to remember just how to place our biasing polarities. The letters which designate the type of transistor can be used to designate the polarities for proper bias. Let *n* stand for negative and *p* for positive. Think of the first letter as the emitter and the second letter as the collector. Thus, for the *npn,* forward bias is provided across the emitter-base junction by making the emitter negative with respect to the base. Reverse bias across the collector-base junction is provided by making the collector positive with respect to the base. This is illustrated for both types of transistor in Fig. 3-1. It should also be noted that the first and second letters indicate the polarities of emitter and collector respectively.

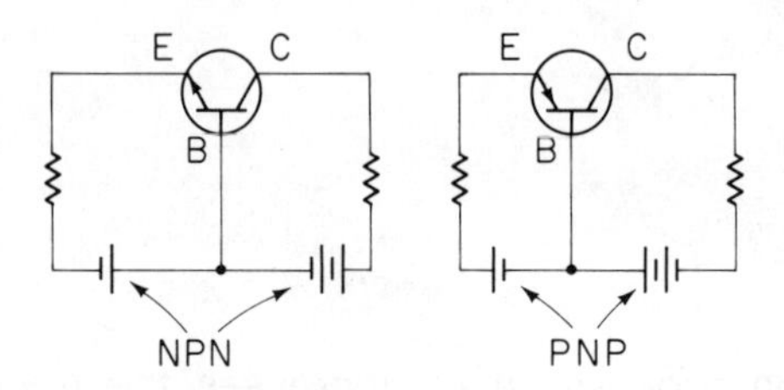


**Fig. 3-1.** Polarities for Proper Bias.

A word about the direction of current. For our purposes, current is the flow of electrons. In all cases, the emitter is represented by the arrow and carries the total transistor current. The direction of current, when the emitter-base junction is forward biased, is always against the arrow. With the *npn,* electrons flow from the emitter into the transistor and divide between the base and the collector. Electrons for the *pnp* flow in through both base and collector and out through the emitter.

When reverse bias is applied to the emitter-base junction, the transistor is cut off. We will discount the small amount of leakage current which occurs under these conditions. Amplification is accomplished by using a small signal to control the emitter-base bias. This signal should not be large enough to cause a reverse bias condition. The small change in bias alters the emitter current, and the resultant change in collector current causes a larger change in the collector voltage.

### *types of amplifiers*

Amplifiers are named according to the job they do. We have audio amplifiers to amplify audio frequencies and video amplifiers to amplify video frequencies. There are radio frequency amplifiers, intermediate frequency amplifiers, and pulse amplifiers. An amplifier is designated as a current amplifier, a voltage amplifier, or a power amplifier depending upon which component of the signal it concentrates on.

The current amplifier produces a gain in current; that is, the output current is larger than the input current. Current gain is expressed as a number which describes the ratio of output current to input current. Actually, this is ac, and it is the *change in output* current divided by the *change in input* current. Current gain equals change in output current divided by the change in input current. Even when the ratio is less than unity, it is still referred to as current gain. In some transistors, current gain ranges from 0.95 to 0.99 which means that the output is smaller than the input. Depending upon the transistor and the circuit configuration, some amplifiers have a current gain much greater than unity. Perhaps a better way to express gain is this: gain is a number which expresses how many times the input has been taken as a factor in order to produce the output. When the formula is transposed it expresses the same idea. Change in output equals change in input multiplied by gain.

The voltage amplifier concentrates on the voltage aspects of the signal. Voltage gain is a ratio between output voltage and input voltage. Again, we are dealing with alternating voltage. So, it is the change ratio that we are interested in. Voltage gain equals change in voltage output divided by the change in voltage input. Nearly all amplifiers have a voltage gain greater than unity. This can be true of an amplifier without it being a voltage amplifier. Most current and power amplifiers have output voltage larger than the input voltage. In order to qualify as a voltage amplifier, the primary purpose of the stage must be to enlarge the signal voltage.

The power amplifier is designed to concentrate on the power component of the signal. Input power is input voltage times input current (ac effective). The output power is output voltage times output current (ac effective). The power gain is the change ratio between output power and input power. Power gain equals change in output power divided by the change in input power.

### classes of amplifiers

Amplifiers are classed according to the percentage of time that the transistor is conducting. Notice the distinction here between amplifier and transistor. The transistor is our amplifying vehicle, but the whole stage is an amplifier. A stage is the transistor and all its associated circuits.

The transistor in a class A amplifier is biased so that it conducts all the time (100%). The signals into this amplifier never drive it beyond either extreme of the conducting limits.

The class B amplifier uses a transistor that is biased to cut off about 50 percent of the time. A sine wave applied to this amplifier would have one alternation amplified and the other eliminated.

The class A–B amplifier has the transistor biased in such a manner that it will cut off, but it is cut off less than 50 per cent of the time.

The transistor in a class C amplifier will be cut off more than 50 percent of the time. Figure 3-2 illustrates the effect of each class on a sine wave input.

Which class is best? That depends on what the amplifier is expected to do. In some cases, class A is best; at other times, only class C can do the job.

## *the input signal*

We have been referring to the input signal to an amplifier. Perhaps we should clarify the meaning of the term *signal*. It means a wave variation in respect to time. When not otherwise specified the signal will be a sine wave. Figure 3-3 illustrates some waveforms that could be used as signals.

Where do these signals come from? Only one place. The input signal always comes from the previous stage. That previous stage may be anything. It could be an ac generator, a trigger circuit, an

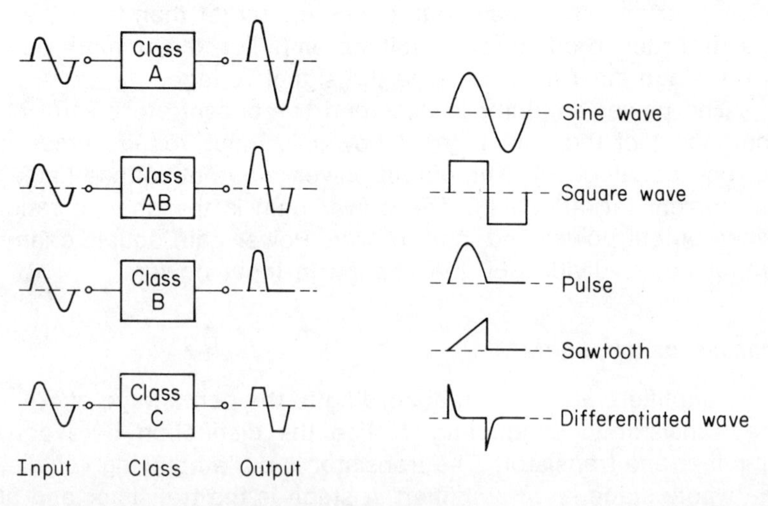


**Fig. 3-2.** Classes of Operation. **Fig. 3-3.** Some Common Signals.

antenna, or some other device. It is convenient to imagine that the previous stage is a generator which produces our input signal. One amplifier stage becomes a generator to the next. Now let's examine some amplifier circuit arrangements.

## CIRCUIT CONFIGURATION

There are many possible ways to arrange an amplifier circuit, but all fall into three categories: common base, common emitter, and common collector. The word common refers to both the input and the output section of the circuit. The word ground is sometimes used instead of common.

### *common base*

When the base of the transistor is common to both the input circuit and the output circuit, the amplifier is using the common base configuration. Either type of transistor may be used. For normal operation, the input junction is forward biased, and the output junction is reverse biased. Figure 3-4 shows proper bias for both types of transistors.

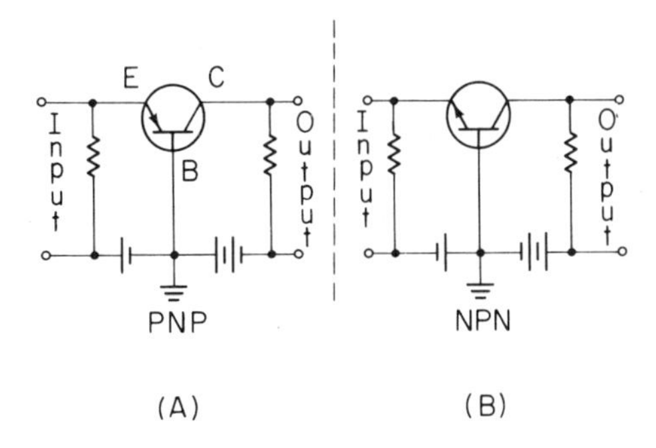


**Fig. 3-4.** Proper Biasing for Common Base.

Thus far we have been using batteries to illustrate bias on our transistors. Obviously this is not done in most equipment. It is time to drop these batteries out of the picture and move a little closer to an operational circuit. Remember that the power and voltage control circuits are designed to give us the type of voltage we need and deliver it to the spot where it will be used. The voltage can be large or small, regulated or unregulated, and positive or negative. Figure 3-5 is the *pnp* configuration without the bias batteries.

$R_1$, $R_2$, and $R_3$ are part of the voltage divider in the power supply. If the taps on the voltage divider are not sufficient to supply all our voltage needs, it is a simple matter to put in a resistive network and divide it some more. Compare the relative potentials between the various elements of the transistor in Fig. 3-4a against the transistor in Fig. 3-5. In both cases, the input section is forward biased, and the output section is reverse biased. In both cases the emitter is positive with respect to the base, and the collector is negative with respect to the base. Hereafter, we will use neither batteries nor voltage dividers. A dc voltage will be designated by $V$. $V_{BB}$ will mean the base supply voltage, $V_{CC}$ will be the collector supply voltage, and $V_{EE}$ will be the emitter supply voltage. The above symbols will be preceded by a minus sign when the supply voltage is negative. When it is essential, the quantity of dc will be specified. Otherwise, we will just assume that the proper quantity is present.

After all the trouble we have had providing a reasonably smooth dc, we can't afford to mix ac signals into our dc power supply. This is prevented by adding ac bypass capacitors to the circuit. For every frequency of ac, there is a capacitor which will act as a short. At every tap on the voltage divider, a by-pass capacitor is connected from that point to ground. Sometimes these capacitors are part of the visible circuit; sometimes they are hidden in the power supply, but they are always there. These capacitors are direct shorts to ground for the ac signal. Points 1 and 2 in Fig. 3-5 are ac grounds because of bypass capacitors which are not shown in the circuit.

Now let's redraw the common base configuration, and omit all unnecessary circuits. It should be similar to Fig. 3-6.

The input signal is applied between point 1 and $V_{EE}$ ($V_{EE}$ is also ac ground). As this signal swings positive, it increases forward bias and emitter current. The change in current goes into ac ground at

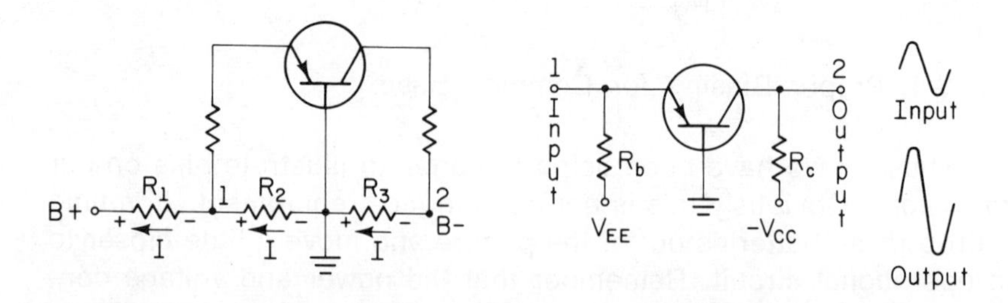


**Fig. 3-5.** Voltage Divider Bias. **Fig. 3-6.** Simplified Common Base.

$V_{EE}$. A small part of this current re-enters the transistor base. Most of the change comes out of ac ground at $V_{CC}$, and goes upward through the collector resistor ($R_C$) to the collector. Therefore, the small change in base current produces a large change in collector current. The output is taken between point 2 and ac ground at $V_{CC}$. The output is the same polarity and phase as the input, but the output is much larger. Typical values for this circuit are:

1. Input resistance (internal) 30 to 150 Ω.
2. Output resistance (internal) 300 to 500 kΩ.
3. Voltage gain 300 to 1500.
4. Current gain less than unity.
5. Power gain 20 to 30 dB. (dB is an abbreviation for decibel; for our purposes, the power level doubles for every 3-dB increase.)

Therefore, this common base configuration provides excellent voltage gain, a fair gain in power, but no actual gain in current.

The transistor manufacturer supplies static characteristic curves for the common base configuration. The curves for each type of transistor will be unique in some of the details, but all of them will be similar in most respects. These curves show the relationship between collector current and emitter current for all reasonable values of collector voltage. Figure 3-7 is an illustration of such a family of curves. These charts may be obtained any place

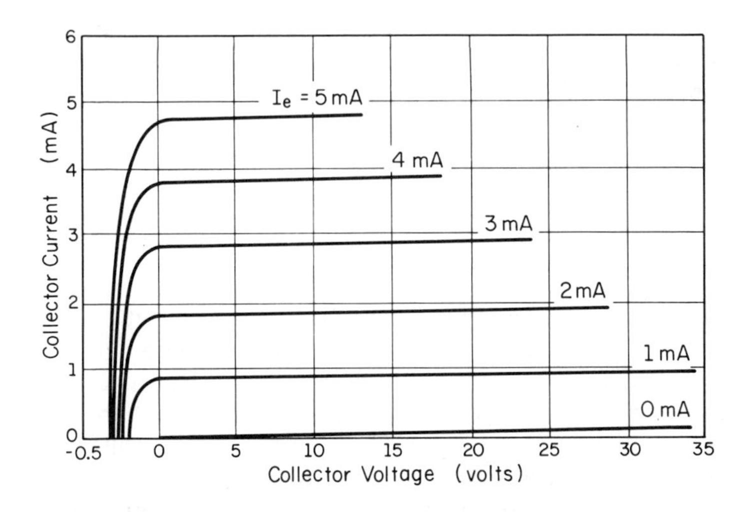


**Fig. 3-7.** Static Characteristics for a Common Base.

where transistors are sold. This drawing is intended to be representative of all common base configurations.

Another bit of useful information which is sometimes supplied along with the characteristic chart is the maximum current gain. In the common base circuit, this gain is symbolized by the Greek letter alpha ($\alpha$).

$$\alpha = \Delta I_c / \Delta I_e \text{ (with collector voltage constant)}$$

where $\alpha$ is maximum current gain, $\Delta$ is a small change, $I_c$ is collector current, and $I_e$ is emitter current.

The average value for $\alpha$ is about 0.97, and as previously stated, it is always less than unity. The static characteristic curves in Fig. 3-7 are not completely accurate, but they are close enough to demonstrate this fact. Select a value of collector voltage. Let's try 15 V. Move up the 15-V line until it intersects $I_e$ of 1 mA. This is our starting point. Note the collector current. It is about 0.85 mA. Move up the 15-V line again until you intersect the $I_e$ 2 mA-line. $\Delta I_e$ is 1 mA. What is $\Delta I_c$? About 0.95 mA; at any rate, it is more than 0.9 and less than unity. $0.95/1 = 0.95$. Examine several other similar small changes and notice the consistency of the results.

There are two resistive components inside a transistor which are associated with junction. The input resistance refers to the internal resistance across the input junction. The output resistance (in most configurations) is the resistance across the output junction. These are characteristic resistances which are apart from the external resistors. In the common base configuration, the input resistance is between emitter and base. The output resistance is between collector and base. These are measured with the external leads open.

The gain in both voltage and power makes the common base configuration a very popular amplifier circuit. Another popular configuration is the common emitter.

### *common emitter*

This configuration is so named because the emitter is common to both input and output. It is also called a grounded emitter. Figure 3-8 illustrates the proper bias for both types of transistors.

In the *pnp* circuit, the emitter is the most positive point. Bias across the emitter-base junction is provided by a small dc from the negative power supply. This is input bias. A larger negative dc is used for output bias from collector to emitter. Electrons flow into the emitter from both base and collector. Base current ($I_b$) is the input current. The output current is collector current ($I_c$). The path for total circuit current is out through the emitter to ground.

In the *npn* circuit, ground is the most negative point. Electrons which comprise the total circuit current leave ground and enter the emitter. Inside the transistor, they divide. Input current is base current; output current is collector current.

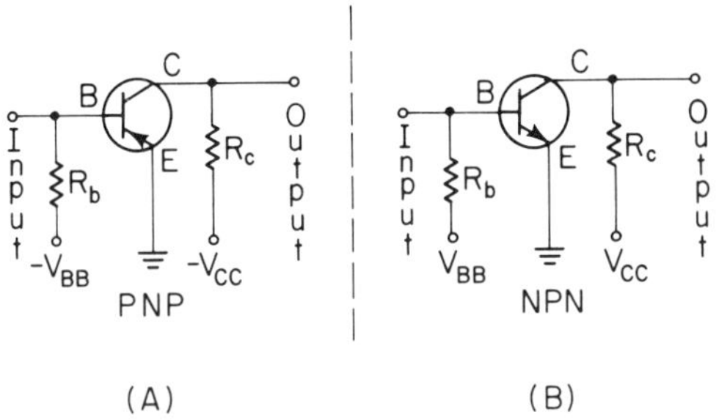

**Fig. 3-8.** Proper Biasing for Common Emitter.

In both circuits, the input junction is forward biased, and the output junction is reverse biased. In the *pnp* circuit, a positive going input signal opposes the forward bias by decreasing base current. This causes a decrease in collector current. Less current upward through $R_c$ makes the output more negative with resepct to ac ground ($-V_{CC}$). This means that there is a 180° phase difference between input and output signals. Figure 3-9 shows the action for an entire sine wave input and output. Notice that both signals are entirely below ground (0) potential.

The action in the *npn* circuit is exactly opposite, but the results are the same. In this case, a positive going input aids the forward bias, increases both input and output currents, and drives the output in a negative direction. More current downward through $R_c$ makes the output swing negative with respect to ac ground ($V_{CC}$). This is shown for the entire sine wave in Fig. 3-10.

Notice, in this case, that both signals are entirely above ground potential (0). Compare these signals with those in Fig. 3-9. As far as the ac component is concerned, there is no difference. Typical values for a common emitter configuration are:

1. Input resistance 500 to 1500 Ω.
2. Output resistance 30 to 50 kΩ.

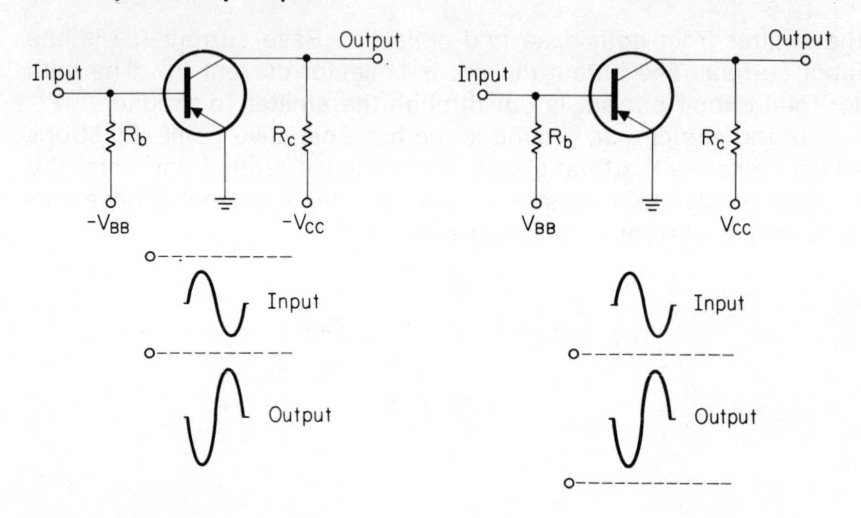


**Fig. 3-9.** Relation of Input and Output Using *pnp.*

**Fig. 3-10.** Relation of Input and Output Using *npn.*

3. Voltage gain 300 to 1000.
4. Current gain 25 to 50.
5. Power gain 25 to 40 dB.

Here we have a fair voltage gain with an excellent gain in both current and power.

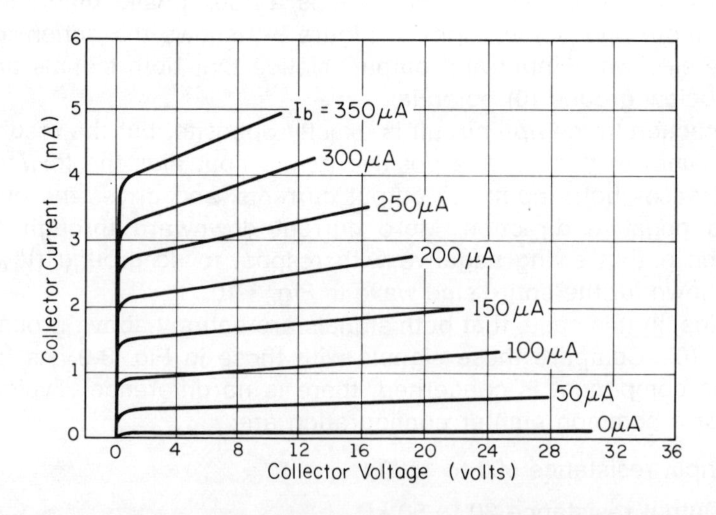


**Fig. 3-11.** Static Characteristics for a Common Emitter.

Static characteristics curves are also available for the common emitter configuration. Figure 3-11 illustrates this family of curves.

Maximum current gain for the common emitter configuration is symbolized by the Greek letter beta ($\beta$). This is sometimes provided along with the chart. If not, it is easily calculated using the following formula:

$$\beta = \frac{\Delta I_c}{\Delta I_b} \qquad \text{(with collector voltage constant)}$$

where $\beta$ is maximum current gain, $\Delta$ is a small change, $I_c$ is collector current, and $I_b$ is base current.

Let's try an example using 16 V as the collector voltage and starting with $I_b$ of 100 $\mu$A. Moving up the graph to the 150 $\mu$A $I_b$ line, we have a $\Delta I_b$ of 50 $\mu$A. What is $\Delta I_c$ for this change? About 0.7 mA. Substituting into the formula we have:

$$\beta = \frac{0.7 \times 10^{-3}}{50 \times 10^{-6}}$$
$$= 0.014 \times 10^{3}$$
$$= 14$$

This figure is short of our typical value of current gain, but remember that this chart is only a sample drawing.

The input resistance in the common base configuration is between the base and the emitter. The output resistance is between the collector and the emitter. Both of these, of course, are measured with the external leads open.

The strong gain in current, voltage, and power enables this configuration to be adapted to almost any amplifier problem. This is probably the most popular of the three configurations, but there is another possible arrangement; the common collector.

### *common collector*

The common collector configuration is not a very popular circuit where the chief concern is amplification. It does, however, have its special application. This circuit is frequently called an emitter follower as well as a grounded collector. The common collector is a more descriptive name because the collector is common to both input and output. Figure 3-12 illustrates proper bias for both types of transistors.

In both circuits the input is applied across the collector-base junction, and the output is from the emitter. The input junctions are forward biased and the output circuits are reverse biased.

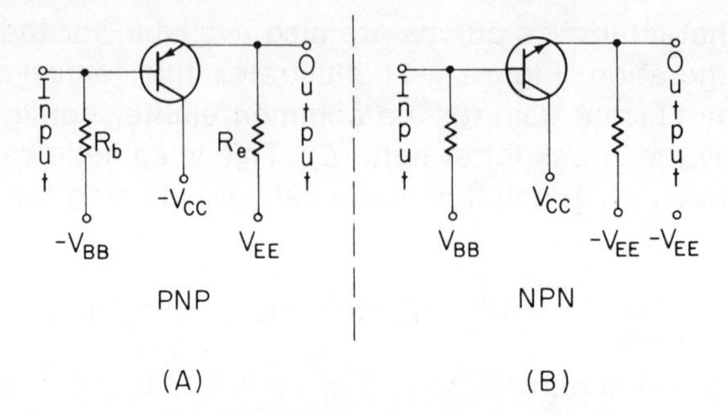


**Fig. 3-12.** Proper Biasing for Common Collector.

With the *pnp* transistor, when the input signal swings positive, it opposes the forward bias. This causes a decrease in both input and output current. Less current downward through $R_e$ allows the emitter to become more positive in respect to ac ground ($V_{EE}$). When the input swings negative, it aids forward bias and increases both input and output currents. More current downward through $R_e$ causes the emitter to swing negative with respect to ac ground ($V_{EE}$).

The *npn* produces identical results for opposite reasons. All polarities are reversed. Therefore, a positive input aids input bias and increases currents. More current upward through $R_e$ makes the emitter swing in a positive direction. The negative swing of the input reduces currents and allows the emitter to swing in a negative direction. Phase and relative size of inputs and outputs are shown in Fig. 3-13.

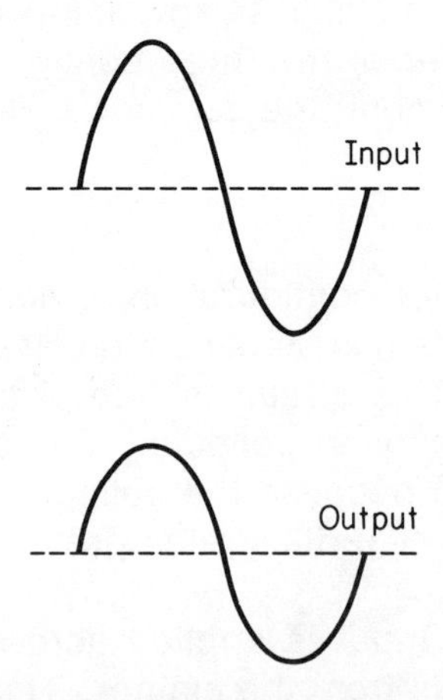


**Fig. 3-13.** Relative Amplitude and Phase.

Typical values for a common collector configuration are:

1. Input resistance 20 to 500 kΩ.
2. Output resistance 50 to 1000 Ω.
3. Voltage gain less than unity.
4. Current gain 25 to 50.
5. Power gain 10 to 20 dB.

The current gain is excellent; the power gain is fair; but the voltage is less than one. The common collector may be used as a current amplifier, and it can be used as a power amplifier. Its most important application, however, is a direct result of the internal characteristics. Take another look at the typical input and output resistances. What do you see? It has a high input resistance and a low output resistance. That's what makes it an important circuit.

Remember that we get maximum transfer of power when the impedance is matched. Signals are usually amplified before they enter a cable to transfer from one chassis to another. The input impedance to a cable is low, but the output impedance of most amplifiers is high. How can this be resolved? Insert an extra stage. Let this stage be a common collector. Its high input impedance will match the high output of the previous amplifier. This will enable maximum power transfer into the common collector. At the same time, its low output impedance will match the low input impedance of the cable. This will enable maximum power to transfer from the common collector into the cable. The ability to match two widely different impedances, and produce a small power gain at the same time, makes the common collector valuable in many situations.

There is no separate family of static characteristic curves for the common collector. However, the chart for the common emitter has all the necessary information.

There is no symbol for maximum current gain in this configuration. But it is equivalent to $\Delta I_e/\Delta I_b$ at a constant collector voltage.

$$\text{Max current gain} = \frac{\Delta I_e}{\Delta I_b} \qquad \text{(with } V_c \text{ constant)}$$

Looking at the common emitter graph in Fig. 3-11, you will find that $I_e$ is not given. This fact is a small problem, however, because $I_e$ is always the sum of $I_b$ and $I_c$. The previous formula may be restated as:

$$\text{Max current gain} = \frac{\Delta I_c + \Delta I_b}{\Delta I_b}$$

**Example:** Moving up the 8-V collector voltage line from collector current 2 mA to collector current 3 mA, what is the current gain?

Solution:

$\Delta I_c$ is 1 mA, $\Delta I_b$ is 72 $\mu$A. Therefore:

$$\text{Max current gain} = \frac{(1 \times 10^{-3}) + (72 \times 10^{-6})}{(72 \times 10^{-6})}$$

$$= \frac{1072}{72} = 14.9$$

Thus far, we have considered all three configurations, but we have been speaking primarily of static conditions. The characteristic curves are strictly static, and the gains we have calculated have been maximum gains. These factors tell a great deal about the capabilities of a transistor, but not how it will react in a given circuit. It is possible to convert these static characteristics into dynamic characteristics.

## PARAMETERS

Equivalent circuits are useful tools when analyzing amplifier circuits. The values in these equivalent circuits are called parameters. When we consider the transistor as an unknown object, we have four variable quantities: input current, output current, input voltage, and output voltage. This is the black box concept as illustrated in Fig. 3-14.

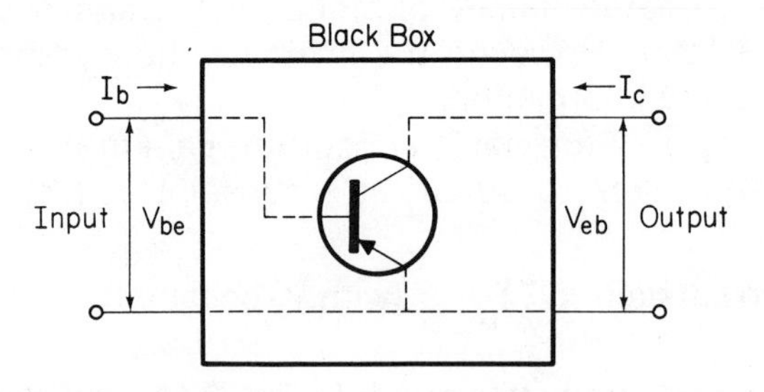


**Fig. 3-14.** Four Variable Quantities.

We may choose any two values to be the independent variables. In either case, the remaining two become dependent variables. This type of arrangement is used to measure values and construct equivalent circuits. Parameters can be calculated from these values to predict the behavior of a transistor in any of the three basic circuit configurations.

If we choose the input current and the output voltage as the independent variables, the resulting parameters are called *hybrid* (or mixed) parameters. When the two currents are used as independent variables, our parameters are called *open circuit* parameters. Using the two voltages as independent variables results in *short circuit* parameters.

We also call these the small signal parameters (or characteristics). The test circuits place the transistor in the most linear portion of its characteristic curve, and the changes are made in very small increments.

## Hybrid Parameters

We have four hybrid parameters which normally appear on the manufacturer's electrical characteristic chart. These are:

1. Input resistance.
2. Forward current gain.
3. Output conductance.
4. Reverse Voltage.

These parameters differ according to the circuit configuration we use. If we use the letter *h* to mean hybrid, *i* for resistance, *f* for forward current gain, *o* for output conductance, and *r* for reverse voltage gain, we can shorten our notations for hybrid parameters as follows:

$h_i$ = hybrid input resistance.

$h_f$ = hybrid forward current gain.

$h_o$ = hybrid output conductance.

$h_r$ = hybrid reverse voltage gain.

Now we need a symbol to designate the circuit configuration. We will use these:

$b$ = common base.

$e$ = common emitter.

$c$ = common collector.

Complete designations for all three configurations then become those in Fig. 3-15.

| Hybrid Parameters | Common Base | Common Emitter | Common Collector |
|---|---|---|---|
| Input Resistance | hib | hie | hic |
| Forward Current Gain | hfb | hfe | hfc |
| Output Conductance | hob | hoe | hoc |
| Reverse Voltage Gain | hrb | hre | hrc |

**Fig. 3-15.** Hybrid Parameter Designations.

In some texts and in some manufacturer's charts, you will find the following variations in the designations of hybrid parameters:

$h_f$ may be $\propto_f$
$h_r$ may be $\mu_r$

The meaning is still the same as in Fig. 3-15, and the circuit designator is still present.

The hybrid parameter values for a given configuration will vary over a wide range. There is no such thing as a typical set of parameters. However, Fig. 3-16 shows the relative magnitudes of parameters among the three circuit configurations for one particular transistor.

| Common Base | Common Emitter | Common Collector |
|---|---|---|
| hib = 39Ω | hie = 1950 Ω | hic = 1950Ω |
| hfb = -0.98 | hfe = 49 | hfc = -50 |
| hob = 0.49μV | hoe = 24.5μV | hoc = 24.5 μV |
| hrb = $380 \times 10^{-6}$ | hre = $575 \times 10^{-6}$ | hrc = 1 |

**Fig. 3-16.** Relative Parameter Values.

Hybrid parameters are calculated from values obtained from the graph of characteristic curves. We will use a hypothetical transistor and calculate the parameters for a common emitter configuration.

### *input resistance ($h_{ie}$)*

The value of input resistance is a small change in base voltage divided by a small change in base current. The collector voltage is held constant for these changes.

$$h_{ie} = \frac{\Delta V_{be}}{\Delta I_b} \qquad (V_{ce}\ \text{constant})$$

The graph in Fig. 3-17 will provide our values of voltage and current.

First we establish bias by specifying the values of base current and collector to emitter voltage. In this case, we set $I_b$ to 150 $\mu$A and $V_{ce}$ to 7.5 V. This point is marked with an *X* on the graph.

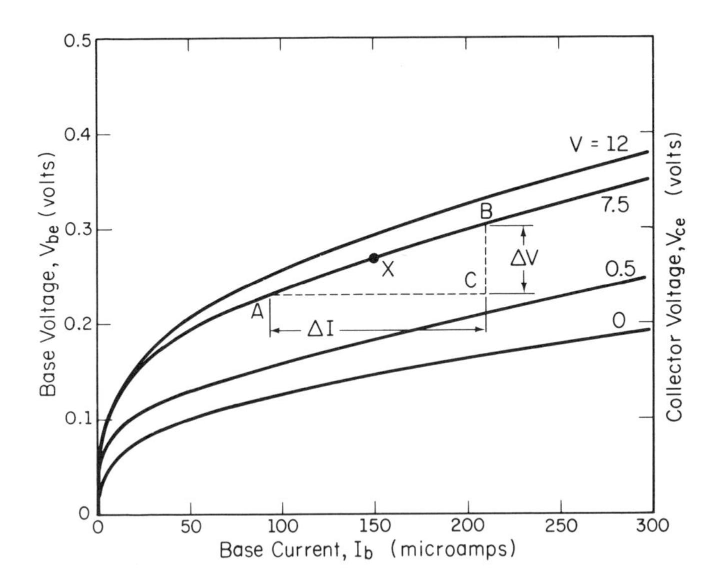


**Fig. 3-17.** Values for Calculating $h_{ie}$.

$\Delta I_b$ is the value of $I_b$ at point *A* subtracted from the value of $I_b$ at point *C*.

$$\Delta I_b = 210\ \mu\text{A} - 95\ \mu\text{A} = 115\ \mu\text{A}$$

$\Delta V_{be}$ is determined by subtracting $V_{be}$ at point *C* from $V_{be}$ at point *B*.

$$\Delta V_{be} = 0.3\ \text{V} - 0.23\ \text{V} = 0.07\ \text{V}$$

Solve for input resistance:

$$h_{ie} = \frac{\Delta V_{be}}{\Delta I_b}$$

$$= \frac{0.07\ \text{V}}{115\ \mu\text{A}}$$

$$= \frac{0.07 \times 10^6}{115} = 608\ \Omega$$

### *forward current gain ($h_{fe}$)*

We calculate forward current gain by dividing a small change of collector current by a small change of base current while holding the collector voltage constant.

$$h_{fe} = \frac{\Delta I_c}{\Delta I_b} \qquad (V_{ce} \text{ constant})$$

We will use the graph in Fig. 3-18 to obtain our value.

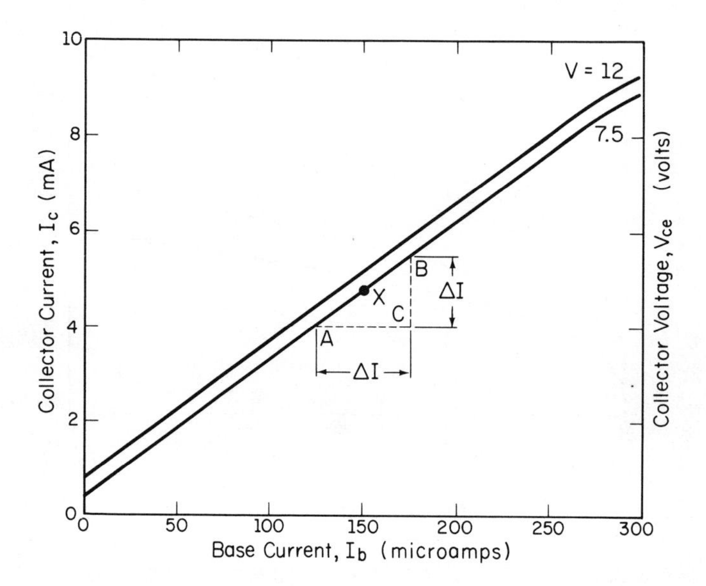


**Fig. 3-18.** Values for Calculating $h_{fe}$.

Let's use the same bias conditions as before; $I_b = 150\ \mu A$ and $V_{ce} = 7.5$ V. This places us at point *X* on the graph.

$\Delta I_b$ is obtained by subtracting the value of $I_b$ at point *A* from the value of $I_b$ at point *C*.

$$\Delta I_b = 175\ \mu A - 120\ \mu A = 55\ \mu A$$

$\Delta I_c$ is $I_c$ at point *C* subtracted from $I_c$ at point *B*.

$$\Delta I_c = 5.5 \text{ mA} - 4 \text{ mA} = 1.5 \text{ mA}$$

$$h_{fe} = \frac{\Delta I_c}{\Delta I_b} = \frac{1.5 \text{ mA}}{55\ \mu A} = 27$$

***output conductance ($h_{oe}$)***

Output conductance is the reciprocal of output resistance. Since resistance is *E/I*, conductance is *I/E*. In this case, we are interested in $V_{ce}$ and $I_c$ with base current held constant.

$$h_{oe} = \frac{\Delta I_c}{\Delta V_{ce}} \qquad (I_b \text{ constant})$$

We will obtain our values from the graph in Fig. 3-19.

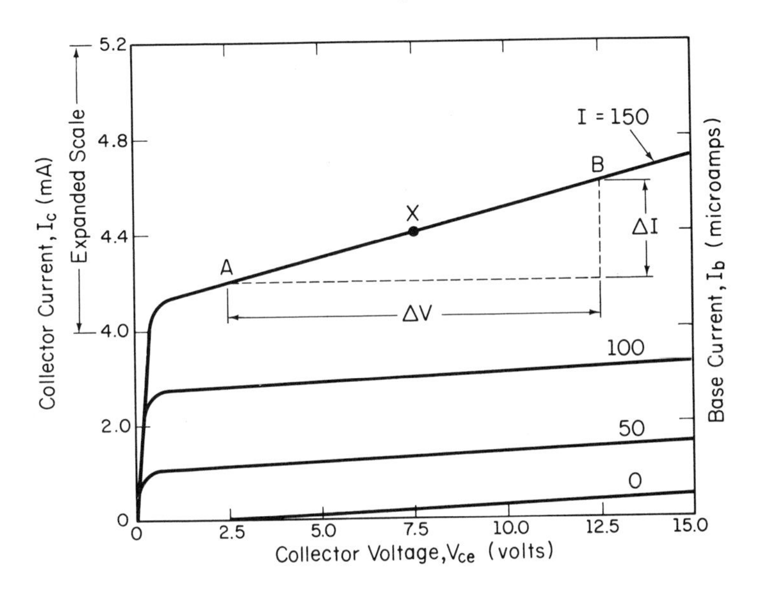


**Fig. 3-19.** Values for Calculating $h_{oe}$.

We will assume the same bias as before ($I_b = 150\ \mu A$ and $V_{ce} = 7.5$ V) and start at point *X*.

$\Delta I_c$ is the value of $I_c$ at point *C* subtracted from the value of $I_c$ at point *B*.

$$\Delta I_c = 4.6 \text{ mA} - 4.2 \text{ mA} = 0.4 \text{ mA}$$

$\Delta V_{ce}$ is $V_{ce}$ at point *A* subtracted from $V_{ce}$ at point *C*.

$$\Delta V_{ce} = 12.5 \text{ V} - 2.5 \text{ V} = 10 \text{ V}$$

Calculate $h_{oe}$:

$$h_{oe} = \frac{I_c}{V_{ce}}$$

$$= \frac{0.4 \text{ mA}}{10 \text{ V}} = 40 \times 10^{-6} = 40\ \mu\mho$$

This conductance is expressed in mhos (℧) the same as any other conductance. Since it is such a small value with transistors, the general term is micromhos (μ℧).

***reverse voltage gain ($h_{re}$)***

Reverse voltage is a reciprocal of forward voltage. Forward voltage gain is $E_{out}/E_{in}$; so, reverse voltage gain is $E_{in}/E_{out}$. We determine the value of reverse voltage gain by dividing a small change of base voltage by a small change of collector voltage. The base current will be held constant.

$$h_{re} = \frac{\Delta V_{be}}{\Delta V_{ce}} \qquad (I_b \text{ constant})$$

The graph in Fig. 3-20 will provide our values, and we will again start at point $X$ with $I_b$ of 150 μA and $V_{ce}$ of 7.5 V.

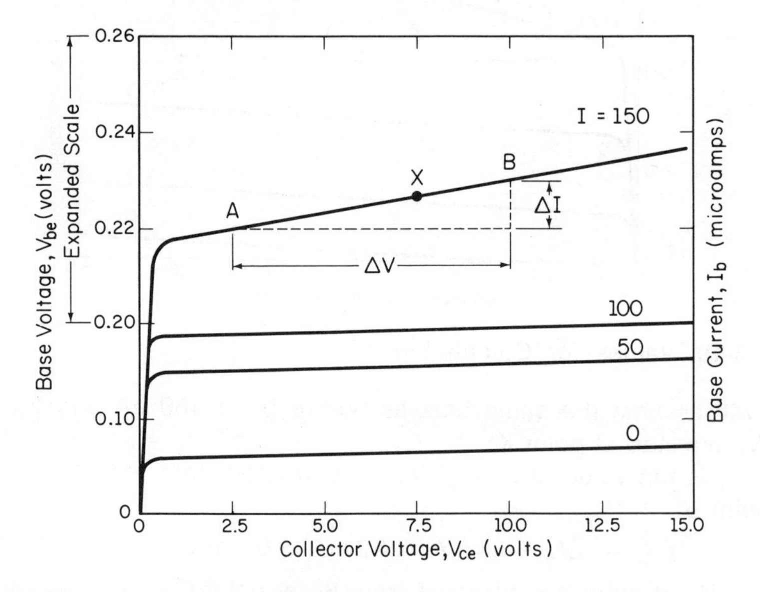


**Fig. 3-20.** Values for Calculating $h_{re}$.

We obtain $\Delta V_{be}$ by subtracting the value of $V_{be}$ at point $C$ from the value of $V_{be}$ at point $B$.

$$\Delta V_{be} = 0.23 \text{ V} - 0.22 \text{ V} = 0.01 \text{ V}$$

$\Delta V_{ce}$ is $V_{ce}$ at point $A$ subtracted from $V_{ce}$ at point $C$.

$$\Delta V_{ce} = 10 \text{ V} - 3 \text{ V} = 7 \text{ V}$$

Solve for $h_{re}$:

$$h_{re} = \frac{\Delta V_{be}}{\Delta V_{ce}}$$

$$= \frac{0.01\ \text{V}}{7\ \text{V}} = 0.0014 = 14 \times 10^{-4}$$

Why do we wish to know the reverse voltage gain? It is important because it shows the ratio of feedback voltage inside the transistor. This is an adverse action which tends to reduce the forward voltage gain. Once we have $h_{re}$, it is a very easy matter to calculate forward voltage gain.

$$\text{Forward voltage gain} = \frac{1}{h_{re}}$$

$$= \frac{1}{0.0014} = 714$$

***hybrid parameters for other configurations***

We will not repeat the procedure for calculating the hybrid parameters for each configuration. The procedure is much the same in each case. Here are the formulas for the common base configuration.

$$h_{ib} = \frac{\Delta V_{be}}{\Delta I_e} \qquad (V_{cb}\ \text{constant})$$

$$h_{fb} = \frac{\Delta I_c}{\Delta I_e} \qquad (V_{cb}\ \text{constant})$$

$$h_{ob} = \frac{\Delta I_c}{\Delta V_{cb}} \qquad (I_e\ \text{constant})$$

$$h_{rb} = \frac{\Delta V_{eb}}{\Delta V_{cb}} \qquad (I_e\ \text{constant})$$

The following hybrid parameter formulas apply to the common collector.

$$h_{ic} = \frac{\Delta V_{bc}}{\Delta I_b} \qquad (V_{ec}\ \text{constant})$$

$$h_{fc} = \frac{\Delta I_e}{\Delta I_b} \qquad (V_{ec}\ \text{constant})$$

$$h_{oc} = \frac{\Delta I_e}{\Delta V_{ec}} \qquad (I_b\ \text{constant})$$

$$h_{rc} = \frac{\Delta V_{bc}}{\Delta V_{ec}} \qquad (I_b\ \text{constant})$$

The manufacturer's characteristic charts normally reveal the hybrid parameters for only one circuit configuration. This is either the common emitter or the common base configuration. The chart in Fig. 3-21 provides formulas for converting the given parameters to the other configurations.

| From CE to CB | From CE to CC | From CB to CE | From CB to CC |
|---|---|---|---|
| $h_{ib} = \frac{h_{ie}}{h_{fe}+1}$ | $h_{ic} = h_{ie}$ | $h_{ie} = \frac{h_{ib}}{h_{fb}+1}$ | $h_{ic} = \frac{h_{ib}}{h_{fb}+1}$ |
| $h_{fb} = \frac{-h_{fe}}{h_{fe}+1}$ | $h_{fc} = -(h_{fe}+1)$ | $h_{fe} = \frac{-h_{fb}}{h_{fb}+1}$ | $h_{fc} = \frac{-1}{h_{fb}+1}$ |
| $h_{ob} = \frac{h_{oe}}{h_{fe}+1}$ | $h_{oc} = h_{oe}$ | $h_{oe} = \frac{h_{ob}}{h_{fb}+1}$ | $h_{oc} = \frac{h_{ob}}{h_{fb}+1}$ |
| $h_{rb} = \frac{(h_{ie})(h_{oe})}{h_{fe}+1} - h_{re}$ | $h_{rc} = 1 - h_{re}$ | $h_{re} = \frac{(h_{ib})(h_{ob})}{h_{fb}+1} - h_{rb}$ | $h_{rc} = \frac{(h_{ib})(h_{ob})}{h_{fb}+1} - h_{rb}$ |

**Fig. 3-21.** Hybrid Conversion Formulas.

**Problem:** Let's use values from the table in Fig. 3-16 and convert some hybrid parameters from one configuration to another. We can use the same table to check the accuracy of our answer.

1. Convert $h_{ie}$ to $h_{ib}$.

$$h_{ib} = \frac{h_{ie}}{(h_{fe}+1)} = \frac{1950\ \Omega}{(49+1)} = \frac{1950}{50} = 39\ \Omega$$

2. Change $h_{re}$ to $h_{rc}$.

$$h_{rc} = 1 - h_{re} = 1 - 575 \times 10^{-6} = 0.999425$$

3. Change $h_{ob}$ to $h_{oc}$.

$$h_{oc} = \frac{h_{ob}}{h_{fb}+1} = \frac{0.49\ \mu\mho}{-0.98+1} = \frac{0.49 \times 10^{-6}}{0.02} = 24.4\ \mu\mho$$

4. Convert $h_{fe}$ to $h_{fc}$.

$$h_{fc} = -(h_{fe} + 1)$$
$$= -(49 + 1) = -50$$

5. Convert $h_{fb}$ to $h_{fe}$.

$$h_{fe} = \frac{-h_{fb}}{(h_{fb} + 1)}$$
$$= \frac{-(-0.98)}{(-0.98 + 1)} = 49$$

## OPEN CIRCUIT PARAMETERS

When we choose the two currents as our independent variables, our parameters become *resistance* values. They are called *open circuit* parameters because the values for calculating the parameters are measured under specified open circuit conditions. We will use a small *r* to designate resistance followed by *i, r, f,* or *o* to designate input, reverse, forward, and output respectively. The parameters are:

$r_i$ = input resistance with output open.

$r_r$ = reverse transfer resistance with input open.

$r_f$ = forward transfer resistance with output open.

$r_o$ = output resistance with input open.

We will add a third letter into each parameter to indicate the circuit configuration. These will be *b, e,* or *c* with the same meaning as in our hybrid parameters.

### *measuring values*

We measure our values for calculating the open circuit formulas as indicated in Fig. 3-22.

### *calculating parameters*

We have assumed the common emitter configuration, but other configurations are the same. We can now use the values measured under the depicted conditions and construct these formulas:

$$r_{ie} = \frac{\Delta V_{be}}{\Delta I_b}$$

$$r_{re} = \frac{\Delta V_{be}}{\Delta I_c}$$

$$r_{fe} = \frac{\Delta V_{ce}}{\Delta I_b}$$

$$r_{oe} = \frac{\Delta V_{ce}}{\Delta I_c}$$

In each case, the open circuit parameter is obtained by a simple application of Ohm's law; $R = E/I$. We obtain the values of $E$ and $I$ by measuring, and use these values to calculate $R$.

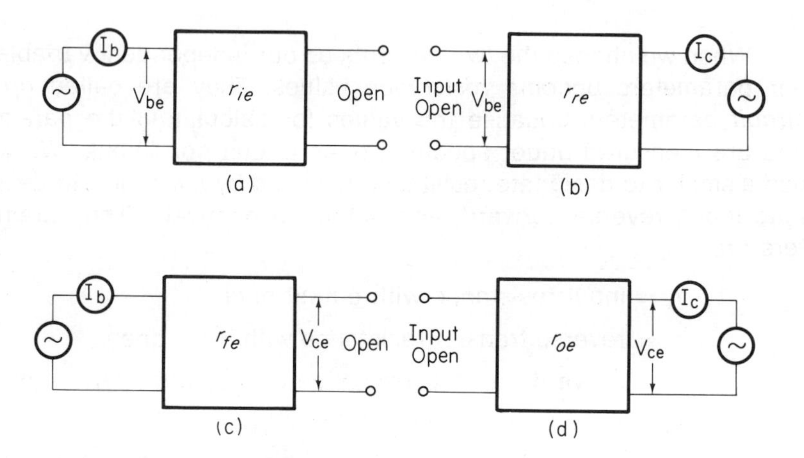


**Fig. 3-22.** Test Circuits for Open Circuit Parameters.

### Short Circuit Parameters

When we use the two voltages as the independent variables, our parameters will be values of *conductance*. We call these *short circuit* parameters because of the method used to take the measurements. These parameters are:

$g_i$ = input conductance with output shorted.

$g_r$ = reverse transfer conductance with input shorted.

$g_f$ = forward transfer conductance with output shorted.

$g_o$ = output conductance with input shorted.

Again, we will add a third letter (*b, e,* or *c*) to designate the circuit configuration.

### *measuring values*

The test circuits are illustrated in Fig. 3-23.

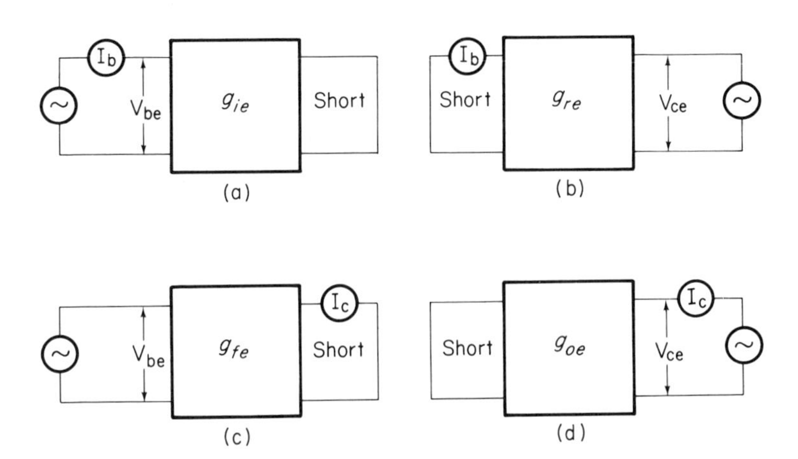


**Fig. 3-23.** Test Circuits for Short Circuit Parameters.

### *calculating parameters*

Again, we have assumed the common emitter configuration, and the other configurations differ only in the values measured. These are our formulas:

$$g_{ie} = \frac{\Delta I_b}{\Delta V_{be}}$$
$$g_{re} = \frac{\Delta - I_b}{\Delta V_{ce}}$$
$$g_{fe} = \frac{\Delta I_c}{\Delta V_{be}}$$
$$g_{oe} = \frac{\Delta I_c}{\Delta V_{ce}}$$

As you have probably noticed, the short and open circuit parameters are exact reciprocals of each other. In this case, we use the conductance formula, $G = I/E$. We measure the values of $I$ and $E$, and use them to calculate $G$.

## INTERRELATIONSHIP OF PARAMETERS

We have already seen the reciprocal relationship between short and open circuit parameters. There is also a relationship between each of these and our hybrid parameters.

### *parameter conversion*

The formulas in Fig. 3-24 are used for parameter conversion.

| From Hybrid to Open Circuit | From Hybrid to Short Circuit | From Open Circuit to Hybrid |
|---|---|---|
| $r_{ie} = h_{ie} - \frac{(h_{re})(h_{fe})}{h_{oe}}$ | $g_{ie} = \frac{1}{h_{ie}}$ | $h_{ie} = r_{ie} - \frac{(r_{re})(r_{fe})}{r_{oe}}$ |
| $r_{re} = \frac{h_{re}}{h_{oe}}$ | $g_{re} = \frac{-h_{re}}{h_{ie}}$ | $h_{re} = \frac{r_{re}}{r_{oe}}$ |
| $r_{fe} = \frac{-h_{fe}}{h_{oe}}$ | $g_{fe} = \frac{h_{fe}}{h_{ie}}$ | $h_{fe} = -\frac{r_{fe}}{r_{oe}}$ |
| $r_{oe} = \frac{1}{h_{oe}}$ | $g_{oe} = h_{oe} - \frac{(h_{re})(h_{fe})}{h_{ie}}$ | $h_{oe} = \frac{1}{r_{oe}}$ |

**Fig. 3-24.** Parameter Conversion Formulas.

Obviously the table in Fig. 3-24 relates directly to the common emitter configuration. The formulas are exactly the same for all three configurations; only the values are different. If we reconstructed the table and replaced each *e* with *b*, the table would be for the common base. If we replaced each *e* with *c*, the table would be for the common collector.

### *conversion exercise*

Let's use the sample hybrid parameters from Fig. 3-16 and convert the common base values to open circuit and short circuit values.

1. $r_{ib} = \frac{h_{ib} - (h_{rb})(h_{fb})}{h_{ob}}$

$$= \frac{39\,\Omega - (380 \times 10^{-6})(-0.98)}{0.49 \times 10^{-6}}$$

$$= 39 + 760 = 799\,\Omega$$

2. $r_{rb} = \dfrac{h_{rb}}{h_{ob}}$

$$= \frac{380 \times 10^{-6}}{0.49 \times 10^{-6}}$$

$$= 775.51\ \Omega$$

3. $r_{fb} = \dfrac{-h_{fb}}{h_{ob}}$

$$= \frac{0.98}{0.49 \times 10^{-6}}$$

$$= 2\ \text{M}\Omega$$

4. $r_{ob} = \dfrac{1}{h_{ob}}$

$$= \frac{1}{0.49 \times 10^{-6}}$$

$$= 2.04\ \text{M}\Omega$$

Let's check our results by converting the values back to hybrid parameters.

5. $h_{ib} = \dfrac{r_{ib} - (r_{rb})\,(r_{fb})}{r_{ob}}$

$$= \frac{799\ \Omega - (775.51)\ (2\ \text{M}\Omega)}{2.04\ \text{M}\Omega}$$

$$= 799 - 760 = 39\ \Omega$$

6. $h_{rb} = \dfrac{r_{rb}}{r_{ob}}$

$$= \frac{775.51\ \Omega}{2.04\ \text{M}\Omega}$$

$$= 380 \times 10^{-6}$$

7. $h_{fb} = \dfrac{-r_{fb}}{r_{oe}}$

$$= \frac{-(2\ \text{M}\Omega)}{2.04\ \text{M}\Omega}$$

$$= -0.98$$

8. $h_{ob} = \dfrac{1}{r_{ob}}$

$$= \frac{1}{2.04\ \mathrm{M\Omega}}$$

$$= 0.49\mu\mho$$

## APPLICATIONS FOR PARAMETERS

### *equivalent circuits*

Figure 3-25 is an equivalent circuit for a transistor amplifier in a common emitter configuration.

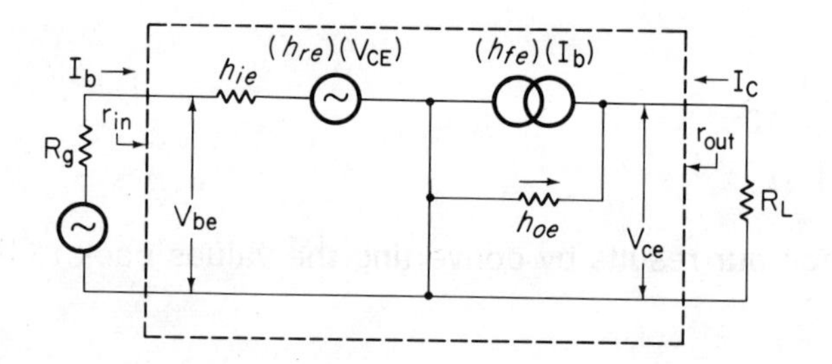


**Fig. 3-25.** Equivalent Circuit for CE Amplifier.

### *current gain*

Current gain ($A_i$) is always the output current divided by the input current. In this case, we have:

$$A_i = I_c / I_b$$

Expressed in hybrid parameters from the equivalent circuit, it becomes:

$$A_i = \frac{h_{fe}}{(h_{oe})(R_L) + 1}$$

### *voltage gain*

Voltage gain ($A_v$) is the output voltage divided by the input voltage. With our CE amplifier, it is:

$$A_v = \frac{V_{ce}}{V_{be}}$$

When we substitute hybrid terms from our equivalent circuit, it becomes:

$$A_v = \frac{(-h_{fe})(R_L)}{[(h_{ie})(h_{oe}) - (h_{fe})(h_{re})][R_L + (h_{ie})]}$$

We have a negative result which indicates a voltage phase inversion through the amplifier.

### *power gain*

Power gain ($G$) is power out divided by power in. With our CE amplifier, this is:

$$G = \frac{(I_c V_{ce})}{(I_b V_{cb})}$$

Power gain may also be expressed as the product of current gain times voltage gain. Placing hybrid values into our formula produces:

$$G = \frac{(h_{fe})^2 R_L}{(h_{oe} R_L + 1)[(h_{ie})(h_{oe}) - (h_{fe})(h_{re})][R_L + h_{ie}]}$$

### *input resistance*

The input resistance is input voltage divided by input current. With the CE amplifier, it is:

$$r_{\text{in}} = \frac{V_{be}}{I_b}$$

Substituting values:

$$r_{\text{in}} = \frac{h_{ie} + [(h_{oe})(h_{ie}) - (h_{fe})(h_{re})][R_L]}{(1 + h_{oe}) R_L}$$

### *output resistance*

The output resistance is the reciprocal of output conductance, and with the CE amplifier it is:

$$r_{\text{out}} = \frac{I_c}{I_{ce}}$$

The hybrid parameters change this to:

$$r_{\text{out}} = \frac{(h_{ie} + R_g)}{[(h_{oe})(h_{ie}) - (h_{re})(h_{fe})] + (h_{oe} R_g)}$$

This formula is derived by moving the signal source to the output circuit, but the resistance of the source ($R_g$) is still considered to be in the input circuit. The voltage output of the generator then becomes:

$$\frac{(h_{fe})\,(I_b)}{h_{oe}}$$

### *exercise in calculating values*

Suppose that we have a transistor with the following hybrid parameters.

$$h_{ie} = 1500\ \Omega$$

$$h_{re} = 5 \times 10^{-4}$$

$$h_{fe} = 50$$

$$h_{oe} = 20\ \mu\mho$$

Let's construct an equivalent circuit and calculate the values of current gain, voltage gain, power gain, input resistance, and output resistance. Our parameters are for the CE configuration, and we will further assume these values:

$$R_L = 15\ \text{k}\Omega$$

$$R_g = 1.5\ \text{k}\Omega$$

Our equivalent circuit should be the same as that in Figure 3-26.

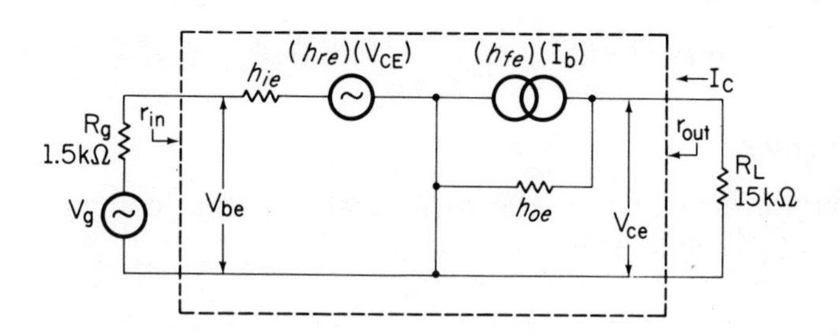


**Fig. 3-26.** Equivalent of Specified Circuit.

1. $$A_i = \frac{(h_{oe})(R_L) + 1}{h_{oe}}$$

$$= \frac{50}{20 \times 10^{-6} \times 15 \times 10^3 + 1}$$

$$= \frac{50}{1.3}$$

$$= 38.46$$

2. $$A_v = \frac{h_{fe} R_L}{[(h_{ie})(h_{oe}) - (h_{fe})(h_{re})][R_L + h_{ie}]}$$

$$= \frac{50 \times 15 \times 10^3}{[(1.5 \times 10^3 \times 20 \times 10^{-6} - 50 \times 5 \times 10^{-4}) \times 15 \times 10^3 + 1.5 \times 10^3]}$$

$$= \frac{750 \times 10^3}{[75 + 1500]}$$

$$= 476$$

3. $$G = \frac{(h_{fe})^2 R_L}{(h_{oe} R_L + 1)[(h_{ie} h_{oe} - h_{fe} h_{re}) R_L + h_{ie}]}$$

$$= \frac{50 \times 50 \times 15 \times 10^3}{(20 \times 10^{-6} \times 15 \times 10^3 + 1)[(1.5 \times 10^3 \times 20 + 10^{-6} - 50 \times 5 \times 10^{-4}) 15 \times 10^3 + 1.5 \times 10^3]}$$

$$= \frac{37.5 \times 10^6}{1.3 \times 1575}$$

$$= 18{,}314.8$$

We could have calculated the power gain by:

$$G = A_i \times A_v$$

$$= 38.46 \times 476$$

$$= 18{,}306.96$$

Notice that the difference is very slight, but the first method has a greater degree of accuracy. We rounded off the result for both $A_i$ and $A_v$ which gave us an error factor when we multiplied $A_i \times A_v$.

4. $$r_{in} = \frac{[h_{ie} + (h_{oe} h_{ie} - h_{fe} h_{re}) R_L]}{(1 + h_{oe} R_L)}$$

$$= \frac{[1.5 \times 10^3 + (20 \times 10^{-6} \times 1.5 \times 10^3 - 50 \times 5 \times 10^{-4}) 15 \times 10^3]}{(1 + 20 \times 10^{-6} \times 15 \times 10^3)}$$

$$= \frac{[(1.5 \times 10^3) + 75]}{(1 + 0.3)}$$

$$= 1211.5\ \Omega$$

5. $$r_{out} = \frac{(h_{ie} + R_g)}{[(h_{oe}\,h_{ie}) - (h_{re}\,h_{fe}) + (h_{oe}\,R_g)]}$$

$$= \frac{(1.5 \times 10^3 + 1.5 \times 10^3)}{(20 \times 10^{-6} \times 1.5 \times 10^3 - 5 \times 10^{-4} \times 50 \times 20 \times 10^{-6} \times 1.5 \times 10^3)}$$

$$= \frac{3 \times 10^3}{[(30 \times 10^{-3}) - (25 \times 10^{-3}) + (30 \times 10^{-3})]}$$

$$= \frac{3 \times 10^3}{35 \times 10^{-3}} = 85.7\ \text{k}\Omega$$

Assume that our amplifier has an input signal of 5 mV peak to peak which causes a change of 100 $\mu$A in the input current. Let's use the results of the previous calculations to determine the output current and voltage.

6. $$I_{out} = I_{in} \times A_i$$
$$= 100\ \mu\text{A} \times 38.46$$
$$= 3846 \times 10^{-6}$$
$$= 3.846\ \text{mA} \qquad \text{(peak to peak)}$$

7. $$E_{out} = E_{in} \times A_v$$
$$= 5\ \text{mV} \times 476$$
$$= 2380 \times 10^{-3}$$
$$= 2.38\ \text{V} \qquad \text{(peak to peak)}$$

## CHAPTER 3 REVIEW EXERCISES

1. Draw a *pnp* transistor in a common base circuit showing the proper bias circuits.
2. Name three types of amlifiers.
3. Name four classes of amplifiers.
4. What is the conduction time for a class AB amplifier?
5. Name the four variables in determining parameters.
6. What parameters are determined when input current and output voltage are made independent variables?
7. Draw a black box test circuit and label the four variables for a common base circuit.
8. The short circuit parameters are obtained when under what conditions?
9. Name the four hybrid parameters.

10. We have calculated $h_{ie} = 10\ k\Omega$.
    (a) What is the circuit configuration?
    (b) What value was held constant during the test?
    (c) A 2-V change of $V_{be}$ causes how much change in $I_b$?
    (d) Is this a typical value of hie for this type of circuit? Explain.
11. A transistor has a $h_{ie}$ of 2 kΩ and an $h_{fe}$ of 49 when used in a CE configuration. What is its $h_{ib}$ in a CB circuit?
12. A transistor has a $h_{rc}$ of $600 \times 10^{-6}$. What is its $h_{rc}$?
13. Explain the meaning of these terms:
    (a) $r_{ie}$.
    (b) $r_{rb}$.
    (c) $r_{oc}$.
14. Draw a test circuit for obtaining rfe and write the formula.
15. Draw the equivalent circuit for Fig. 3-27.

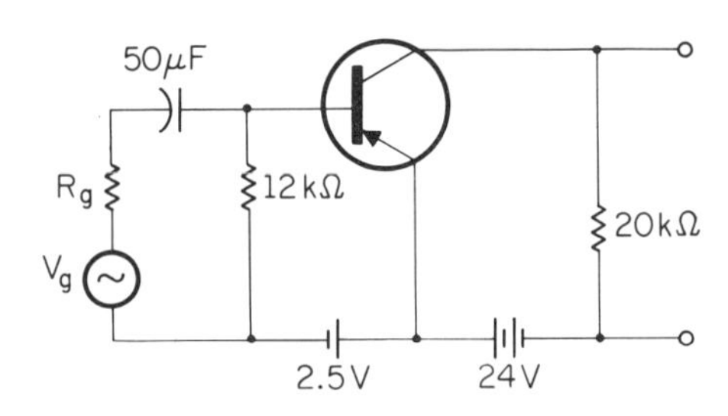


**Fig. 3-27.** Amplifier Circuit.

16. Referring to Fig. 3-27, express the following factors in terms of the hybrid parameters:
    (a) Current gain.
    (b) Voltage gain.
    (c) Power gain.
17. The hybrid parameters for the circuit in Fig. 3-27 are:

$$h_{ie} = 200\ \Omega$$

$$h_{re} = 600 \times 10^{-6}$$

$$h_{fe} = 50$$

$$h_{oe} = 25\mu\mho$$

    Calculate $A_i$, $A_v$ and $G$.
18. Convert the hybrid parameters of item 17 into open circuit parameters.

# 4

# dynamic analysis of amplifiers

## DYNAMIC CHARACTERISTICS

The dynamic characteristics of a transistor enable an accurate prediction of how the transistor will react to a specified set of conditions. The circuit designer works from dynamic characteristics. Those characteristics are different for every different circuit. Therefore, the designer must create his own. The first step is to construct a load line.

### *plotting the load line*

The load line for a particular circuit is constructed by drawing a diagonal line on a family of characteristic curves. The chart to be used, of course, must match the circuit configuration. The exact position of the line is determined by two factors:

1. The size of the load resistor.
2. The amplitude of the output bias voltage.

We now need a particular circuit. So let's use the one in Fig. 4-1.

This is a common emitter configuration using an *npn* transistor. The matching static characteristics are given in Fig. 4-2. Notice that two points have been marked on this graph. They are 20 V of collector voltage ($V_C$) and 4 mA of collector current ($I_c$).

These two points represent the two extreme conditions for the

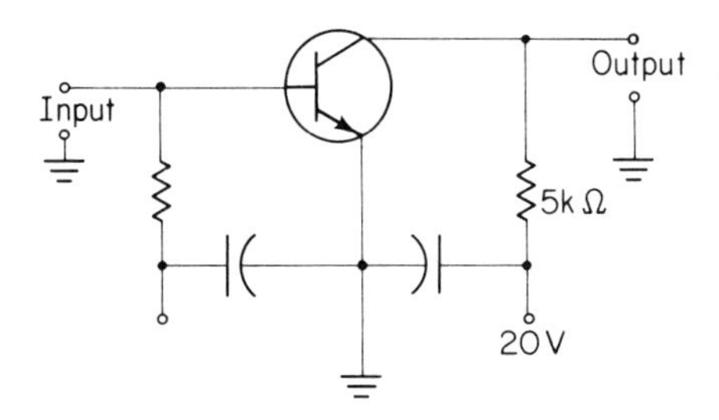


**Fig. 4-1.** Circuit with Load Resistor and Output Bias.

circuit in Fig. 4-1. If the transistor is cut off, the full 20 V will be on the collector. If the transistor is saturated, nearly all the voltage will be dropped across the resistor. In the first case, we have 20 V of collector voltage and 0 collector current. This is point *A* on the graph. Under the second condition, we have the maximum current of 4 mA. This is determined by dividing the 20 V by the load resistance of 5 kΩ. This gives us point *B*. To summarize, point *A* is maximum volts $V_c$ and zero $I_c$; point *B* is maximum $I_c$ and zero $V_c$. ($V_c$ is dc voltage measured from collector to ground.)

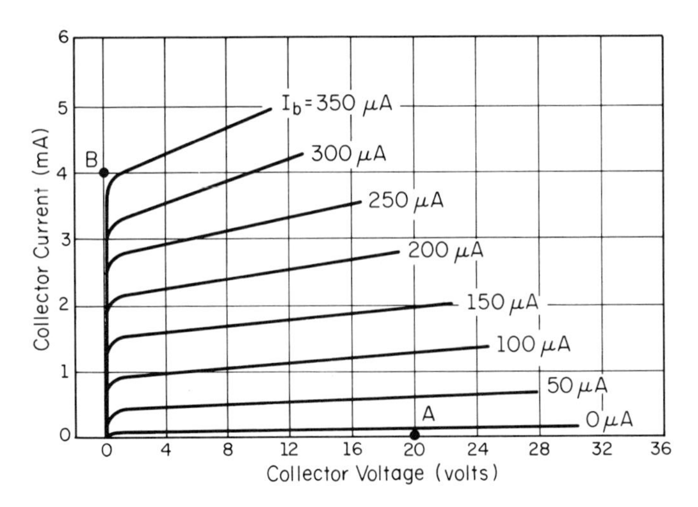


**Fig. 4-2.** Maximum Voltage and Maximum Current.

When the points *A* and *B* of Fig. 4-2 are connected together by a straight line, we have a load line. This is illustrated in Fig. 4-3.

This load line indicates all the possible values of collector voltage, collector current, and base current. Remember now, we are discussing a particular transistor in a particular circuit; specifically

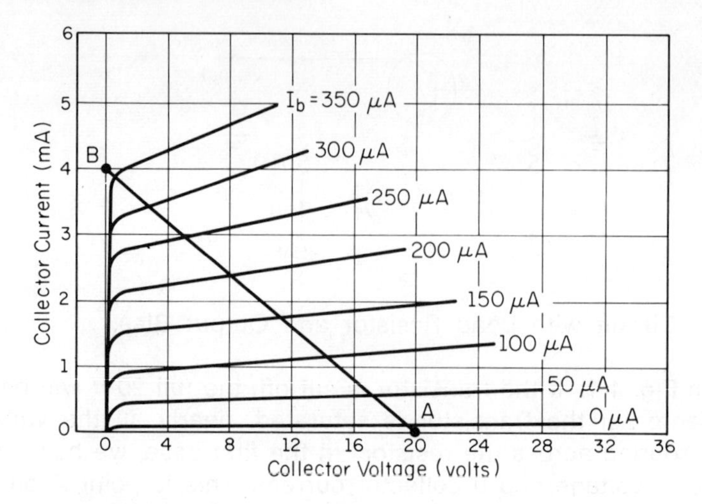


**Fig. 4-3.** Load Line for Fig. 4-1.

the circuit of Fig. 4-1. Notice that the value of the base resistor ($R_b$) has not been specified. The size of $R_b$ in conjunction with the magnitude of input bias, determines the base current. By manipulating these two factors, we can establish $I_b$ at just about any point we desire. If we can establish an $I_b$ of 200 μA we will be near the center of the load line.

What size resistor will give us 200 μA of $I_b$ when we use 20 V for bias? $R_b = 20\ \text{V}/200\ \mu\text{A} = 100\ \text{k}\Omega$ as shown in Fig. 4-4.

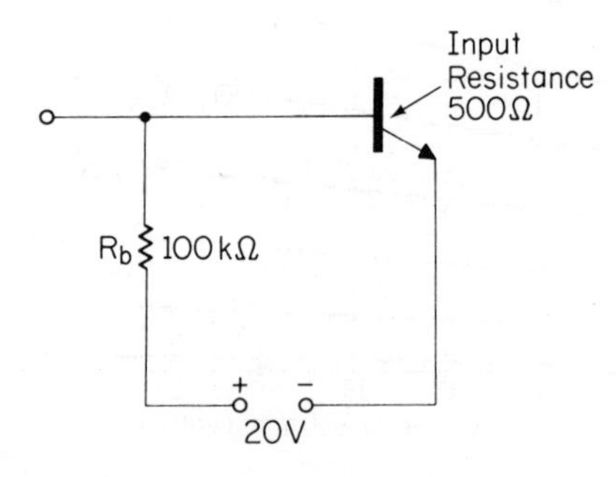


**Fig. 4-4.** Relation of Input Resistance to $R_b$ in Bias Circuit.

In this calculation, we ignored the input resistance which appears across the emitter-base junction. This is possible because $R_b$ is in series with the input resistance, and $R_b$ is very large when compared to input resistance. In this case, input resistance is 500 to 1500 Ω according to typical common emitter values. We'll use 500 Ω as our value. This combination is illustrated in Fig. 4-4. Let's insert this resistor and connect it to a 20-V bias. The circuit of Fig. 4-1 then becomes that in Fig. 4-5. Now, when we mark the point where the load

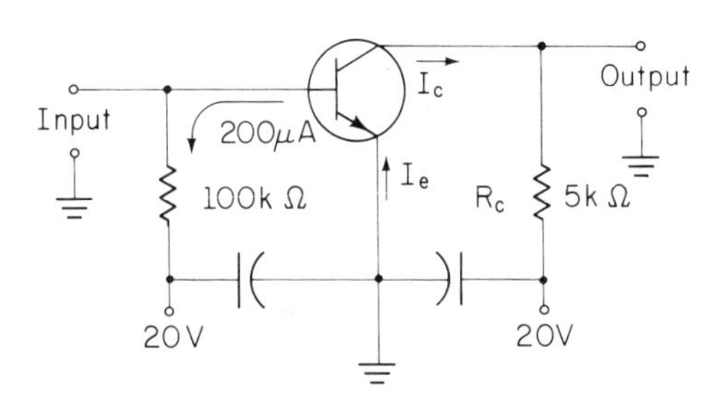


**Fig. 4-5.** Establishing Base Current.

line intersects the 200-μA line, the graph of Fig. 4-3 becomes that in Fig. 4-6. Point *C* has just been established.

The graph in Fig. 4-6 will tell us all we need to know about the circuit in Fig. 4-5. Point *C,* which we established when we selected

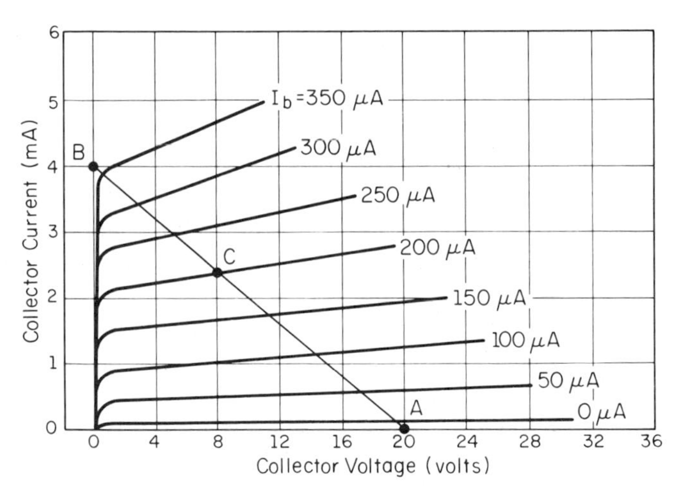


**Fig. 4-6.** Establishing the Operating Point.

$R_b$ and bias, is the *operating point* for this circuit. The operating point indicates the value of $I_b$, $I_c$, and $V_c$ when we have no input signal. Drop a vertical line from the operating point to determine the collector voltage. It seems to be 8.2 V. Project a horizontal line to the left to determine collector current. 2.3 mA is about right. The operating point has provided:

$$I_b = 200\ \mu A$$
$$I_c = 2.3\ mA$$
$$V_c = 8.2\ V$$

***using the load line to calculate values***

An incoming signal will vary the bias and set up ac variations around the operating point. A small amplitude input signal will cause a large change in output current. To fully understand this operation, we need to take another look at the input resistance. Let's assume that we have an input signal that is 0.05 V peak to peak, and apply this into our bias circuit. This is illustrated in Fig. 4-7.

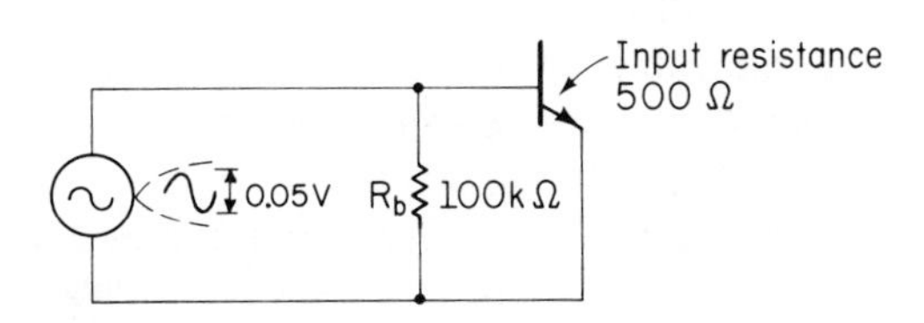


**Fig. 4-7.** Signal Effect on Bias.

The dc source has been omitted for sake of clarity; we are interested in ac change. The signal source is an ac generator with a peak to peak output of 0.05 V. This ac signal is applied across both $R_b$ and the input resistance. As far as the signal is concerned, these two resistors are in parallel. The full 0.05 V appears across each resistor. Across the 100 kΩ resistance of $R_b$, the effect is negligible. Across the 500 Ω of internal resistance, the effect is significant. It will increase the collector to base current by 0.05 V/500 Ω which is 0.0001 A or 100 μA.

Let's go back to the circuit and load line (Figs. 4-5 and 4-6), and determine the effect of a 100-μA change in $I_b$. From the operating point (*C*), move up the load line to the 250-μA $I_b$ line. Back at the operating point, move down the load line to the 150-μA $I_b$ line. Mark these two points. Extend horizontal lines from each point to the left margin. The difference in $I_c$ between these two lines indicate the change in collector current. It seems to have changed by 1.1 mA from 1.8 to 2.9 mA. The 0.05 V, 100-μA signal input has caused a collector current change of 1.1 mA. This is a current gain of 11. This is considerably less than the typical gain of a common emitter amplifier. Our circuit must not be in the typical circuit category.

How about the change in collector voltage? Dropping a vertical line from 250-$\mu$A $I_b$ point on the load line, we find that the minimum collector voltage is 5.5 V. A second vertical line down from the 150-$\mu$A $I_b$ point shows the maximum collector voltage to be 11.2 V. The change in $V_c$ is 11.2 V — 5.5 V which is 5.7 V. The voltage gain is Δ ouput/Δ input. This is 5.7 V/0.05 V which is a gain of 114.

The power gain is equivalent to the current gain times the voltage gain. This is 11 × 114 or 1254.

The gain through a particular transistor is greatly affected by the size of the load resistor ($R_c$). Figure 4-8 shows three load lines. Each line corresponds to a specific value of $R_c$. All other circuit values are the same for all three.

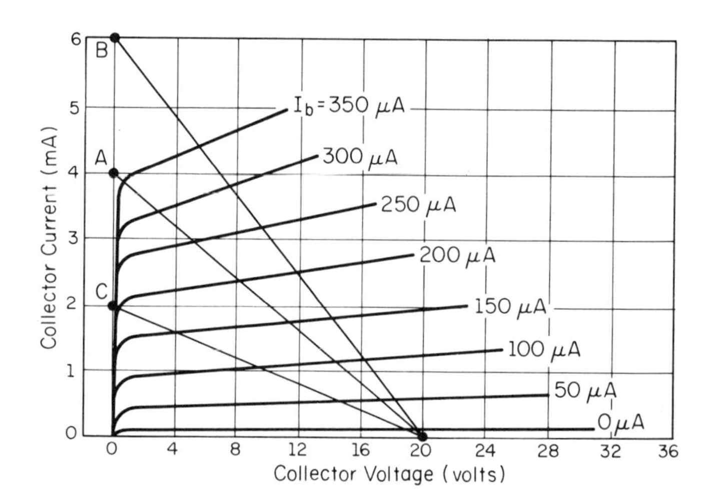


**Fig. 4-8.** Effect of $R_c$ on Gain.

Line *A* is the same as our previous load line. It represents the circuit with an $R_c$ of 5 kΩ. Line *B* is for a 3.3 kΩ $R_c$, and line *C* is for an $R_c$ of 10 kΩ.

For a given input signal, line *B* provides the greatest change in $I_c$ while line *C* indicates the greatest change in voltage. From this, we may conclude that current gain is inversely proportional to the ohmic value of $R_c$, and the voltage gain is directly proportional to the same value. We have already calculated the gains in the circuit represented by line *A*. So, it should not be necessary to repeat all the details with the other two lines.

Changing the output bias affects the gain only slightly. Raising the value of bias, moves the load line to the right; lowering it moves the line to the left. Figure 4-9 shows load lines for three values of collector bias. All circuit components are the same in each case.

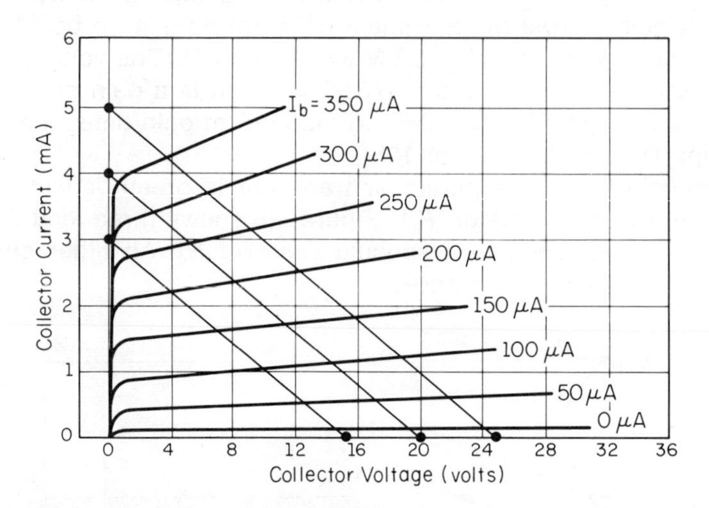


**Fig. 4-9.** Effect of Collector Bias on Gain.

Line *A* is our familiar load line with a $V_c$ of 20 V. The other two lines are for the same circuit with different values of $V_c$. Notice that the load lines are parallel to one another, while the base current ($I_b$) lines fan out slightly on the right side. This fanning out of $I_b$ lines accounts for the slight effect $V_c$ has on gain. As $V_c$ increases, the load line moves from the lower left toward the upper right. Voltage, current, and power gains all increase slightly with an increase in $V_c$.

The usefulness of the load line can be extended greatly by projecting it onto another chart. This chart is called a dynamic transfer characteristic curve.

## DYNAMIC TRANSFER CHARACTERISTIC CURVE

Figure 4-10 illustrates the method of constructing this curve.

Chart B is a load line for a particular common emitter amplifier circuit. Eleven points have been emphasized along the load line. Each point represents a specific value of base current and collector current. Chart A is a projection of these eleven points. When these projected points are connected in series by a smooth curve, we

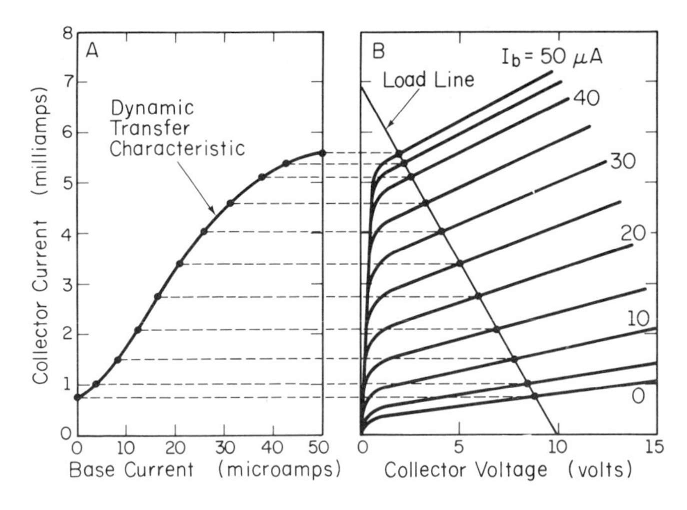


**Fig. 4-10.** Projecting the Dynamic Characteristics.

have the dynamic transfer characteristic curve. This curve represents all possible combinations of input current ($I_b$) and output current ($I_c$).

Chart A can be used to predict the exact effect that this amplifier will have on any input signal. Remember that we can move the operating point to any spot along the load line. How did we do that? We adjusted the input bias. An operating point has not been established in this amplifier circuit. So, we are free to set it where we like. The position of the operating point, as projected onto the dynamic curve, will establish the ac ground level for both input and output signals. Let's examine the use of this dynamic transfer characteristic curve.

### *predicting the output*

As previously stated, the dynamic transfer characteristic curve can be used to predict the exact effect on any input signal. By that we mean that it is possible to determine the amplitude and shape of the output with respect to the input.

Figure 4-11 is the same dynamic curve that we saw before. This time it has the eleven points projected both horizontally and vertically. The vertical projections represent inputs ($I_b$) while the horizontal projections represent outputs ($I_c$).

The point of projection is determined by establishing the operating point. The input line (vertical) to this point is the ac ground for the input. The output line (horizontal) from the operating point is the ac ground for the output signal. On the input, the position of the chart to the left of the operating point is negative; to the right is positive. On the output, negative is below the operating point; above this point is positive. Let's analyze some inputs and outputs. Figure 4-12 shows an input with a distorted output.

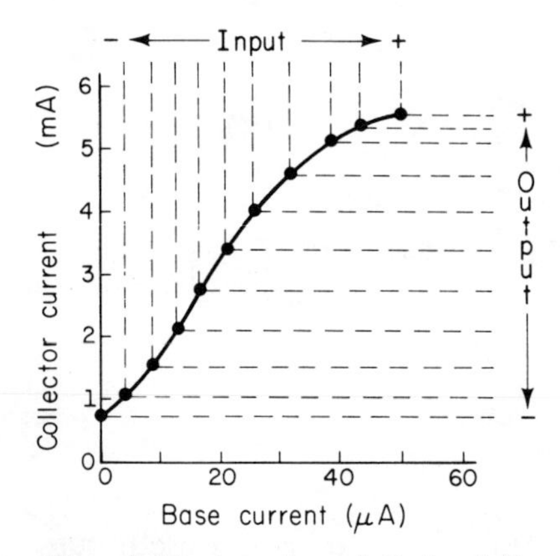


**Fig. 4-11.** Input and Output Projections.

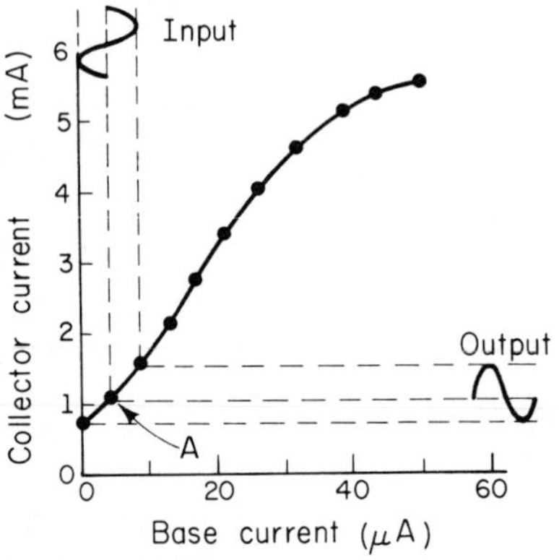

**Fig. 4-12.** Operating Point Too Low.

Point *A* is the operating point, and it is very low on the curve. The positive alternation of the input gets a fair amplification, but the negative alternation is distorted. The negative alternation of the input drives the transistor to near cutoff which limits the ampitude of the output. Also, there is a nonlinear change ratio between $I_b$ and $I_c$ in this area near cutoff. This nonlinearity flattens the negative alternation of the output. A considerable overall amplification has occurred despite these facts, but we have a poor reproduction of the input signal.

Moving the operating point to the other end of the curve produces the result shown in Fig. 4-13.

The operating point (*A*) is very near the point of transistor saturation. Amplification (gain) is very limited because the change in input bias has little effect on the output current.

These two examples should emphasize the fact that distortion occurs when the operating point is near either curved end of the line. For a good reproduction of the input signal, the operating point should be near the center of the linear portion of the curve. Not only will this give good fidelity, but it will result in maximum gain at the same time. (Electronically, fidelity is the degree of exactness to which the input is reproduced in the output. High fidelity is a good reproduction. Low fidelity is a poor reproduction.) Figure 4-14 illustrates the input and output of a high fidelity circuit.

Here the operating point (*A*) has been set for the exact center of the linear portion of the curve. The signal input is 10 μA peak

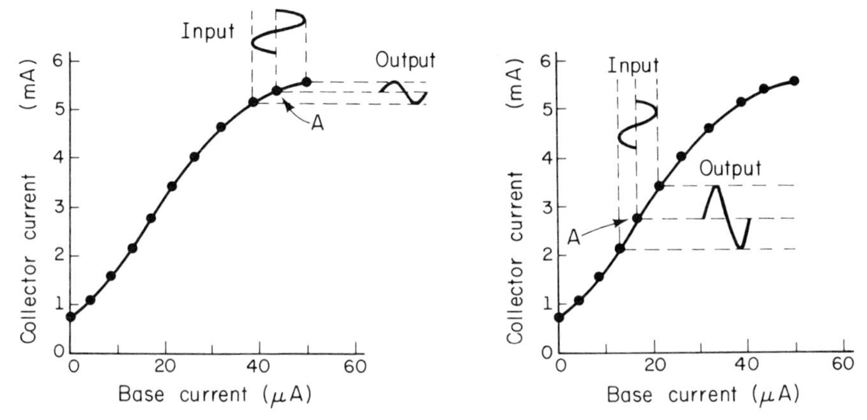


**Fig. 4-13.** Operating Point Too High.

**Fig. 4-14.** High Fidelity.

to peak. If this is the maximum input signal, any one of the three projection points could be the operating point and still obtain high fidelity. With the point as shown, this amplifier can handle signals twice this large with no noticeable distortion.

Transistors are very effective and reliable amplification devices, but they have certain limitations that must be considered. One of these is frequency.

## FREQUENCY LIMITATIONS

Transistors are fairly broad band devices, but no given transistor can be expected to handle all frequencies. The reason for this is the internal impedance between elements. We have already discussed an input resistance as well as an output resistance. Actually, there is an internal *RC* impedance network between any two of the elements. This is illustrated in Fig. 4-15.

Impedance *A* is across the emitter-base junction and represents the input impedance. In this common emitter, impedance *B* is the output impedance. Impedance *C* is across the collector-base junction. At low frequencies, the impedances are largely resistive with the capacitors acting as open circuits. Transistors which are designed to operate in the power and audio frequency ranges, have rather large internal capacitances.

As frequency increases, the impedances become more capacitive in nature. The capacitive reactance grows smaller, and allows

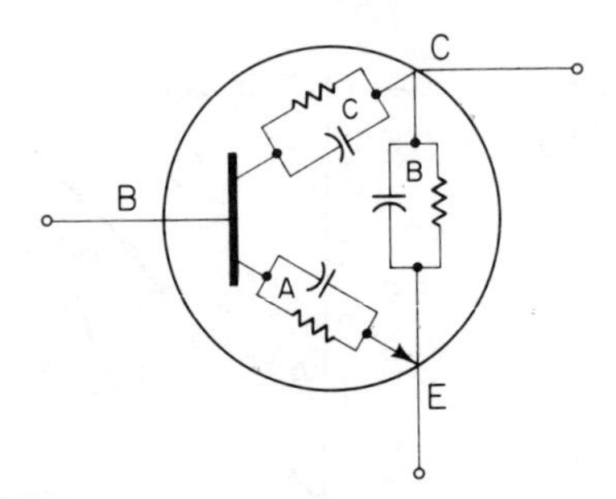


**Fig. 4-15.** Interelement Impedance.

more of the signal to be shunted past the resistor. Radio frequency transistors are manufactured with very small interelement capacitances. The smaller capacitance increases $X_C$ and enables the transistor to handle higher frequencies.

The interelement capacitance varies to some extent in accordance with the current and voltage. That is another reason why it is important to observe the recommended voltages. The collector supply voltage specified by the manufacturer should be kept within a close tolerance.

## POWER LIMITATION

Transistors are rated as to the maximum collector power. This rated power should not be exceeded for extended periods of time. In order to make sure that this rated value is not exceeded, a constant power line should be drawn on the static characteristic curves. This line should connect all the values where voltage times current are equal to the rated power. This line should be drawn before the load line is constructed. Figure 4-16 shows such a line.

Here we have assumed that the maximum rated power is 18 mW. The curved line on the chart connects all the points where power is 18 mW. This is a boundary that the load line must not cross. If you will examine the load line in Figs. 4-3 and 4-6, you will see that we have not exceeded the maximum constant power rating.

Our primary concern in some amplifiers is the power gain. If

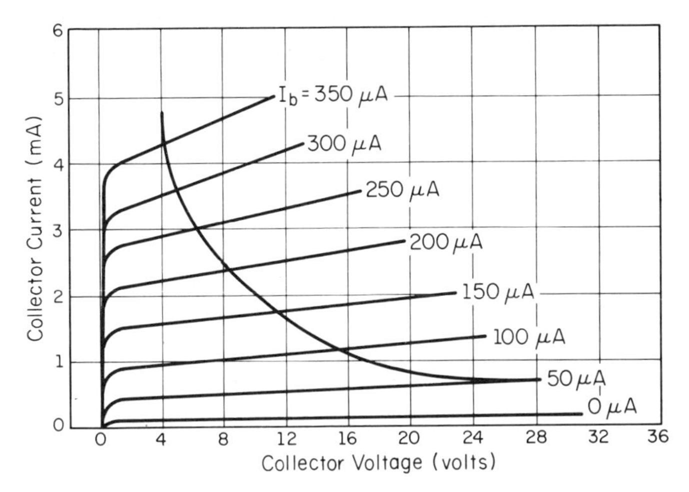


**Fig. 4-16.** Constant Power Line.

the load line is constructed so that it is tangent to the constant power line, it will allow maximum power amplification without exceeding the limits. This is illustrated in Fig. 4-17.

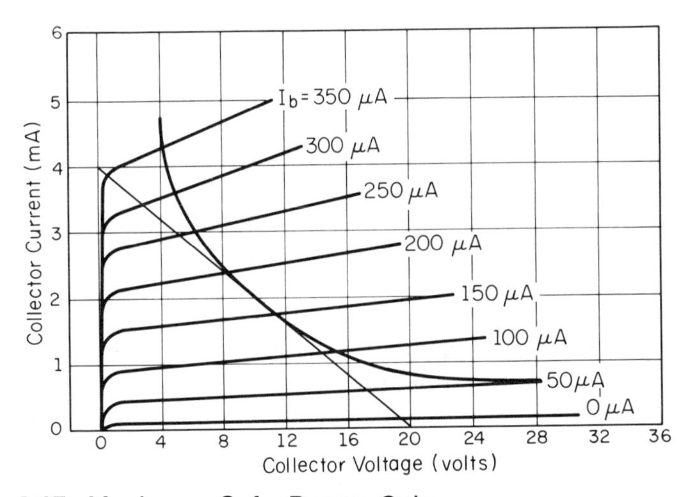


**Fig. 4-17.** Maximum Safe Power Gain.

The chart in Fig. 4-17 was constructed by superimposing Figs. 4-3 and 4-16. Remember that the circuit represented by this load line had a power gain of over 1200. This chart shows that this was the most that circuit could do without exceeding the rated power.

Remember, that the power line is called a constant power line.

This is the amount of power that the collector can handle on a continuous basis. A short surge of power beyond the safe limits should do no harm. However, if the surge is too high, or lasts too long, the transistor will be damaged.

Even a properly biased transistor can be overdriven by signals that are too large. When this happens, limiting occurs in the amplifier.

## OVERDRIVEN AMPLIFIERS

We have seen the distortion of signals which resulted from the misplaced operating point. In those cases, the transistor was never completely cut off. All of our previous amplifiers have been operated class A. It is possible for an incoming signal to stop the current. When this happens, the class of operation becomes AB, B, or C. The class is determined by the percentage of signal time represented by the cutoff period. This is called cutoff limiting, and it is similar to the action in a half wave rectifier. The portion of the signal below the cutoff point will be eliminated from the output.

Another type of limiting occurs when the transistor is driven into saturation. After saturation is reached, the changing bias has no effect on the collector current. The portion of the signal which occurs during saturation is also lost from the output wave shape. This action is known as saturation limiting.

It sometimes happens that a signal will be strong enough to overdrive the amplifier in both directions. This results in cutoff limiting during part of one alternation and saturation limiting during part of the other alternation. Fig. 4-18 illustrates the action in an overdriven amplifier.

Here, the operating point is near the center of the curve where base current is 25 $\mu$A. The amplifier will operate class A for all input signals with peak to peak amplitudes that do not exceed 50 $\mu$A. But we are applying a signal which has a peak to peak amplitude of 70 $\mu$A.

The first 25 $\mu$A of the positive alternation of the input is amplified and passes through to the output. The transistor is then saturated. The input continues to rise for another 10 $\mu$A but it causes no change in collector current. The shaded area A indicates the portion of the input that takes place during saturation. The output signal is squared during the saturation period. Shaded area *A* is eliminated from the output signal.

The input passes the positive peak and starts to decrease.

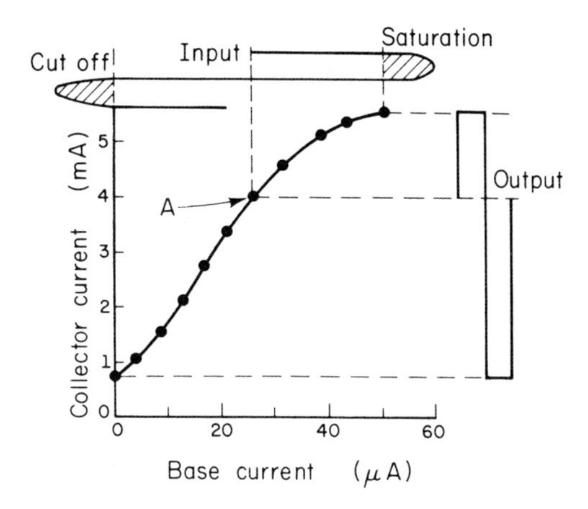


**Fig. 4-18.** Cutoff and Saturation Limiting.

When it reaches an amplitude of 25 μA, it regains control of the collector current. Input and output currents drop together until cutoff is reached. Cutoff occurs when the negative alternation of the input reaches an amplitude of —25 μA. Collector current ceases at this time. The output remains at zero while the input signal passes through the changes indicated by shaded area *B*. During this time the negative alternation increases from —25 μA to —35 μA, passes the negative peak, and decreases back to —25 μA. At this point collector current starts again. The remainder of the input is reproduced in the output.

An overdriven amplifier is not necessarily a bad thing. It depends upon the purpose of the amplifier stage. If the purpose was high fidelity amplification of a sine wave, this is not the way to do it. If our purpose is to amplify a sine wave and transform it into a square wave, that job is done very well.

Up to this point our amplifiers have had fixed bias. Let's take a look at some other possible ways of arranging bias.

## SELF-BIASING

Input bias for an amplifier can be provided in part by the circuit. One method of doing so is shown in Fig. 4-19.

Notice that $R_b$ is connected into the collector circuit instead of to a separate power source. With this arrangement the collector voltage becomes the power source. The base current is controlled by the level of voltage at the collector. The split resistor for $R_b$ in

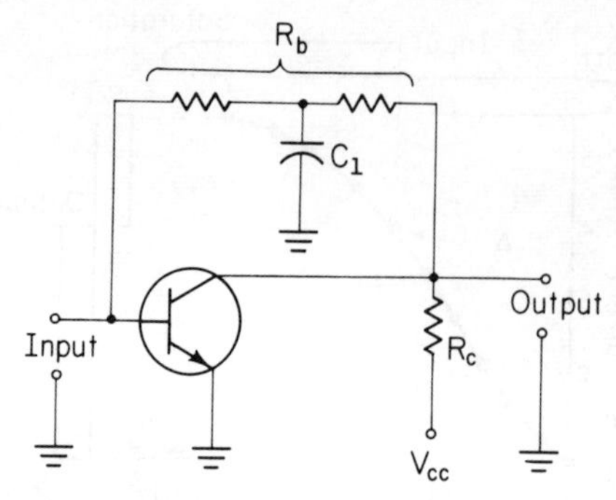


**Fig. 4-19.** Self-Biased Common Emitter.

conjunction with capacitor $C_1$ prevents the output ac signal from coupling back to the input.

Biasing a transistor in this fashion offers a certain amount of thermal stabilization. Bias on a transistor may become unsatisfactory when the transistor reaches a certain temperature. With a self-bias arrangement, the change in temperature adjusts the bias accordingly. This provides a more stable operation.

We have examined the amplifier stage in considerable detail. What happens to the output when it leaves our amplifier? How does it pass from one stage to the next? This is a job performed by the coupling network.

## COUPLING NETWORKS

All coupling networks can be placed into one of four categories: direct coupling, *RC* coupling, impedance coupling, and transformer coupling. The type of coupling to use in a particular case is generally dictated by the frequency of the signal being coupled.

### *direct coupling*

This is essentially a direct wire from the output of one stage to the input of another. This is a satisfactory method of connecting dc and very low frequency signals. A sample direct coupling arrangement is shown in Fig. 4-20.

In this arrangement, the collector current of $Q_1$ is limited to the base current of $Q_2$. This does not have to be the case. A resistor can be connected from point *A* to either ground or B+. This will provide a shunt path and relieve $Q_2$ from carrying the collector current for $Q_1$.

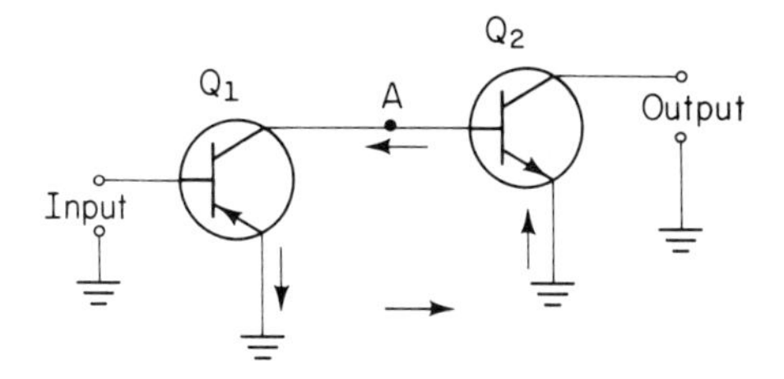


**Fig. 4-20.** Direct Coupling.

There is a limit to the number of stages that can be connected by direct coupling. The main reason for the limitation is thermal instability. When temperature varies the bias in the first stage, the variation in current is amplified through all the stages.

### *RC coupling*

Figure 4-21 illustrates *RC* coupling between two amplifier stages.

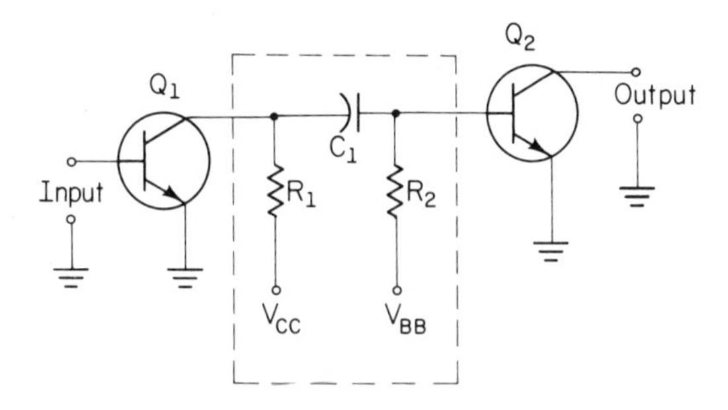


**Fig. 4-21.** *RC* Coupling.

The coupling circuit is composed of $R_1$, $C_1$, and $R_2$. But as frequently happens, some elements of the filter are performing a dual role, $R_1$ is the collector resistor for $Q_1$, and $R_2$ is the base resistor for $Q_2$.

The signal from $Q_1$ is developed across $R_1$ then coupled through $C_1$ to the input of $Q_2$. This type of coupling is popular in audio amplifiers, is sometimes used in power amplifiers, but it is not widely used otherwise. As frequencies decrease $X_C$ of $C_1$ increases. Thus, the very low frequencies are attenuated. Raising the frequency decreases $X_C$, and there comes a point when the capacitor acts as a short.

### *impedance coupling*

Figure 4-22 shows two stages connected by impedance coupling.

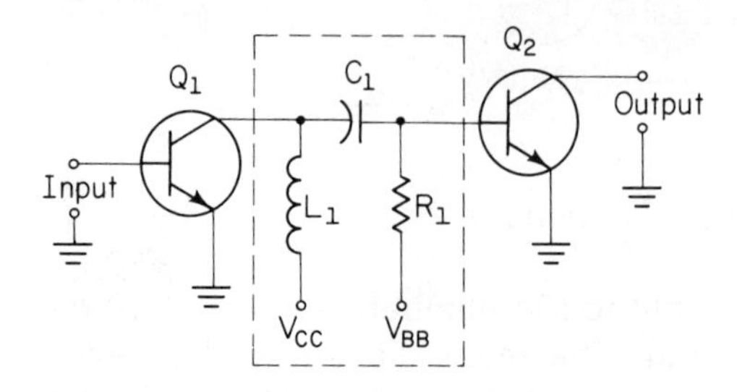


**Fig. 4-22.** Impedance Coupling.

This coupling is similar in appearance to RC except for the fact that one or more resistors have been replaced by inductors. In this diagram, $L_1$ is the collector load for $Q_1$. This coupling is good for high audio through low radio frequencies. At low frequencies the capacitor tends to block the signal, and the inductor tends to short it out. At very high frequencies the inductor becomes an open while the capacitor acts as a short.

### *transformer coupling*

Figure 4-23 illustrates the use of transformer coupling.

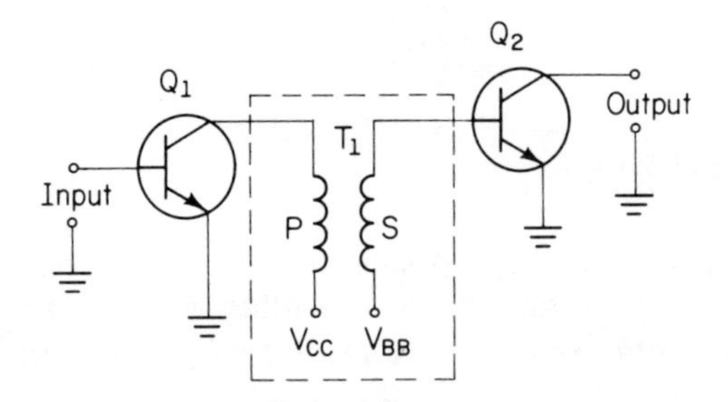


**Fig. 4-23.** Transformer Coupling.

The primary of transformer $T_1$ is the collector load for $Q_1$. The signal is developed here and electromagnetically coupled to the secondary. The secondary has replaced the base resistor for $Q_2$. With some variations, this type of coupling is suitable for a wide range of frequencies from high audio to very high radio. At the lower frequencies, the transformer has an iron core. As frequency increases, the iron core gives way to powdered iron, and finally, to air core.

Maximum power transfer between stages occurs when the coupling matches the output impedance of one to the input impedance of the other. This matching is accomplished easily with transformer coupling. It is largely a matter of selecting a transformer with the proper primary to secondary turns ratio.

## *link coupling*

Link coupling is special application of transformer coupling. It is a transformer link between two tuned circuits as shown in Fig. 4-24.

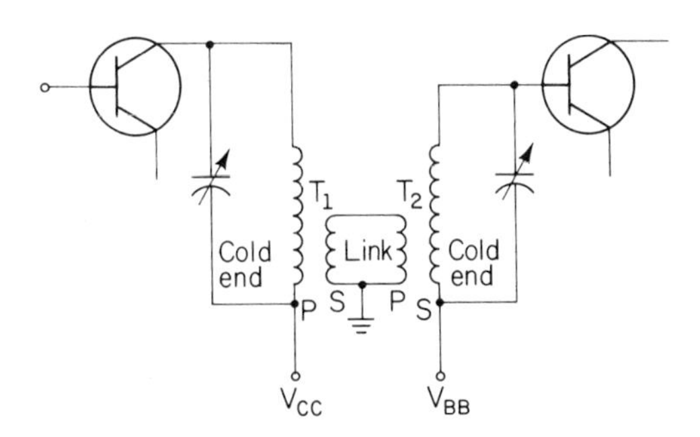


**Fig. 4-24.** Link Coupling.

Tuned tank circuits of the type shown are generally unbalanced. This causes one end of the tank to be at a higher signal potential than the other. The high end of the tank is called the hot end; the low end is the cold end. The transformer link balances the impedances and provides maximum coupling from the cold end of one tank circuit to the cold end of another tank circuit.

This is a versatile coupling and finds its greatest use in coupling stages of RF amplifiers. It is found most often in transmitter circuits.

The link is a low impedance circuit that generally couples two widely separated tank circuits. The link bridges this gap with a minimum of power loss. The coupling between the link and either tuned circuit can be made adjustable without complex mechanical problems. One side of the link is normally grounded to eliminate stray capacitive coupling. This also eliminates harmonics which could cause distortion feedback.

## INTERNAL FEEDBACK

One problem that is inherent to transistors is caused by the fact that they couple signals in both directions. This coupling takes place through the interelement impedance between input and output.

The portion of the output signal that couples back to the input is called feedback. This feedback will cause trouble in most amplifiers if it is not neutralized. If the feedback aids the incoming signal (same phase), it is called positive or regenerative feedback. Regenerative feedback tends to set up oscillations. When the feedback is out of phase with the input, it is called degenerative or negative feedback. Degenerative feedback reduces the amplitude of the incoming signal which results in a lower output. The oscillations from regenerative feedback upset the stabilization of the amplifier. Attenuation from degenerative feedback results in a loss of power. As frequency increases this undesirable feedback becomes more pronounced.

The best way to eliminate the effect of feedback is to construct a coupling circuit which transforms the transistor into a unilateral device. In order to do this, an external circuit must be designed to neutralize both the resistance and capacitance of the internal impedance. In other words, an external feedback circuit which exactly cancels the effect of the internal feedback circuit. Figure 4-25 shows such an arrangement.

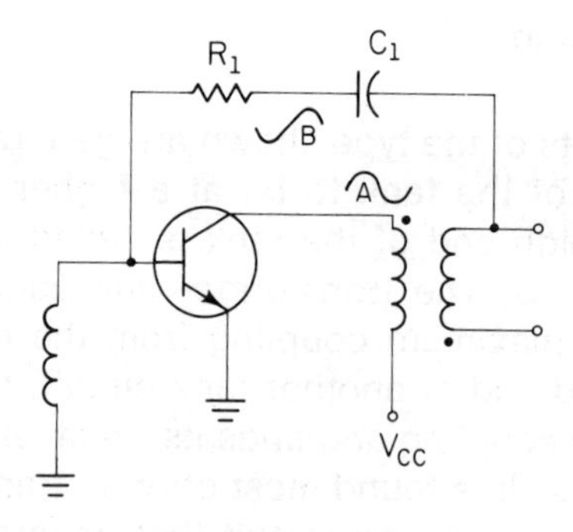


**Fig. 4-25.** External Feedback.

$R_1$ and $C_1$ form the external feedback circuit. These components should match the values of the internal components between base and collector.

Signal *A* represents the phase and amplitude of the portion of the signal which is coupled back internally from collector to base. Signal *B* represents the phase and amplitude of the signal which is coupled back to the base through the external circuit. These two feedback signals should be of the same amplitude and 180° out of phase.

When the effect of both internal components are cancelled in this fashion, the transistor can pass a signal in only one direction. It is, in effect, a unilateral device.

It was mentioned at the beginning of this chapter that amplifiers are named according to the job they do. Let's consider the circuits of one that is designed to amplify audio frequencies.

## AUDIO AMPLIFIER

These amplifiers are intended to amplify frequencies that are classified in the audible range. The extreme limits of the audio range are from 16 Hz to 20 kHz. It is not likely that we will have an amplifier that can equally amplify all the frequencies in this band. However, the audio amplifier must be able to efficiently amplify the majority of these signals. Therefore, the audio amplifier is a wide band amplifier.

### *circuit considerations*

Any of the transistors that we have considered are suitable for use in an audio amplifier. Let's use an *npn* junction. The largest part of our problem is deciding what circuit components to use with it. As to circuit configuration, the common emitter will do fine.

Audio signals on the order of $\mu$V must be separated from noise and amplified. These frequencies range over a wide band, and the very low frequency signals need amplification as much as those near 20 kHz. The circuit then must contain a device for noise elimination and a means of enhancing the gain for frequencies on the low end of the band. The high end needs no special attention.

We are interested in a high fidelity amplification. This means that the bias must be arranged for an operating point near the center of our load line. Furthermore, this bias must be stabilized for variations in temperature. As to coupling, *RC* coupling does a good job. Let's put these together and see what form the circuit will take.

### *analyzing the circuit*

Figure 4-26 shows two stages of audio amplification.

In our first stage of amplification, we need the maximum power input. This is accomplished by transformer coupling from the previous stage. The primary of $T_1$ matches the high output impedance of the previous stage. The secondary matches the low input impedance of $Q_1$. This gives us the maximum obtainable input power.

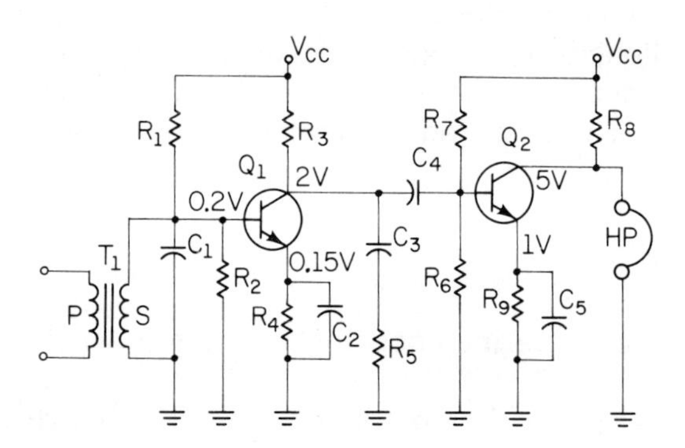


**Fig. 4-26.** Audio Amplification.

The noise accompanying our audio signal is almost as large as the signal itself. Most of the noise signals, however, are of relatively high frequencies. Capacitor $C_1$ decreases this noise level by shunting its high frequency components to ground. Care must be exercised in selecting this capacitor. If it is too large, it will shunt the high audio signals as well as the noise.

$R_1$and $R_2$ form a bleeder network for the input bias. This bias is a fixed value of approximately 0.2 V at the base of $Q_1$. The fixed bias is both aided and stabilized by the *RC* network in the emitter circuit of $Q_1$. $R_4$ develops a positive potential of 0.15 V at the emitter. Slow current changes caused from temperature will vary this potential. $C_2$ is a direct short for audio frequency signals. In effect $C_2$ places ac ground directly on the emitter. This prevents $R_4$ from attenuating the signal.

The audio frequencies are amplified through the transistor and are coupled out on the collector. Ordinarily there would be a low amplification of very low frequency signals, but this is prevented by $C_3$ and $R_5$. The impedance of this shunting network is inversely proportional to frequency. This forces the low frequencies to be amplified almost as much as the high frequencies.

Our next coupling network is $C_4$ and $R_6$. $C_4$ should be as large as possible to avoid undue attenuation of the low frequency components. $R_6$ should be large in order to couple maximum signal to the input resistance of $Q_2$.

$R_6$ and $R_7$ also form a voltage divider network for the base bias on $Q_2$. This bias is enhanced and stabilized by the emitter self-bias furnished by $R_9$ and $C_5$.

The audio signal from the collector of $Q_2$ is directly coupled to a pair of head phones.

After audio amplifiers we have amplifiers designed for narrow-band intermediate and radio frequencies. Let's bypass these and move on to the video amplifier.

## VIDEO AMPLIFIER

The video band of frequencies not only demands specific circuit design, it also puts a bit of a strain on the imagination. These frequencies begin at 0 and go up to about 4 MHz. Before we get into the amplifier, let's consider the nature of a video signal.

### *the video signal*

The video signal is partially composed square waves. These square waves are composed of a fundamental frequency and an infinite number of odd harmonics. (The fundamental frequency is known as the first harmonic. Other harmonics are integral multiples of the fundamental frequency. For instance, for a fundamental of 10 Hz, the 2nd harmonic is 20 Hz, the 5th harmonic is 50 Hz, the 100th harmonic is 1 kHz, etc.) Figure 4-27 illustrates the approximate result of combining the 1st, 3rd, and 5th harmonics.

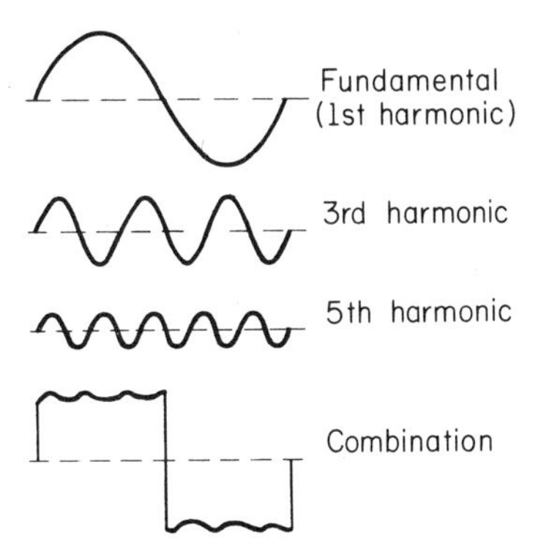


**Fig. 4-27.** Start of a Square Wave.

The resultant of combining just three harmonics is an exaggeration. Actually, it takes many more harmonics to produce a wave this close to a square wave.

The leading and trailing edges of each alternation of the square wave are primarily composed of very high frequencies. The flat top of a square wave is dependent upon low frequencies. Failure to properly amplify the high frequency components results in a distorted leading edge with rounded corners. If the low frequencies are *not* properly amplified, the flat top will be *distorted.* Failure to properly amplify any frequencies in the band will result in some variation of the true square wave.

### *circuit considerations*

We first need a transistor capable of a wide frequency response. Let's choose an *npn* and place it in a common emitter circuit. The *npn* is a matter of personal preference. The common emitter will give us the necessary amplification of current, voltage, and power.

Next we need bias stabilization. This has been amply covered previously. It is just mentioned here as a reminder.

The interelement impedance will likely cause a little feedback at high frequencies. The transistor will need an external feedback circuit to cancel this effect.

If we can manage proper amplification of both low and high frequencies, the center frequencies will be adequately handled. Therefore, we will need special circuits to boost the amplification of both low and high frequencies. If we can get all of these features into one stage of amplification, we will have a good video amplifier.

### *analyzing the circuit*

Figure 4-28 illustrates a video amplifier circuit. The ordinary biasing circuits have been omitted for the sake of clarity.

There are two invisible capacitors in this circuit. The first is the collector capacitance of $Q_1$. It is a combination of all distributive capacitances between point *A* and ground. We'll refer to it as $C_0$. The other is the input capacitance to the next stage. It exists between point *B* and ground. We will call it $C_1$.

Take a look at $R_1$. This is our degenerative feedback circuit. It functions like this: the potential at the top of $R_1$ always opposes the change caused by the signal. Not much, understand, but enough to reduce gain somewhat. This is caused by not using an ac bypass capacitor. What do we get in return for this loss of gain? We get

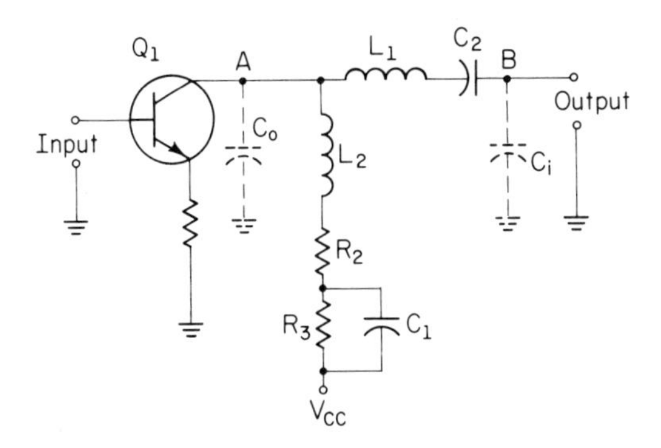


**Fig. 4-28.** Video Amplifier Circuit.

high fidelity, a broader frequency response curve, and thermal bias stabilization.

The coupling capacitor ($C_2$) has a tendency to attenuate low frequencies. This is countered in two ways. $R_3$ and $C_1$ form a high-load resistance for low frequencies. At zero frequency the entire resistance of $R_3$ is in series with the collector load resistor which is $R_2$. This provides maximum amplification of the low frequencies. This gets the low frequencies off to a good start, and $L_1$ helps them on the way. $L_1$ in series with $C_2$ cancels the tendency of $C_2$ to attenuate the low frequencies. Therefore, our low frequencies have received maximum amplification and optimum coupling.

The distributive capacitance represented by $C_0$ and $C_1$ have no effect on low frequencies. As the frequencies increase, their capacitive reactances decrease. If we don't counteract this tendency, the high frequencies will be shorted out. That is the purpose of $L_2$. This inductor has no effect on the low frequencies, but it becomes a strong factor to the high frequencies. The principle circuit components to the high frequencies are $C_0$, $L_2$, and $C_1$; and these three are in parallel. They form a low *Q,* wide band, tank circuit which eventually becomes the collector load impedance. Not only does this provide a considerable gain for these frequencies, but it also passes maximum signal to the next stage.

At the high frequencies, $C_1$ becomes a short and removes $R_3$ from the circuit.

Care must be exercised not to overcompensate. Coils, capacitors, and resistors can be adjustable if necessary. The effect of a fixed coil can be made adjustable by placing an adjustable resistor in parallel with it. One or more of these measures may be necessary if our circuit produces less than a perfect video signal.

There is another type of amplifier which may appear at most any frequency. It is primarily a power amplifier and is frequently called a driver. Drivers are used to produce sufficient power to send signals through long transmission lines, operate speakers, or trigger several circuits at the same time. We will analyze one type of driver without regard to frequency.

## PUSH–PULL AMPLIFIER

The amplifiers that we have seen previously are called single-ended amplifiers. This means that one amplifying device handles the entire signal. The push–pull amplifier uses two amplifying devices with each device handling half of the signal. This enables maximum power with maximum fidelity which is very important to many driver stages.

### *the basic circuit*

Figure 4-29 illustrates a push–pull amplifier in its simplest form. We are using two perfectly matched *npn* transistors with zero input bias and transformer coupling. Each transistor is in a common emitter configuration.

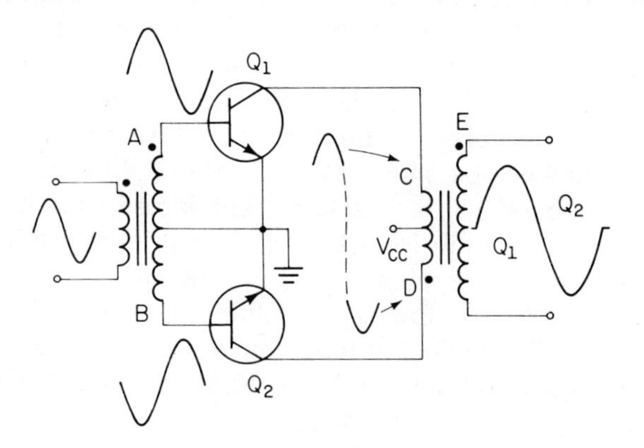


**Fig. 4-29.** Basic Push–Pull Amplifier.

The wave shapes on this diagram represent voltages. The small signal sine wave is stepped up slightly across each half of the transformer secondary. Point *A* goes positive at the same instant that point *B* goes negative. This causes $Q_1$ to conduct and $Q_2$ to cutoff.

The collector current from $Q_1$ causes point $C$ to swing in a negative direction. This negative at point $C$ is inverted across the output transformer which causes point $E$ to become positive with respect to point $F$. The section of the output that is labeled, $Q_1$, represents the positive alternation of the input and the conduction time of $Q_1$.

On the negative alternation of the input, point $A$ becomes negative, and point $B$ becomes positive. This causes $Q_2$ to conduct and cuts off $Q_1$. The collector current of $Q_2$ causes point $D$ to swing in a negative direction. The negative potential at point $D$ causes point $E$ to become negative with respect to point $F$. The output transformer has recombined the two halves of the signal.

### *signal distortion*

This output appears to be a well-formed signal, but with the circuit configuration used, there would be some distortion. Let's analyze the action on a dynamic transfer characteristic curve. Figure 4-30 is the combined curve for the two transistors.

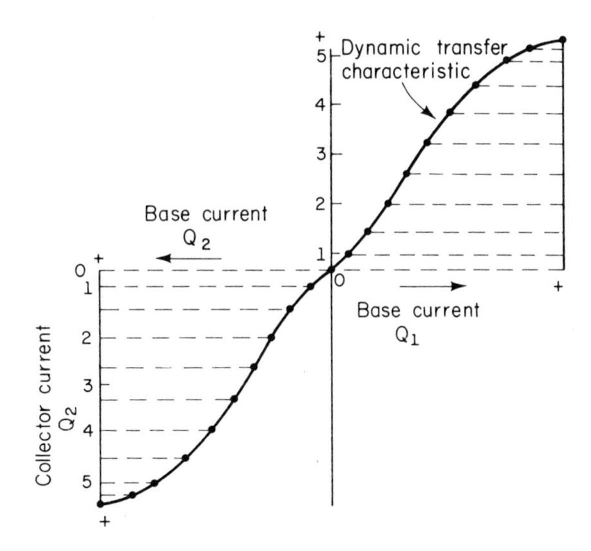


**Fig. 4-30.** Dynamic Characteristics of $Q_1$ and $Q_2$.

Notice that collector currents, as well as base currents, for the two transistors are in opposite directions. This curve is formed by taking two curves, inverting one, and placing them back to back. Figure 4-31 illustrates the current distortion caused from the nonlinearity near the zero cross over point.

### *improving fidelity*

Since high fidelity is one feature especially desirable in the driver, this distortion must be eliminated. Let's examine the problem. If we could overlap the conduction time of the two transistors, that zero cross over point would be eliminated. This fact is illustrated in Fig. 4-32.

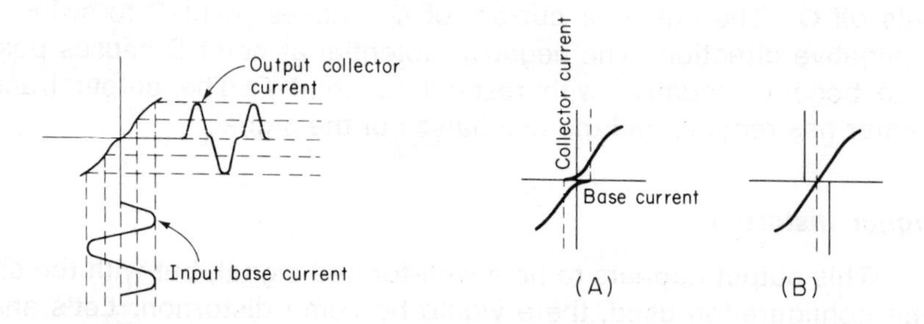


**Fig. 4-31.** Distortion With Zero Bias.

**Fig. 4-32.** Eliminating Zero Cross Over.

Part A of the drawing shows the amount of overlap required. Part B shows the ideal dynamic curve that results from the overlap.

### *the improved circuit*

How do we obtain such an overlap? Simple. We just add a little forward bias to both transistors. All we need to do is add a voltage divider to the circuit of Fig. 4-28. The result is shown in Fig. 4-33.

The small positive potential at point *A* provides forward bias

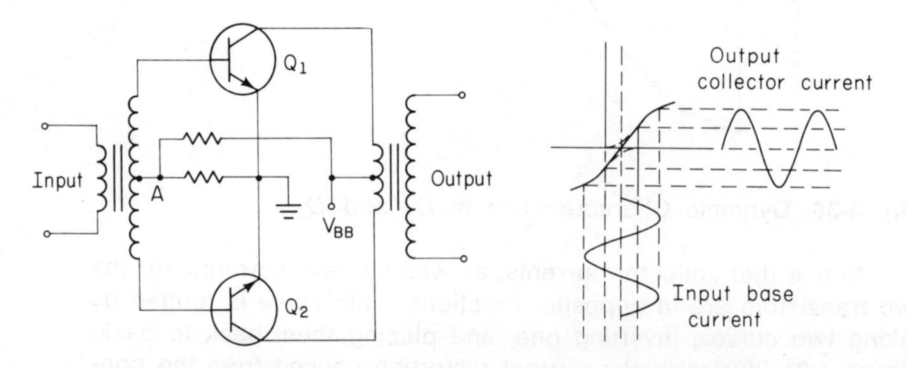


**Fig. 4-33.** Push-Pull With Forward Bias.

**Fig. 4-34.** A Faithful Reproduction.

across the emitter-base junction of both transistors. With no signal input, both transistors are conducting. On the arrival of a signal, one transistor gradually cuts off while the other increases conduction. This action reverses on the next alternation. This arrangement provides maximum current and power gain along with a high degree of fidelity. Figure 4-34 illustrates the input and output current waveshapes.

## RADIO FREQUENCY AMPLIFIERS

Radio frequency (RF) as used here is a broad term. It is meant to cover the transmitted and received frequencies which carry intelligence. This puts us most anywhere in the frequency spectrum. The exact spot to be determined by the type of equipment being used.

### *uses of* RF *amplifiers*

These amplifiers fall into two general classes, the untuned and the tuned. The untuned is for amplification only, and it handles a relatively wide band of frequencies. The tuned amplifier provides both amplification and selectivity. The tuned RF amplifier features high amplification over a very narrow frequency band.

Intermediate frequency (IF) amplifiers are actually RF amplifiers tuned to a fixed frequency. They are found in the IF section of receivers to provide high gain and maximum selectivity. Most receivers contain a cluster of several IF stages. The relative position of these amplifiers is shown in Fig. 4-35.

This is a block diagram of a typical superheterodyne receiver. The carrier frequency is picked up by the antenna and is immedi-

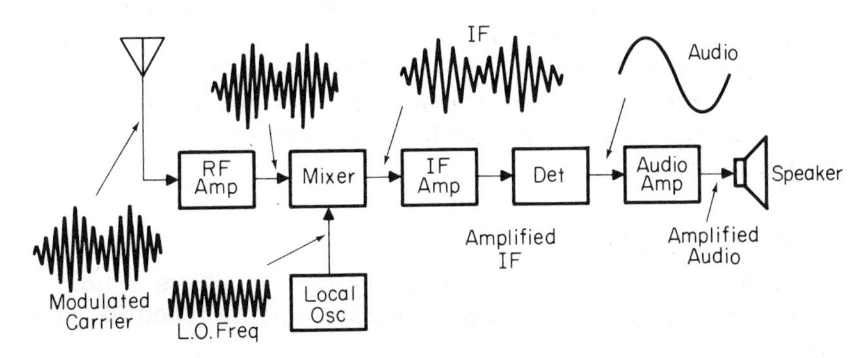


**Fig. 4-35.** Amplifiers in a Receiver.

ately amplified by one or more stages. This improves the signal to noise ratio of the weak incoming signal while building it up to a workable level. The amplified RF is mixed with the continuous RF from the local oscillator to produce the IF. This is a lower frequency, easier to work with, and can be handled by conventional circuit components. The detector stage extracts the intelligence by rectifying and filtering. In this case, intelligence consists of an audio signal. The audio is further amplified to attain sufficient power to drive the speaker.

Reversing the process, the transmitter uses the RF amplifier to increase the amplitude of the carrier frequency before feeding it to the antenna. This is shown in Fig. 4-36.

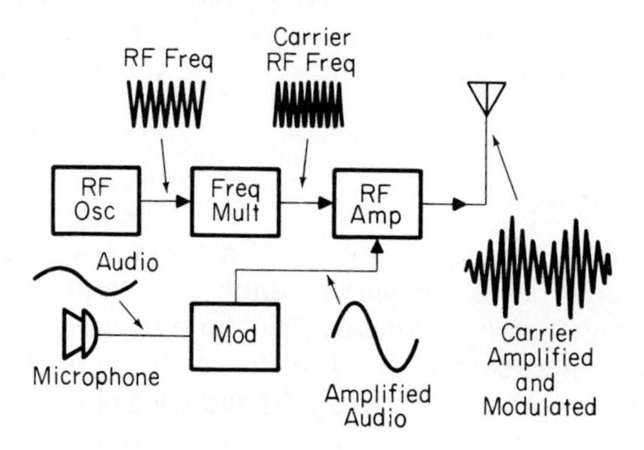


**Fig. 4-36.** Amplifiers in a Transmitter.

The RF oscillator produces a high-frequnecy, continuous-wave signal. The frequency multiplier increases this frequency to the desired broadcast frequency. The resulting RF carrier has sufficient power to drive the antenna but contains no intelligence. The modulator picks up the relatively weak audio signal from the microphone, amplifies it, and uses it to modulate the carrier wave. Modulation is accomplished in the final RF amplifier stage. The audio signal increases and decreases the amplification at an audio rate. As a result the final amplified carrier has the audio signal impressed on its amplitude.

### *tuned* RF *amplifier*

The receiver normally employs tuned RF amplifiers in both the input and IF stages. Figure 4-37 is a representative tuned RF amplifier.

Bias for the transistor is fixed by the voltage divider ($R_1$ and $R_2$). $R_3$ in the emitter circuit establishes temperature compensating bias while $C_3$ is an RF bypass to prevent degeneration. $C_2$ and $C_5$ are also RF bypass to keep the RF away from the dc power supply. $C_4$ neutralizes the interelement coupling.

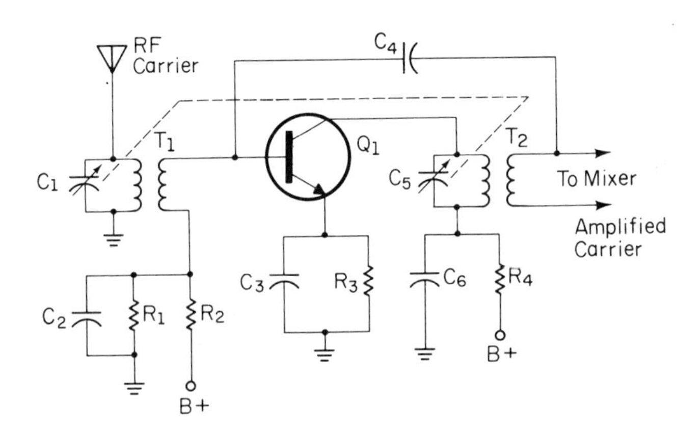


**Fig. 4-37.** RF Amplifier.

The primaries of both transformers are tuned by variable capacitors. The broken line indicates that these two capacitors have a mechanical link. Actually, they are ganged on a common rotor.

This is obviously the first RF stage in a receiver, but it could just as well be an IF stage. The signal on the antenna must match the resonant frequency of $T_1$'s tuned primary. This sets up resonant oscillations and the parallel tank is a maximum impedance for this frequency. The result is maximum signal frequency coupled to the base of $Q_1$. Signals of other frequencies find the tank to be a very low impedance which shorts most of these off frequencies to ground.

The RF signal on the base of $Q_1$ varies the collector current at the same RF rate. The varying collector current sets up and sustains resonant oscillations in the tuned primary of $T_2$. Maximum signal is developed and coupled to the mixer.

Since IF amplifiers operate at somewhat lower frequencies, they may use alloy core transformers. These transformers are frequently tuned by a moveable slug in the transformer core.

### *untuned* RF *amplifier*

The term untuned does not imply that this amplifier will handle any frequency that happens along. It simply means that it has a wider bandpass and consequently lacks a high degree of selectivity. An untuned RF amplifier is shown in Fig. 4-38.

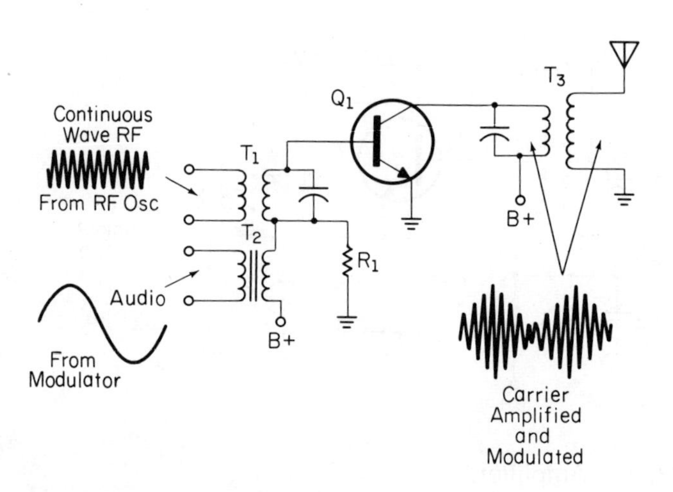


**Fig. 4-38.** Untuned RF Amplifier.

Here we have used the last stage in a transmitter. The principal circuit components are the three transformers. The secondary of $T_1$ and the primary of $T_3$ are resonant tank circuits. They are not sharply tuned in order to allow a reasonable bandwidth. Bias for the transistor is provided by $R_1$ and the secondary of $T_2$.

This stage has two input signals: the continuous RF waves from the oscillator and the audio from the modulator. Both signals will vary the bias on $Q_1$. In this fashion the audio variations are impressed on the amplitude of the RF carrier. The amplitude variations are intelligence. The RF is a means of carrying the intelligence from one antenna to another.

## CHAPTER 4 REVIEW EXERCISES

1. What two extreme points are connected by a load line?
2. The load line represents a locus of all possible values of what three factors?
3. What two factors determine the position of a load line?

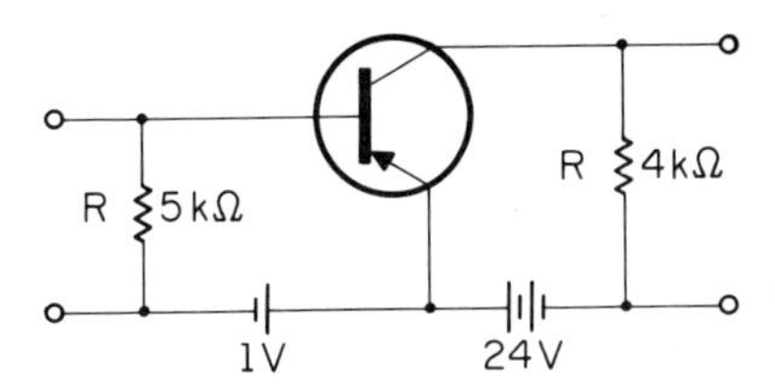


**Fig. 4-39.** Dynamic Circuit.

4. How is the operating point established on a load line?
5. What three values intersect at the operating point?

Items 6 through 9 refer to Fig. 4-39.

6. What is the maximum $I_c$ for this circuit?
7. What is the maximum $V_c$?
8. What is the value of the following at the operating point:
   (a) $I_b$?
   (b) $I_c$?
   (c) $V_c$?
9. Use the graph in Fig. 4-40 to construct a load line for this amplifier.

Items 10 through 16 refer to Figs. 4-39 and 4-40.

10. An input sine wave with 100-$\mu$A peak to peak amplitude will cause $I_c$ to swing between what extremes?

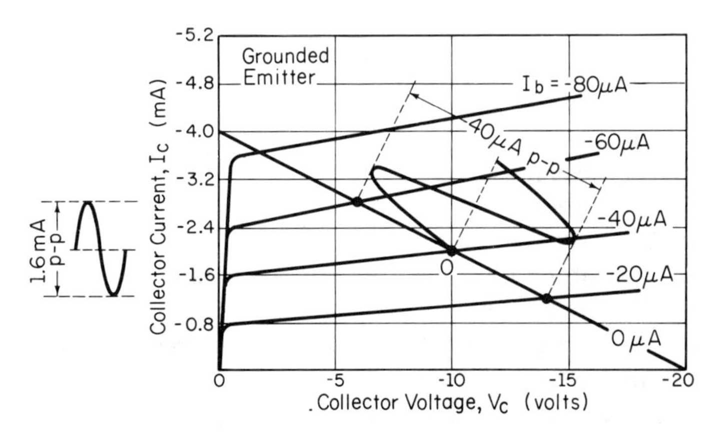


**Fig. 4-40.** Static Curves.

11. What is the value of collector voltage when the signal of item 10 is maximum:
    (a) Negative?
    (b) Positive?
12. The signal of item 10 is 10 mV peak to peak. What is the:
    (a) Voltage amplification?
    (b) Current amplification?
13. What is the output power at the operating point?
14. The maximum safe power for this transistor is 34 mW. Draw a constant power curve on the graph of Fig. 4-40.
15. The maximum safe power is exceeded between what values of collector current?
16. List two ways that the circuit can be changed to avoid excessive power.

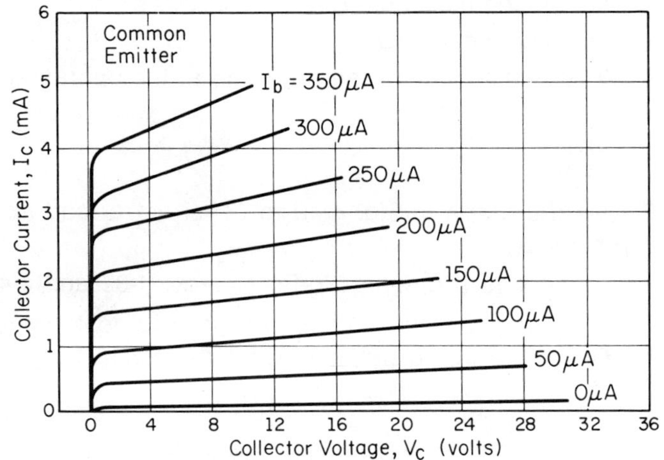


**Fig. 4-41.** Load Line With Signals.

Items 17 through 20 refer to Fig. 4-41.

17. Draw the circuit represented by this load line and label the:
    (a) Type of transistor.
    (b) Resistance of $R_L$.
    (c) Direction of $I_c$ and $I_b$.
    (d) Value of $V_{CC}$.
18. The circuit described by this load line has an $R_b$ of 125 kΩ. What is the input bias?
19. The input signal causes the output voltage to vary between what two extreme points?
20. What is the current amplification?

# 5

# *voltage control*

## POWER CONVERSION

Power is one thing that all electronic systems have in common. Every circuit in every system makes use of electric power in some fashion. The power system has the same importance to electronic equipment that an engine has to an automobile. A vehicle operator may be the best driver on the road, but if he knows nothing about his engine, his knowledge of his car is very limited. A knowledge of power equipment is just as essential to the electronics technician as a knowledge of engines is to an automobile mechanic.

## CONVERSION UNITS

There are many sources of electrical energy, but the two most practical sources are the battery and the generator. Unfortunately the raw output from these two sources is not always suited to our specific needs. The battery has a limited capacity coupled with a tendency to discharge at the most inconvenient times. Generators require a mechanical turning force that is sometimes difficult to provide; they also take up a lot of space. Once a source of voltage is available, it is seldom of the proper type. The equipment seems to require ac when dc is available and vice versa. When the proper type of voltage is available, it seldom, if ever, comes in the proper magnitude. These are a few of the problems that are solved by power conversion. A part of nearly all systems is some type of conversion unit. This conversion unit is designed to perform such tasks as changing ac to dc, changing dc to ac, increasing or decreasing voltage amplitude, and changing the phase of ac.

### *the dynamotor*

This device is a combination motor and generator and is designed to change the amplitude of a direct voltage. Dynamotors are manufactured in many sizes. A small one is a compact device not much larger than a thermos bottle and somewhat resembling one in shape. Figure 5-1 is a drawing of a dynamotor.

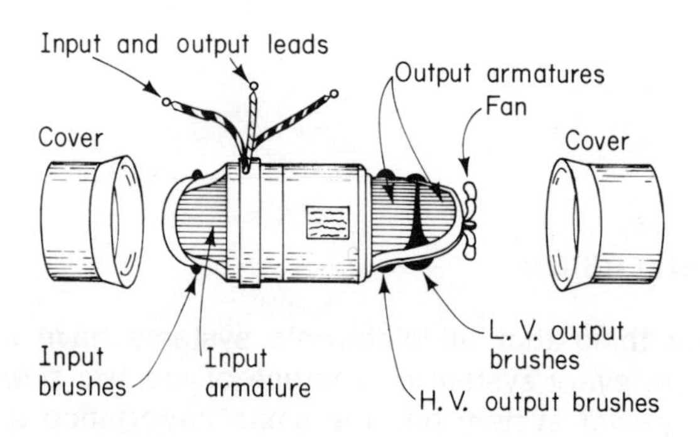


**Fig. 5-1.** A Small Dynamotor.

The dynamotor is operated on dc and provides a dc output. Therefore, it is essentially a dc generator which is turned by a dc motor. The motor side (input) is designed to operate on a particular amplitude of voltage. The proper input causes the motor armature to rotate at a given number of revolutions per minute. The generator armatures are on the same shaft and must rotate at the same speed as the motor. The generator has a low voltage (LV) armature and a high voltage (HV) armature. This provides an output that is either lower or higher than the input voltage. The output lead can be changed to select either the LV or HV output.

### *the rotary converter*

The rotary converter is a combination ac motor and dc generator. Its function is to convert ac to dc. It can be constructed with a single armature. In this case, the ac would be fed into the commutator through slip rings and the output taken from brushes. The single winding converter provides a dc output which is equivalent to the peak ac input. Other ratios of voltage require separate armatures for input and output.

### *the inverter*

This is a converter in reverse. It is a combination dc motor and ac generator. Its function is to change dc to ac. The single armature version of this machine is too unstable for most applications. Therefore, it generally comes with separate input and output armatures.

### *the vibrator*

This device is designed to serve either one of two purposes. One type will change an input dc to an output ac. The other uses a dc input and provides a greater amplitude dc output. The essential parts of a vibrator are shown in Fig. 5-2.

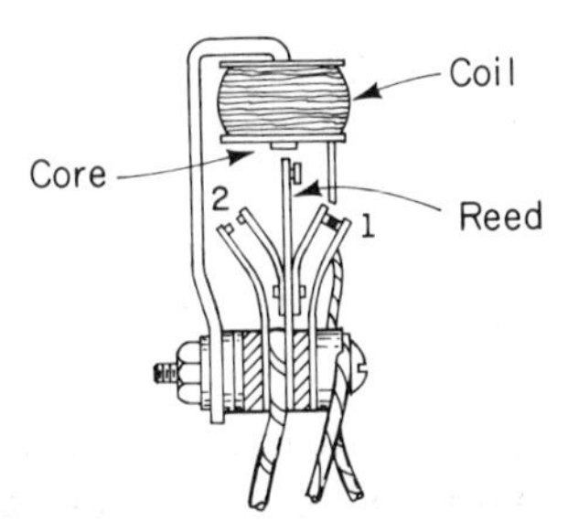


**Fig. 5-2.** Parts of a Vibrator.

The input voltage causes current through contacts 1 and through the coil. The coil energizes, and pulls the reed to the left. This will open contacts 1 and close contacts 2. When contacts 2 are closed, the current bypasses the coil and allows it to deenergize. The deenergized coil allows the reed to swing back to the right which will open contacts 2 and close contacts 1. This completes a full cycle of actions, and this cycle repeats in a periodic fashion. *Not shown* in the drawing is another coil which also energizes by current through these contacts. This second coil is part of the output assembly. If the vibrator is designed for an ac output, the second coil is energized in one direction through contacts 1 and in the opposite direction through contacts 2. If it is designed for a dc output, both sets of contacts provide energizing current in the same direction.

### *multiphase generators*

When we think of an ac generator, we tend to think of the single-phase version. It uses one set of slip rings and provides a single sine wave of output voltage. This is not always the case. Generators are also designed to provide two- and three-phase outputs. Figure 5-3 illustrates a two-phase generator with its two outputs.

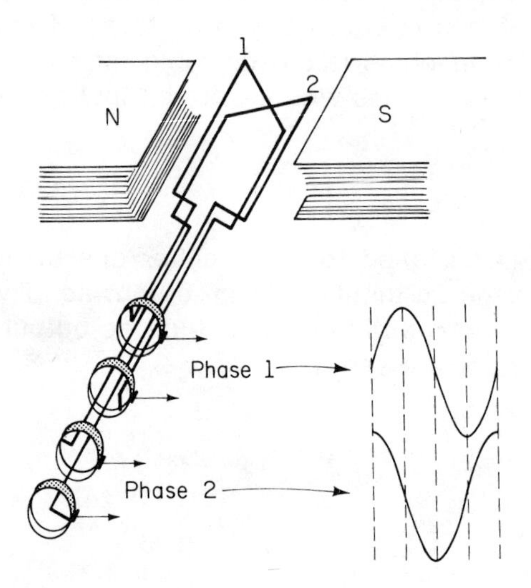


**Fig. 5-3.** Two-Phase Generator.

Here we have represented a pair of slip rings for each output. This is not actually essential. One ring could be common with one additional ring for each output. A three-phase generator produces three voltages 120° apart. When the outputs are perfectly balanced, so that there is always zero current in the common lead, the common lead can be eliminated. This is frequently done in a three-phase system.

The windings in a three-phase generator are represented by either a *Y* or a Δ as shown in Fig. 5-4.

In either case, three outputs are provided, and the sine waves are out of phase with adjacent outputs by 120°. When three leads are used as shown here, an output is taken across any pair of leads (1 and 2, 1 and 3, and 2 and 3). Any given output of the *Y* connection will have an amplitude which is approximately 1.73 times the amplitude of either of its two windings. For instance if each coil has a voltage of 100 V in amplitude, each of the three outputs is 173 V. The three outputs are shown in Fig. 5-5.

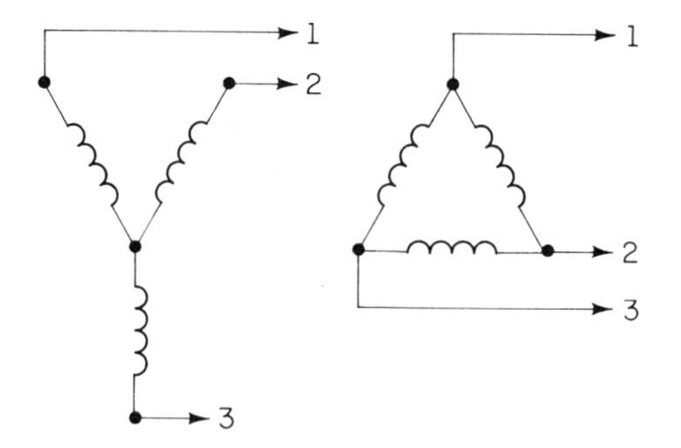


**Fig. 5-4.** *Y* and Δ Connections.

### *summarizing the power problem*

More than 90 per cent of the power in the United States is supplied from ac sources. Generally speaking, then, our problem can be summarized as follows:

1. Obtain a source of ac.
2. Adjust the amplitude to usable proportions.
3. Change the ac to pulsating dc.

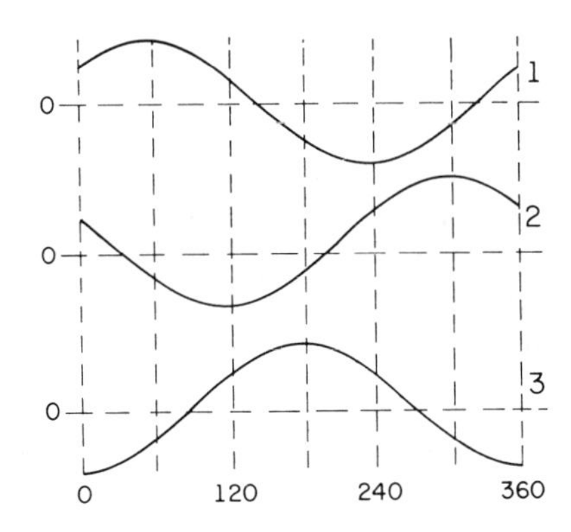


**Fig. 5-5.** Outputs from a Three-Phase Generator.

Obtaining a source of ac is no longer a big problem. To most of us, ac is as near as the closest convenience outlet (wall plug). However, the designer must create a power system to solve the other steps in the problem. The block diagram in Fig. 5-6 indicates the steps and the type of equipment required.

The transformer accepts the ac and changes its amplitude to

meet the need. The rectifier changes the sine wave of the ac to pulses of dc.

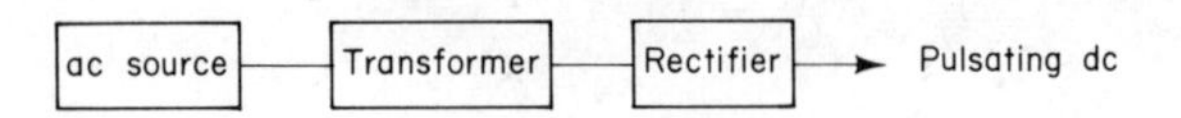


**Fig. 5-6.** Solving the Power Problem.

Delivering the ac is a task performed by specialists in the field of power engineering. Caring for the power equipment, which converts this ac to our particular need, is clearly in the field of electronics. Fortunately, all power supplies have a great deal in common. From one piece of equipment to the next, the principle difference is in the voltage amplitudes. Nearly all power systems utilize some version of the components listed in Fig. 5-6. Once you understand their general function and relationships, you have a working knowledge of most common power systems. The remainder of this chapter is devoted to these components.

## TRANSFORMER PRINCIPLES

A transformer is an electromagnetic device which changes the amplitude of ac. It is composed of an input coil and an output coil which are coupled together by an electromagnetic field. The input section is called a primary, and the output section is the secondary. There are no moving parts, and there is no electrical connection between primary and secondary. The input ac causes a changing flux around the primary. This changing flux cuts across the windings of the secondary and produces an ac voltage in this coil. This is a practical application of the principle of mutual inductance.

### *construction of transformers*

A transformer may be constructed with a magnetic iron alloy core or an air core depending upon its operating frequency. The air core transformer is principally intended for coupling of radio frequency signals from one point to another. We will return to this type a bit later in this chapter. Just now, we are interested in low frequencies which are used to handle our input voltage. This type of transformer invariably contains an iron core. This core is not a simple chunk of molded iron. A core of that type would allow induced currents to circulate freely within the core. This would create heat and greatly increase the transformer losses. These core currents are called eddy currents, and the power loss they cause, is eddy current loss. These losses can be minimized by using laminated cores.

There are two popular designs in core lamination. These are the *E* and *I* combination and the *O*. Both of these are illustrated in Fig. 5-7.

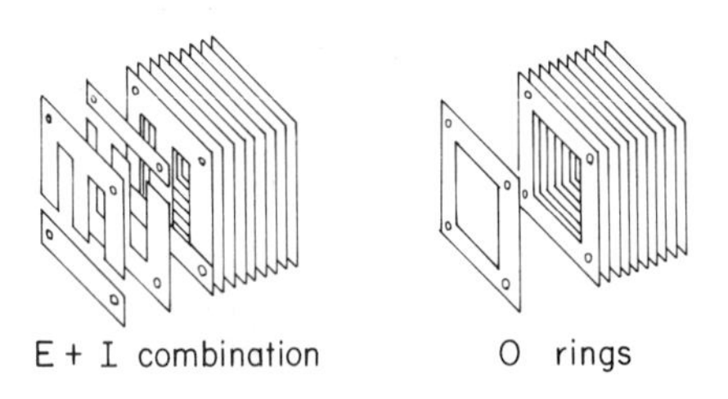


**Fig. 5-7.** Two Lamination Designs.

The *E* and *I* combination has two advantages over the *O* rings. It has fewer losses and provides a more compact transformer. Regardless of the design of the lamination, each slice of the core is coated with an insulating varnish, and the laminations are pressed tightly together. In the final product, small bolts secure the laminations. The windings on the two designs are illustrated in Fig. 5-8.

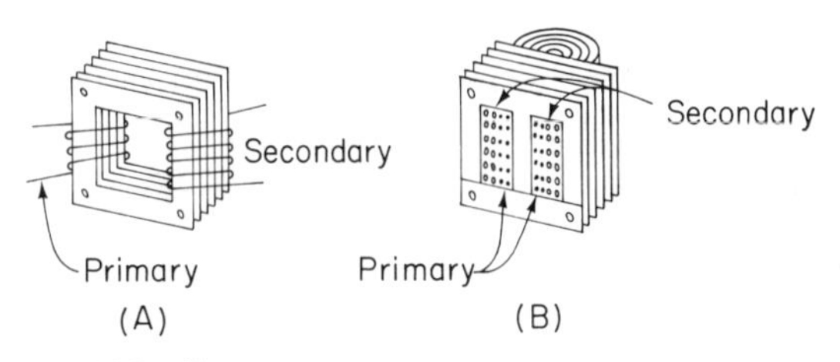


**Fig. 5-8.** Transformer Windings.

Notice in Fig. 5-8b that the secondary is wound directly on top of the primary windings. This places primary and secondary so close together that nearly all flux from the primary cuts all the turns of the secondary. This is a highly efficient arrangement. Also, in this same drawing, the coils are inside the core. This is a space saver, an important consideration in most electronic equipment.

A power transformer may contain more than one primary, and they frequently have several secondaries. Each secondary provides a separate output. All outputs will have a common frequency, but they may have different amplitudes and phases. In addition to having multiple secondary windings, a particular winding may have taps at several places. A secondary winding with a center tap, is equivalent to having two secondaries with outputs 180° out of phase with each other.

The phase relation between input and output is determined by the direction of the windings. Generally these voltages are either in phase or 180° out of phase. When input and output are 180° out of phase, the direction of the windings has caused phase inversion.

Schematic symbols for iron core transformers are shown in Fig. 5-9.

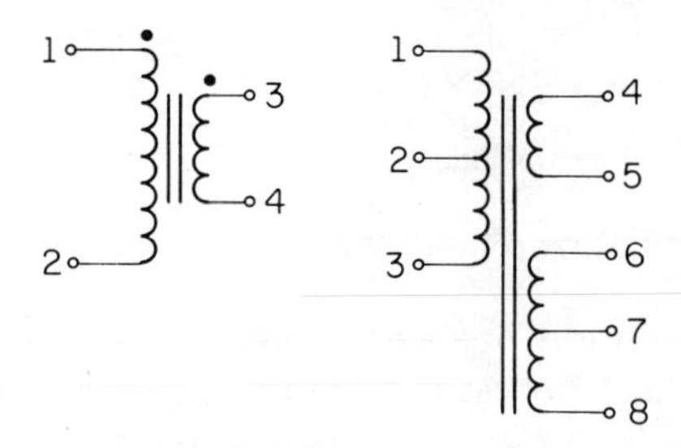


**Fig. 5-9.** Schematic Symbols.

In both drawings, the left side represents the primary. Part B shows a double secondary with taps on both the primary and one secondary. Part A shows a dot on the primary and secondary. Not all schematics use this dot system, but when dots are used, the points that are marked with dots are of the same polarity. For instance, in this drawing, pin 1 of the primary is always of the same polarity as pin 3 of the secondary.

### *voltage ratios*

Apart from the transformer losses which tend to reduce the output voltage, the output is controlled by two factors: the ratio of primary turns to secondary turns and the magnitude of the input voltage. Until further notice, our discussion will be of the ideal transformer which has no losses. When the number of turns on the primary is larger than the number on the secondary, the output will have less amplitude than the input. This is called a *step down* transformer. If the secondary has the greater number of turns, the output will be of a higher magnitude than the input. This is a *step up* transformer. Step up and step down transformers are illustrated in Fig. 5-10.

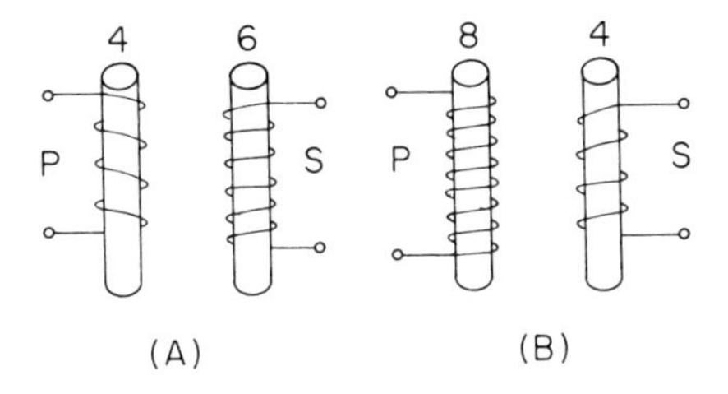


**Fig. 5-10.** Turns Ratio.

The numbers above each coil happen to be the exact count of the turns on that coil. This number is generally a ratio instead of a count. For instance, the primary to secondary turns ratio in Part A is 2:3, and in B it is 2:1. The actual number of turns has no bearing on the amount of output voltage; only the ratio.

Part A is a step up transformer. The ratio of primary to secondary voltage is the same as the turns ratio; 2:3. For every 2 V in, 3 V will be delivered by the secondary. 2 V at the input produces 3 V out; 20 V in gives 30 V out. This same ratio would be true if the primary had 400 turns and the secondary had 600 turns.

Part B is a step down transformer. The ratio is 2:1 because the primary has twice as many turns as the secondary. Any combination of turns that will produce this ratio will give 1 V out for each 2 V in. Mathematically, these ratios are expressed as follows:

$$\frac{E_p}{E_s} = \frac{N_p}{N_s}$$

where $E_p$ is the primary voltage, $E_s$ is the secondary voltage, $N_p$ is the number of turns on the primary, and $N_s$ is the number of turns on the secondary. Actually the formula is a simple comparison of two ratios. It says, "the voltage ratio is equal to the turns ratio."

### *current ratios*

Our perfect transformer has the same amount of power in both primary and secondary. Therefore, any change in voltage must be accompanied by a change of current in the opposite direction. For instance, a step up transformer which increases voltage by a ratio of 1:2 must decrease current by a ratio of 2:1. Such a transformer with 10 V and 2 A in the primary would have a primary power of $10 \times 2$ or 20 W. The secondary would have twice the voltage (20 V), half the current (1 A), and exactly the same power (20 W).

The relation of current ratio to turns ratio is expressed as:

$$\frac{I_s}{I_p} = \frac{N_p}{N_s}$$

Notice that this is an inverse relationship. A step down transformer increases current while a step up transformer decreases current.

***impedance***

The problem of coupling of voltage from primary to secondary of a transformer is similar to many other coupling problems encountered in electronics. The coupling will provide a maximum transfer of power from one circuit to another. The maximum power transfer occurs when the input impedance matches the output impedance. We have already stated that the ideal transformer has the same amount of power in both primary and secondary. This indicates unity coupling between the two coils. The impedance from primary to secondary can be expressed in terms of the turns ratio.

$$\frac{Z_p}{Z_s} = \left(\frac{N_p}{N_s}\right)^2$$

Since transformers can be purchased with a wide range of turns ratios, they are frequently used as coupling devices between two circuits that have widely different impedances. For instance, a 1250-Ω circuit feeding into a 50-Ω circuit would suffer a great loss of power. If the gap between these circuits is bridged by a 5:1 step down transformer, maximum power can still be transferred. The circuit would appear as in Fig. 5-11.

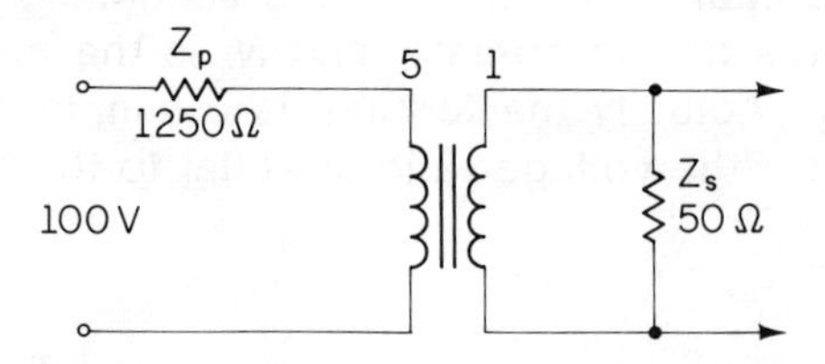


**Fig. 5-11.** Matching Impedance.

With 100 V in the primary, the current is 0.08 A. Power in the primary is $EI = 100\ (0.08) = 8$ W. The voltage in the secondary has been stepped down to 20 V. 20 V across the 50-Ω load produces a current of 0.4 A. The secondary power is $20\ (0.4) = 8$ W.

Of course in actual practice, transformers are not perfect; they do have losses. Some of these losses are discussed in the next section.

### *energy losses*

There are three kinds of losses to be considered in a transformer. These are copper, hysteresis, and eddy current losses. Copper loss is caused by the fact that the windings on both primary and secondary have a certain amount of dc resistance. This can be considerable when one or both windings are composed of many turns of fine wire. This resistance uses energy.

The hysteresis losses are caused by residual magnetism in the core as the voltage passes through zero. A coercive force is required to cancel this residual magnetism. This coercive force is an energy loss.

Eddy currents have already been mentioned as a reason for laminating the transformer core. These eddy currents are induced by the changing flux in the core material. They create heat, which is an energy loss, and also cause a certain amount of shielding of the windings from the flux. Laminating the core material does not eliminate these eddy currents, but it does reduce the losses which they cause. This is attributed to the fact that the insulated laminations shorten the current paths. The eddy current losses are still directly proportional to the square of the thickness of the laminated sections.

### *efficiency*

The efficiency of a transformer is a ratio of power out to power in. In practice, this is always less than unity. The unity figure can be approached but never quite equaled. The efficiency is expressed in terms of percentage and is obtained by multiplying the power out to power in ratio by 100.

$$\%\ \text{efficiency} = \frac{P_{\text{out}}}{P_{\text{in}}} \times 100$$

If a given transformer has an input power of 600 W and an output power of 580 W, it is 96.7% efficient. 20 W are being lost in this transformer because of copper, hysteresis, and eddy currents.

## CLASSIFICATION OF TRANSFORMERS

There are four classes of transformers that concern electronic equipment. They are classified according to the range of frequencies which they are designed to handle. These classes are power frequency, audio frequency, radio frequency, and tuned radio frequency. Only the first of these clearly belongs with power systems, but a brief description of each class is in order at this time.

### *power frequency*

Power frequency transformers are designed to handle one specific frequency in the power frequency range. In broad terms, power frequencies range from 60 to 1600 Hz. Most home equipment and fixed installations use 60-Hz power. Some mobile equipment, especially on aircraft, make use of power from 400 to 1600 Hz. Some devices are designed to use 208 V, 60-Hz, three-phase power. These utilize 60-Hz, three-phase transformers.

### *audio frequency*

The audio frequency band overlaps the power band. It starts at about 16 Hz and goes up to about 20 kHz. This band is supposed to include all frequencies that can be detected by human ears. Not many of us can hear all these frequencies. However, the mid-frequencies of the band should be audible to all of us. Mid-frequencies is a loose term, but it is understood to cover frequencies from 100 to 4000 Hz. A good audio transformer will couple any of the mid-frequencies with very little distortion. This means that it can step up or step down voltages at any of these frequencies by the same amount.

The construction of an audio transformer is similar to that of a power transformer with closer attention to certain details. The core material is more critical, and the capacitance between windings is reduced to a minimum. They are designed with a maximum coupling coefficient which sometimes goes as high as 0.999.

The audio transformer is used in public address systems, broadcast equipment, and communications receivers.

### *radio frequency*

This band of frequencies logically should bridge the gap from the top of the audio range to the bottom of the infrared light band. This is a broad band which starts at about 16 kHz and goes up to 10 THz ($T = 10^{12}$).

At the lower end of this band the transformers are composed of many turns of fine wire wound on a form of insulating material.

The form is sometimes filled with a powdered iron and sometimes left empty. If the form is empty, it constitutes an air core transformer. A schematic symbol for an air core transformer is shown in Fig. 5-12.

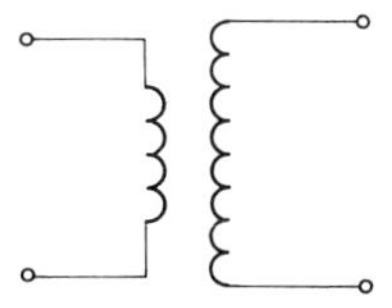

**Fig. 5-12.** Air Core Transformer.

As the frequencies go higher the losses in the iron core transformer increase. When these losses become too great to tolerate, the iron core is abandoned in favor of the air core. Moving on up the frequency band, fewer and fewer turns are used on the transformer. There comes a point when two straight wires become the primary and secondary of a transformer. Beyond this point, other means of coupling must be devised.

Although designed for specific frequencies, the common radio frequency transformer is not highly selective. It passes a band of frequencies above and below the specified frequency. In most cases, this is fine, but there are places that we want to pass a very narrow band. This necessity was the father of the tuned transformer.

### *tuned radio frequency*

Although practiced to some extent in audio transformers, the tuned transformer is mainly used with radio frequencies. They are designed to pass specified frequencies and rigidly discriminate against all others. It is not a difficult process. It simply means the addition of a capacitor to either the primary or the secondary. Sometimes both primary and secondary are tuned. The inductance of the transformer coil and the capacitance of the capacitor are chosen so that they form a resonant circuit for the desired frequency. Maximum response will occur with the resonant frequency. Frequencies to either side of resonance will be attenuated.

That little detour seemed to be necessary, but let's get back to our subject of power conversion. We now have the ac in the equipment and have used the transformer to adjust its amplitude to a manageable level. The next step is rectification.

## RECTIFICATION

Rectification is the process of changing an alternating voltage to a pulsating direct voltage. The device used for this conversion is called a rectifier. The types of rectifiers and their function in the rectification process are an important part of shaping our input power.

### *half-wave*

The half-wave rectifier accomplishes its mission by simply eliminating one alternation from each sine wave of alternating voltage. This leaves the remaining alternations with all the peaks and valleys, but it is direct voltage. It changes amplitude almost constantly, but it is always of a single polarity. Applying this voltage to a circuit would produce a current of very erratic amplitude, but it would be a dc in the sense that it would keep a constant direction. The most common device used for a rectifier is some form of solid-state diode. Fig. 5-13 illustrates a half-wave rectifier.

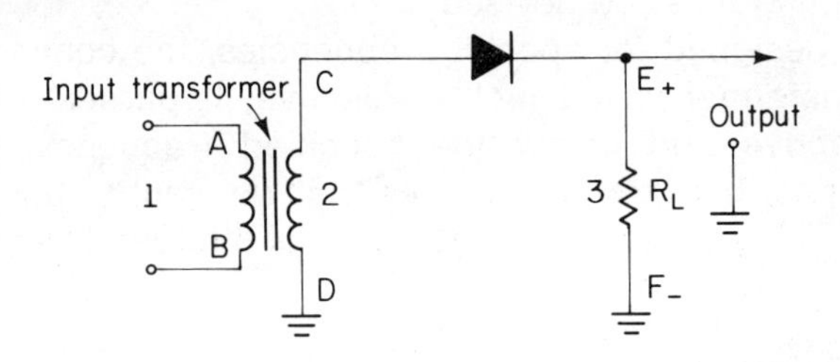


**Fig. 5-13.** Half-Wave Rectifier.

Figure 5-14 illustrates the waveshapes at point 1, 2, and 3 of Fig. 5-13.

Assuming that waveshape 1 is present at point 1 on the input transformer, the other waveshapes will be present at corresponding points on the schematic. From $T_1$ to $T_2$, point *A* of the primary is positive with respect to point *B*. This is inverted across the transformer which makes point *C* negative with respect to point *D* (ground). This negative at point *C* places reverse bias on the diode and holds it cutoff for the duration of this alternation. Since the diode is cutoff, there is no current through $R_L$. This keeps points *E* and *F* at ground potential.

At $T_2$, point *A* of the primary starts negative, and this negative is coupled to point *C* as a positive. As point *C* goes positive, the diode is forward biased which allows current from point *F* to *E* and through the diode to point *C*. The current through $R_L$ will follow the sinusoidal pattern of the positive alternation of the sine wave ($T_2$ to $T_3$). This positive alternation of voltage is fully developed across $R_L$, and it is available at the output.

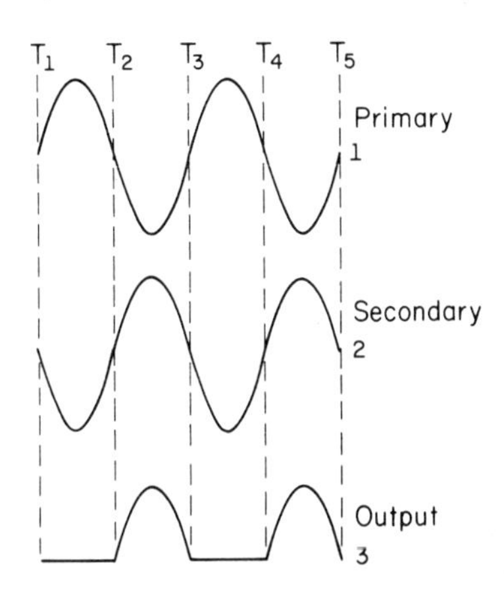


**Fig. 5-14.** Half-Wave Rectifier Waveshapes.

At $T_3$ the cycle of events is complete and another one starts. The action is repetitive as long as the input voltage is present. The result is a string of positive pulses. This output is referred to as pulsating dc. The pulse frequency is the same as the frequency of the input since there is one pulse out for each cycle in. The pulse frequency is also called a ripple frequency.

The amplitude of the pulse is the same as the peak value of the voltage from the transformer. This is discounting the negligible amount of voltage which is dropped across the conducting diode. The average value of voltage during one alternation of a sine wave is 0.637 times the peak value. Since the rectifier output is one alternation of the sine wave, the average output of the rectifier is 0.318 times the peak value. Why 0.318? Because the voltage is present only half the time. The average value when it is present is 0.637, but there is voltage half the time only. Therefore, average output is $\frac{0.637}{2}$ or 0.318 times the peak value.

We have already stated that current from this rectifier would be erratic. The voltage is also difficult to smooth out because of the large gaps between pulses. But this type of rectifier does have its uses. Many types of electronic circuits require a high potential with very little current. The half-wave rectifier is well suited to supplying voltage for such circuits.

Thus far, we have been discussing a positive rectifier; that is, we have eliminated the negative half of the sine wave and provided an output which is positive in respect to ground. When a circuit requires a negative voltage, the diodes in the rectifier would be connected in the opposite direction. This will reverse the current through $R_L$ and produce an output which is negative with respect to ground.

### *full-wave*

The full-wave rectifier has some diodes to conduct on each alternation but still keeps current in a constant direction. This produces an output pulse for each alternation of the input. This circuit is illustrated in Fig. 5-15.

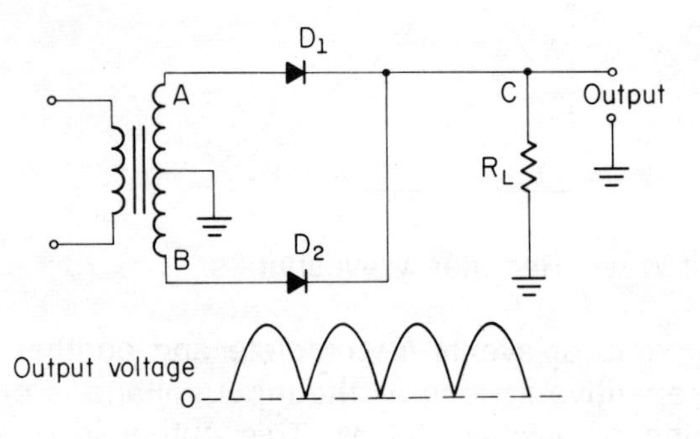


**Fig. 5-15.** Full-Wave Rectifier.

The center tap of the secondary of this transformer is grounded. This allows only half of the secondary peak voltage to be delivered to either diode. For a given size transformer, this means that the maximum amplitude to either diode is only half as much as could be delivered to a half-wave rectifier. Since transformers are manufactured in all sizes, this is not a serious disadvantage. If more voltage is needed, we just select a transformer with more turns on the secondary.

In Fig. 5-15, when point *A* goes positive, $D_1$ conducts. This allows current from ground through $R_L$ and develops a positive alternation between point *C* and ground. At this same time, point *B* is negative which holds $D_2$ cutoff. On the next alternation, point *B* goes positive allowing $D_2$ to conduct. This allows current through $R_L$ in the same direction and develops another positive alternation between point *C* and ground. When point *B* goes positive, point *A* becomes negative and cuts off $D_1$.

Each diode is conducting for one alternation and cutoff for one alternation; one diode is cutoff while the other is conducting. Each pulse of the output is labeled according to the diode that was conducting as it was being developed. These peaks are closer together and the valleys are less pronounced. This voltage is fairly easy to shape into a smooth direct voltage. It's a matter of filling in the gaps between the peaks.

The full-wave rectifier provides a relatively smooth current and can withstand heavy current loads. It is employed with circuits which draw considerable current and need only a moderate amount of voltage. Because each diode uses only half of the transformer secondary, this rectifier is not suitable for high-voltage circuits. They are seldom used with circuits which require more than a 1000 V.

This rectifier has a ripple frequency that is twice the frequency of the input voltage. It produces two peaks for each sine wave of the input.

### *bridge*

The bridge rectifier is a type of full-wave rectifier, but it uses the entire peak voltage of the secondary at all times. This circuit is illustrated in Fig. 5-16.

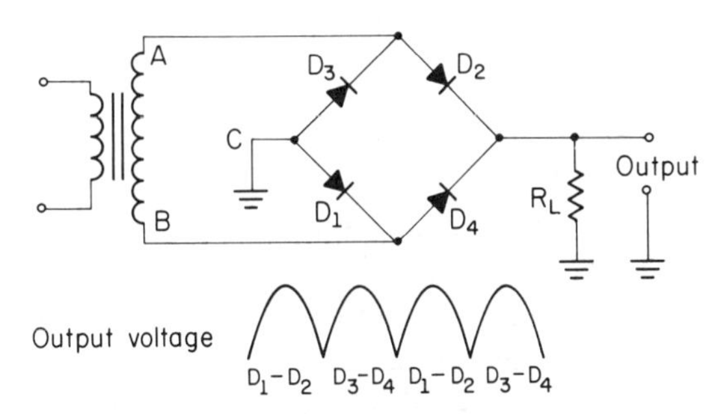


**Fig. 5-16.** Bridge Rectifier.

On one alternation of the input, point *A* goes positive and point *B* goes negative. This places reverse bias on $D_3$ and $D_4$, and forward bias on $D_1$, and $D_2$. $D_3$ and $D_4$ are held cutoff during this alternation, but $D_1$ and $D_2$ will conduct. The path of current is from point *B* through $D_1$ to point *C.* (Point *C* and the bottom of $R_L$ are electrically the same.) The current path is upward through $R_L$ and through $D_2$ back to point *A.* The pulse marked $D_1$–$D_2$ is developed across $R_L$ during this alternation.

On the next alternation, point *B* is positive and point *A* is negative. This places reverse bias on $D_1$ and $D_2$ and forward bias on $D_3$ and $D_4$. $D_1$ and $D_2$ will be held cutoff during this alternation, and $D_3$ and $D_4$ will conduct. The path of current is now from point *A* through $D_3$ to point *C*. From here it is upward through $R_L$ and through $D_4$ back to point *B*. The pulse marked $D_3$–$D_4$ is developed across $R_L$ during this alternation.

The output of the bridge rectifier is identical to the output of the regular full-wave rectifier except for the amplitude. If a negative output is desired, the connections to all four diodes will be reversed.

### *three-phase*

It was stated earlier that three-phase generators are common sources of power. A three-phase generator calls for a three-phase transformer and a three-phase rectifier. Fig. 5-17 illustrates a schematic symbol for a Δ to *Y* transformer.

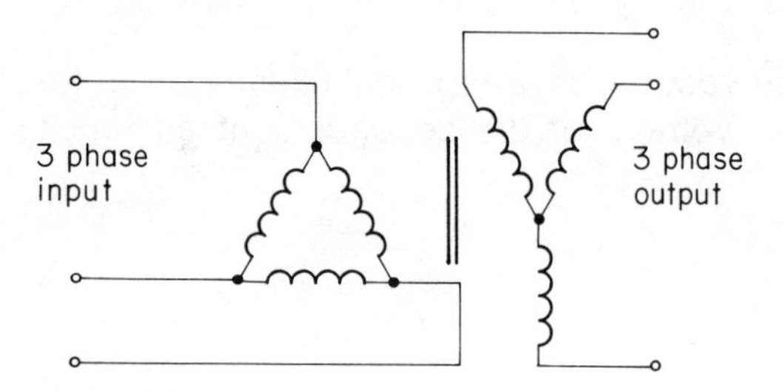


**Fig. 5-17.** Three-Phase Transformer.

Any of the rectifiers mentioned are suitable for use with the three-phase transformer. Let's use a half-wave rectifier. We will need three diodes instead of one, and they will be arranged as shown in Fig. 5-18.

Sine waves *A, B,* and *C* are developed on coils *A, B,* and *C* respectively. One of the diodes is conducting all the time which keeps some current through $R_L$ all the time. The high potential which remains across $R_L$ holds a positive potential on the emitters of the diodes. As a result each diode will conduct for only a short time during the most positive portion of the sine wave at its collector.

The output voltage across $R_L$ remains at a high level with only a slight ripple. This ripple is three times the frequency of any given sine wave input.

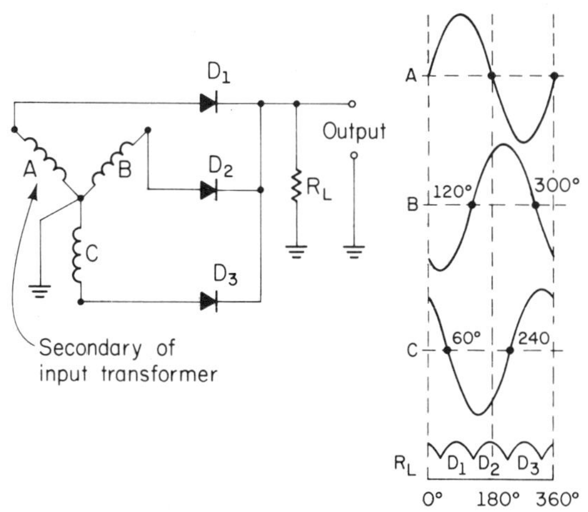


**Fig. 5-18.** Three-Phase Rectifier.

## VOLTAGE CONTROL

Generally, a great deal of experience is required for a person to realize the importance of proper voltage control. Almost every trouble that can occur in electronic equipment involves voltage; that is, the trouble is either caused from improper voltage, or else the trouble causes an improper voltage indication. In this first case, the trouble lies in the voltage control section. In the second case, the voltage control section provides a symptom of an equipment malfunction

The voltage has been delivered and rectified to a pulsating dc. This is not a usable voltage. Let's examine some methods of shaping this voltage to fit the many requirements of our electronic equipment. We will examine methods of removing the ripple, regulating the amplitude, and multiplying and dividing the value.

### *voltage multipliers*

When very large voltages are required, chances are that the load will use very little current. This type of circuit requires that the normal output voltage be doubled, tripled, quadrupled, and so on until it reaches the required level. These increased voltages can be obtained by combining the rectifier with a voltage multiplier. Let's start with a full-wave rectifier and a voltage doubler as illustrated in Fig. 5-19.

On one alternation, point 1 is positive, and point 2 is negative. Electrons leave point 2 and pile onto the curved plate of $C_1$. This drives electrons from the top of $C_1$ and sends them through $D_1$ to point 1. There is practically no opposition to this current. So, the voltage on $C_1$ follows the rise of waveshape *A*. The polarity of the charge is indicated on the schematic.

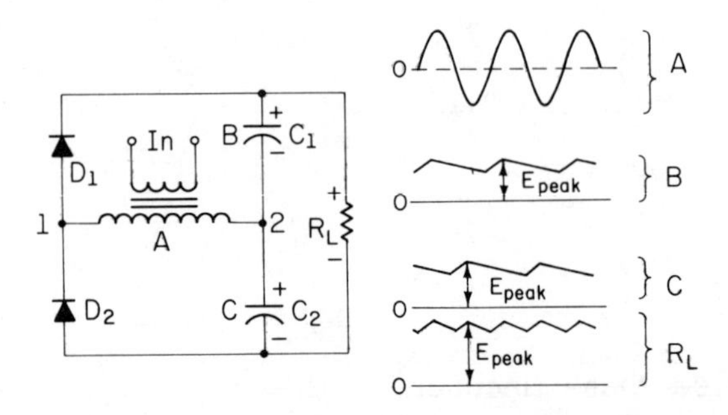


**Fig. 5-19.** Full-Wave Rectifier–Doubler.

On the next alternation, point 2 is positive, and point 1 is negative. Now, electrons leave point 1, go through $D_2$, and pile up on the curved plate of $C_2$. This drives electrons from the top of $C_2$ back to point 2. The voltage on $C_2$ follows the rise to the peak value of waveshape *A*, and in the direction indicated on the schematic.

The only discharge path for the capacitors is in series and upward through $R_L$. $R_L$ is a very large resistor which limits the discharge to a mere trickle of electrons. After the capacitors are once charged, the diodes conduct just enough to replace the charge that leaks off. Waveshapes *B* and *C* represent the charge that is retained on the two capacitors. These two charges in series keep enough current going through $R_L$ to produce waveshape $R_L$. The output has a small ripple, but it is a direct voltage with an amplitude equivalent to twice the peak voltage from the secondary of the transformer.

When the required output voltage is in excess of 10,000 V, multiple capacitors are used. This is necessary because a 10,000-V capacitor is bulky and difficult to find. The same result is obtained by connecting several capacitors as shown in Fig. 5-20.

For the purpose of this voltage doubler, circuits *A* and *B* are the same. The total capacitance is the same; the ratio of resistance per microforad is the same. Dividing the resistor and inserting the shorting bars are necessary in order to assure equal distribution of voltage across the capacitors. Each capacitor in circuit *A* must withstand $\frac{1}{2}$ of the total output voltage. Each capacitor in circuit *B* must withstand only $\frac{1}{6}$ of the output voltage.

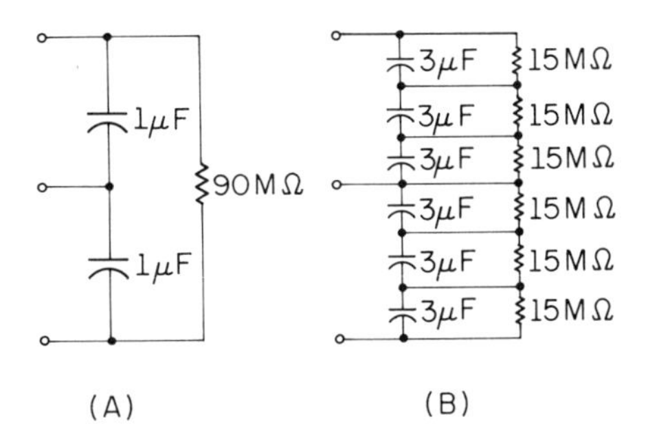


**Fig. 5-20.** Lowering Capacitor Voltage.

With a slight rearrangement of the components in Fig. 5-20, we can change this doubler to a cascade doubler. The cascade doubler is shown in Fig. 5-21.

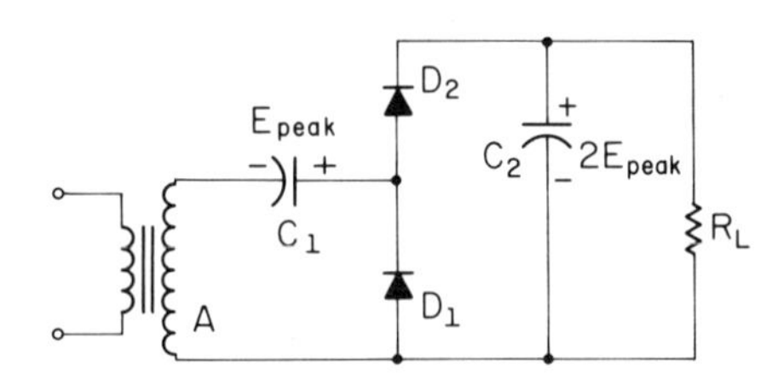


**Fig. 5-21.** Cascade Voltage Doubler.

When point *A* on the transformer goes positive, $D_1$ conducts and charges $C_1$ in the direction indicated. $C_1$ charges to the peak value of the transformer voltage. During the next alternation, $D_2$ conducts, and $C_1$ remains charged. The transformer voltage, and $C_1$ voltage in series, act to impress twice the peak transformer voltage on the anode of $D_2$. $D_2$ conducts and charges $C_2$ in the direction indicated. $C_2$ charges to twice the peak voltage. The charge on $C_2$ is the same as the available output across $R_L$.

This cascade doubler is the first step toward larger multiplications of the voltage. From the basic circuit in Fig. 5-21, diode sections can be added. Each section will increase the output by the peak value of the transformer voltage. Adding one section changes the doubler to a tripler as shown in Fig. 5-22,

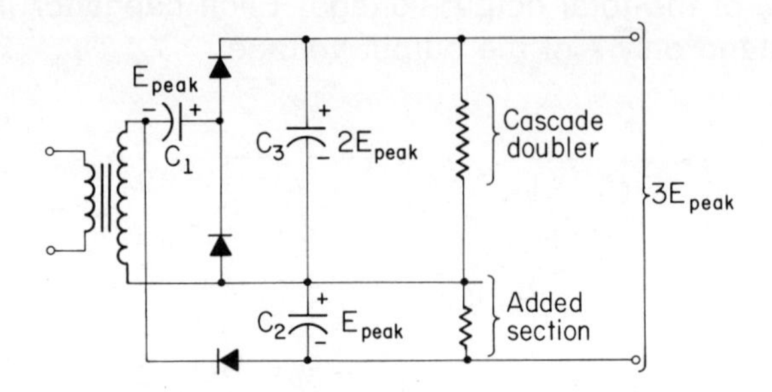


**Fig. 5-22.** Voltage Tripler.

This process of adding sections may continue as long as we like. The limits will be set by the stress which we place on the capacitors. But we have already shown one method of decreasing this stress. Let's add one more section and produce a voltage quadrupler. This is shown in Fig. 5-23.

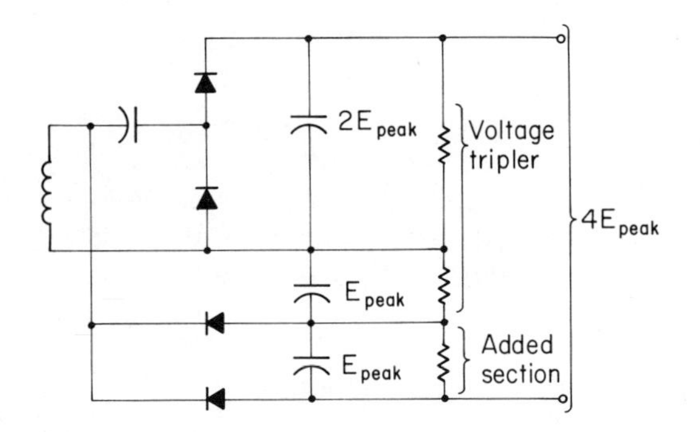


**Fig. 5-23.** Voltage Quadrupler.

No matter which rectifier we choose to change our alternating voltage to direct voltage, we still have a problem. The direct voltage is varying in amplitude. In most cases, this ripple frequency must be removed. That is the purpose of filters.

## FILTERING

Filtering is the process of reducing the ac component of a pulsating dc. This has already been done in the circuits which we called voltage multipliers. We will now examine some common low-frequency filters, but first, let's restate our problem. The ripple frequency is an undesirable ac component. We need a pure direct voltage with a constant amplitude. Our objective is to work out the best compromise that we can manage.

### *capacitor input filters*

The simplest version of this filter has already been shown in Figs. 5–19, 20, 21, 22 and 23. It is a capacitor connected across the load resistor. Figure 5-24 shows this filter again. Here it is filtering the output of a half-wave rectifier. We will call it an *RC* filter because the load resistor is doing part of the work. Let's analyze the action.

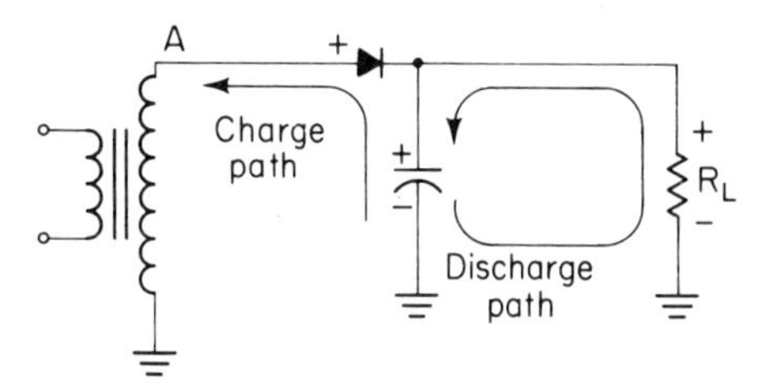


**Fig. 5-24.** *RC* Filter.

When point *A* of the transformer goes positive, the diode is forward biased. Current from ground, through the diode, charges the capacitor as indicated. The capacitor charges to peak value of the transformer voltage. As the voltage at point *A* passes peak and starts to decrease, the diode cuts off, and the capacitor starts to discharge through $R_L$. The values of resistance and capacitance form a very long time constant for the input frequency. As a result, the capacitor loses very little of its charge before point *A* goes positive again.

From this point on, the diode conducts briefly at the peak of each alternation. It conducts just enough to replace the charge that the capacitor has lost. So, the capacitor stores energy at each positive peak and gives up energy for the remainder of the time.

The key to the effectiveness of this filter is the amount of current drawn by the load resistor. Placing a low resistance load across the capacitor will result in a high current. Higher current will rapidly discharge the capacitor. If the capacitor is allowed to discharge, the filtering action is ineffective. This filter is very effective when the load requires a high voltage with very little current.

When an inductor is combined with the capacitor, the filter becomes an *LC* filter. This is necessary if the load is to use any appreciable current. Figure 5-25 illustrates one type of *LC* filter.

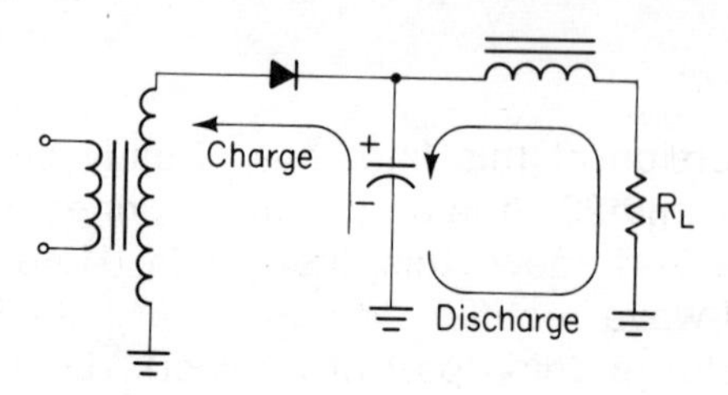


**Fig. 5-25.** *LC* Filter.

This is an L-type, capacitor input, *LC* filter. The capacitor is an open to dc and a virtual short to the ac component. The inductor is a short to dc and a high reactance to the ac component. Since all ripple frequencies in power supplies are low, the filter coils are constructed with iron cores. The inductor action in blocking ac has led to it being called a choke coil.

Another version of the capacitor input, *LC* filter is illustrated in Fig. 5-26.

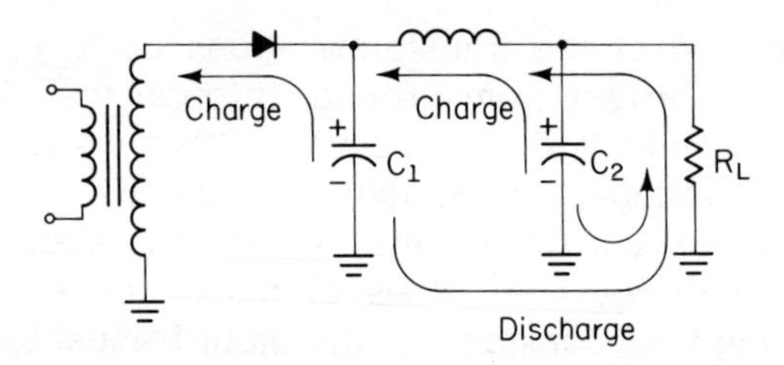


**Fig. 5-26.** Pi-Type *LC* Filter.

This pi- ($\pi$) type filter can handle a bit more current, but when used with a half-wave rectifier as shown here, current is still very limited. There are two distinct actions present in the filtering process. The charge and discharge action of the reactive components smooth out the slow changes. The sudden changes are blocked by the coil and shorted out by the capacitors.

As the requirement for load current increases, we change from a half-wave to a full-wave rectifier. For best filtering action, the configuration of the filter also changes.

The capacitor input filter just described is an effective filter for full-wave rectifiers with a moderate current drain. The capacitors charge twice as often as they do with the half-wave rectifier. This gives them much less time to discharge, and enables more current to be drawn. When still higher currents are needed, we resort to choke input filters.

### *choke input filters*

The choke input filter is an L-type filter. It has an inductor in series with total circuit current and a capacitor paralleling the load resistor. This arrangement is shown in Fig. 5-27.

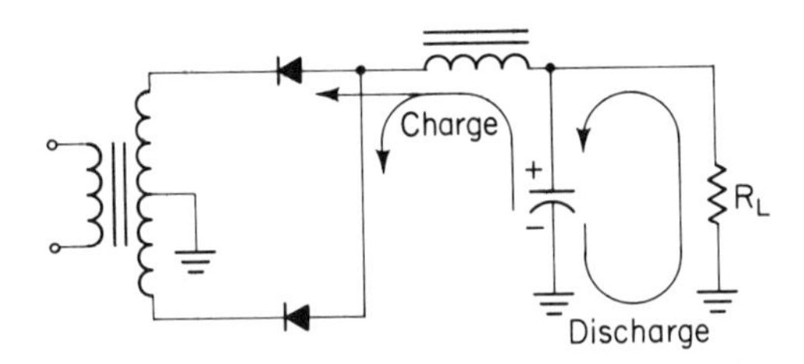


**Fig. 5-27.** Choke Input Filter.

On the peak of each alternation both coil and capacitor charge to the peak ac value. (Since the transformer is center tapped, the peak value of the pulsating dc is $\frac{1}{2}$ the peak transformer ac.) As an alternation passes peak and starts to drop, the capacitor and coil act together to maintain current in the same direction and at the same level. This action does a good job of smoothing out the gradual changes. Sharp changes are suppressed by the coil and shorted by the capacitor.

This filter does a good job on loads requiring relatively heavy current. It also stands up fairly well under a changing load current. It can be improved further by adding other L sections as shown in Fig. 5-28.

This filter will provide a relatively steady direct voltage under

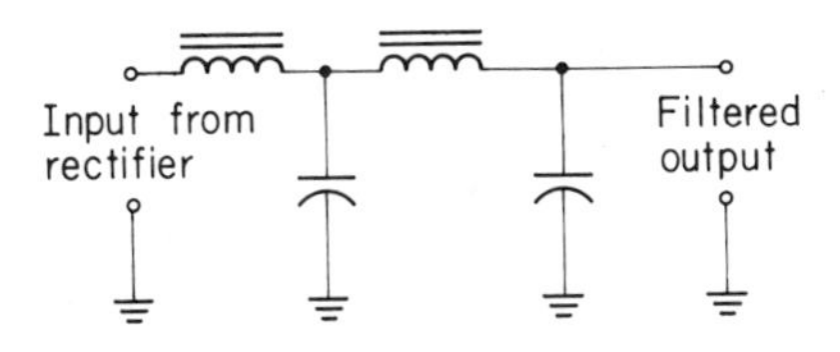


**Fig. 5-28.** Double Filter.

any reasonable load condition. It can withstand heavy current as well as changing current.

In addition to the load resistor, which is sometimes physically removed from the filter circuit, another resistor is normally connected across the output of the filter. This resistor is normally referred to as a bleeder resistor because it provides a discharge path for the capacitor. Such a resistor is shown in Fig. 5-29.

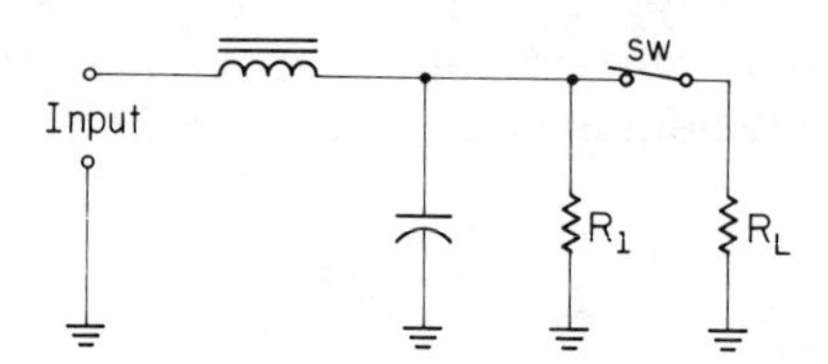


**Fig. 5-29.** Filter With Bleeder.

To illustrate the need for a bleeder resistor, suppose that the following events happen in the order listed.

1. The switch (SW) is opened to remove the load.
2. The filter is disconnected from the input.

This would leave the capacitor fully charged. Without the bleeder resistor ($R_1$) the capacitor would have no discharge path. The charge would eventually leak off, but meanwhile it is a potential hazard to anyone who touches the bare wires of the circuit. Filter capacitors are fairly large, and can deliver a heavy shock. A good capacitor can hold this charge for a long time. It is not unusual to receive a shock from a capacitor which has been without power for several days. The bleeder resistor provides a high-resistance path for this charge to slowly bleed away.

Another useful purpose of the bleeder resistor is to provide a minimum current when the load resistance is disconnected. A steady minimum current prevents surges when loads are connected into the power supply.

Now that we have a steady direct voltage, the next problem is distribution of this voltage to the various loads.

## VOLTAGE DIVIDER

We will not delve into the theory of loaded voltage dividers. They are mentioned here to show their use and position in the system. This relative position is depicted in Fig. 5-30.

The generator converts some form of mechanical energy to electrical energy to produce the alternating voltage. This voltage is delivered to the primary of the transformer. The transformer steps the voltage up or down to provide a manageable level. The rectifier changes the ac to a pulsating dc. The filter removes the ripple frequency from the pulsating dc and provides a steady direct voltage. The voltage divider provides taps to supply the proper voltage, and current, to several different circuits.

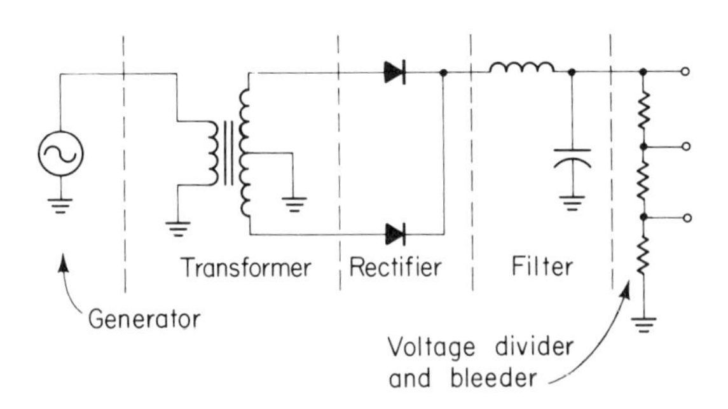


**Fig. 5-30.** Summary of Power System.

Some circuits will require no further refinement of the voltage. We will connect these directly to the voltage divider. Some other circuits are more particular. They are sensitive to any variation of voltage amplitude. They not only demand a smooth dc, they must have a smooth dc that never rises or falls beyond specified limits. These circuits need one more stage in their voltage control system. This stage is the voltage regulator.

## VOLTAGE REGULATION

A voltage regulator is a device which reacts to a voltage change in a manner that keeps the load voltage at a constant level. The filter does a good job of protecting us from sudden changes of short duration. A surge of voltage for a duration of a few milliseconds will be blocked by the inductor and shorted by the capacitor. However, the filter does not react to a slow change. The line voltage from our ac source is not regulated. At certain times of the day it might be as low as 110 V or as high as 130 V. The filter will not adjust this level. We need a device that will adjust the level of voltage which we deliver to our critical circuits.

***principles of shunt regulation***

The simplest form of shunt regulator is a variable resistor in parallel with the load resistor as illustrated in Fig. 5-31.

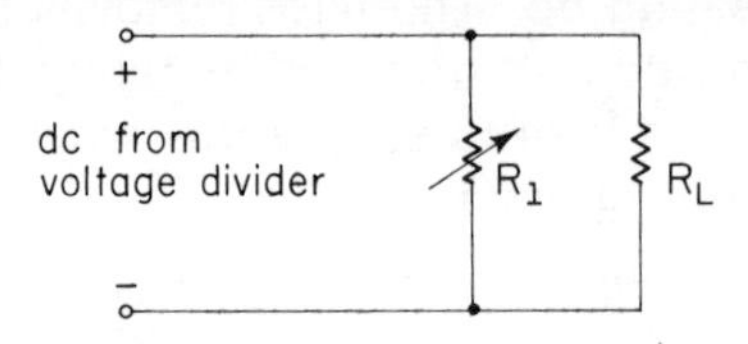


**Fig. 5-31.** Shunt Regulation Principles.

Suppose that the load circuits, represented by $R_L$, require 100 ± 2 V. When the voltage divider delivers more than 102 V, or less than 98 V, some adjustment must be made. The variable resistor, ($R_1$) enables us to make such an adjustment.

When the input voltage is too high, we decrease the resistance of $R_1$. This will cause an increase in total circuit current. The increased current will drop more voltage across the previous circuits. With a careful adjustment, the increased voltage drop can be exactly enough to cancel out the increase of voltage from the ac source. If the adjustment is made instantly, the voltage across $R_L$ will not change.

When the voltage divider delivers a potential that is too low, we increase the resistance of $R_1$. This decreases the total circuit current. The decrease in current causes less voltage drop across the previous components. The decrease in voltage drop offsets the decrease from the ac source, and the voltage across $R_L$ remains constant.

A changing load can also affect the amount of voltage across the load. There are no pure parallel circuits. Our line current is coming to this circuit through a certain amount of opposition. Adding more circuits will cause more line current, increase series voltage drop, and reduce the voltage available to $R_L$. This would be equivalent to reducing the resistance of $R_L$ in Fig. 5-31. Reducing the number of circuits represented by $R_L$ results in decreased line current and less series voltage drop. This increases the voltage level which is delivered to the load resistor. This action is equivalent to increasing the resistance of $R_L$ in the schematic.

The shunt resistor ($R_1$) can compensate for these load changes. When the resistance of $R_L$ increases, the resistance of $R_1$ must decrease at the same time. When the $R_L$ resistance decreases, $R_1$

must increase. This will keep the line current constant and result in a constant voltage level for $R_L$. What we need now is a sensing device which can automatically change the resistance of the shunt resistor.

### *zener diode shunt*

We have already seen a circuit similar to Fig. 5-32 in conjunction with solid-state diodes. Let's examine it again in the light of this new information.

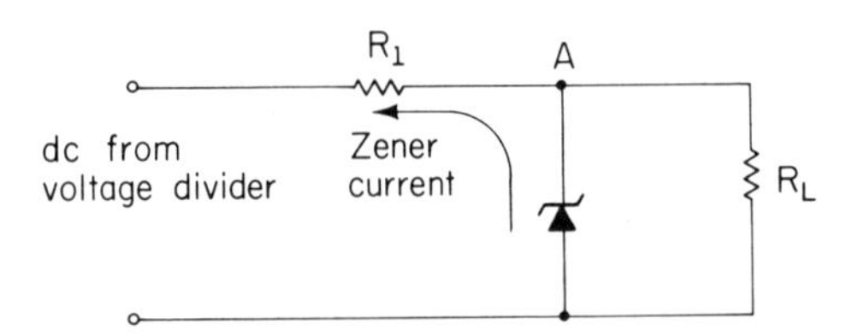


**Fig. 5-32.** Shunt Regulator.

$R_1$ in this circuit may be a physical resistor or it may not be. At any rate, it represents an aggregate of all the series resistance up to point *A*.

The Zener diode has a double role. It senses all changes and adjusts its resistance accordingly. $R_L$ represents all the circuits powered by this regulator.

The voltage at point *A* can be affected by four different actions. The voltage divider can deliver too much or too little voltage, and the load resistance can increase or decrease. The Zener diode can instantly react and compensate for all four of these conditions. Two of these conditions will cause the voltage at point *A* to increase. These are: too much voltage in or an increase in the resistance of $R_L$. The other two conditions result in a decrease in potential at point *A*. These are: not enough voltage delivered or a decrease in $R_L$ resistance.

Regardless of the reason, when point *A* becomes more positive, the reverse bias on the Zener diode is increased. A small increase in this reverse bias results in a sharp increase in Zener current. The increase in Zener current causes more voltage drop across $R_1$. The increased voltage drop across $R_1$ exactly cancels the increase at point *A*.

When point *A* becomes less positive for any reason, the reverse bias on the Zener diode is decreased. This results in a decrease in Zener current which reduces the voltage drop across $R_1$. The reduction in voltage across $R_1$ cancels the decrease in potential at point *A*.

The Zener diode has a very limited current capacity. This can be overcome in two ways: connecting diodes in parallel with each other and placing resistors in the diode branch. A combination of these arrangements is shown in Fig. 5-33.

### *thermistor shunt*

The shunt component does not have to be a Zener diode. A thermistor is another sensitive device frequently employed in shunt regulators. It is capable of handling more current than the Zener diode and offers regulation over a wider range of voltages. Its use is illustrated in Fig. 5-34.

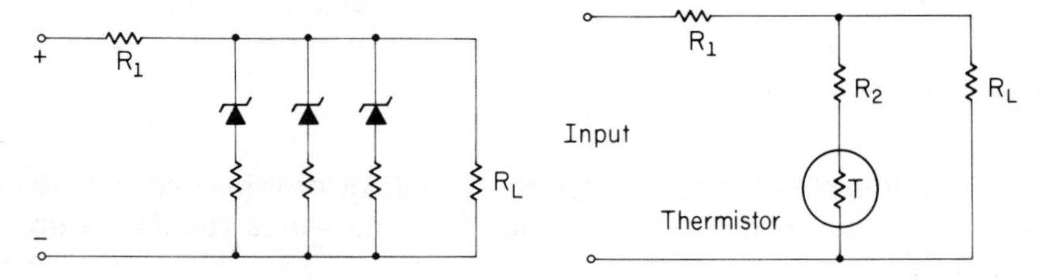


**Fig. 5-33.** Increasing Current Capacity.

**Fig. 5.34.** Thermistor Shunt.

The thermistor is a solid-state device with a negative temperature coefficient. The word thermistor was coined from thermal resistor. In this circuit, $R_2$ is an essential part of the shunt. The thermistor alone cannot compensate adequately for both increases and decreases of voltage. $R_2$ is an ordinary resistor with a constant resistance over a wide temperature range. It broadens the effective range of the thermistor by countering some of its reaction. This is illustrated in Fig. 5-35.

If less sensitivity is required, the shunt may be composed of ordinary crystal diodes. They would be connected opposite to that shown for the Zener diodes because they operate with forward bias. The ordinary diode, with forward bias, carries a very heavy current in comparison to the Zener current. This enables shunting of larger quantities of current.

Many regulators are constructed to allow the regulated voltage level to be adjusted up or down by several volts. This requires a more sophisticated type of shunt.

### *electronic shunt*

The electronic shunt contains one or more transistors with an adjustable bias arrangement. Figure 5-36 is a schematic of an electronic shunt regulator.

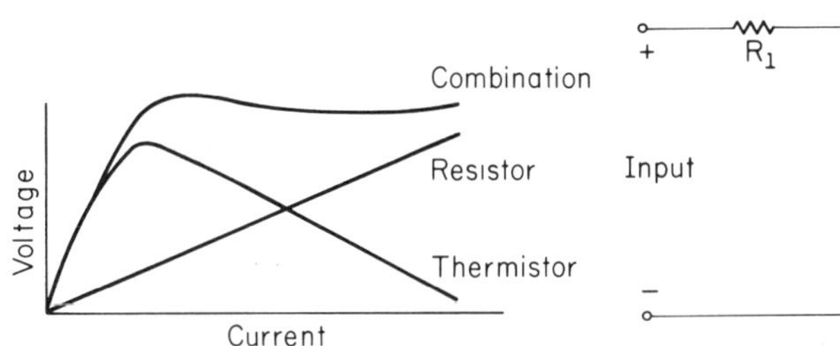


**Fig. 5-35.** Current-Voltage Response.

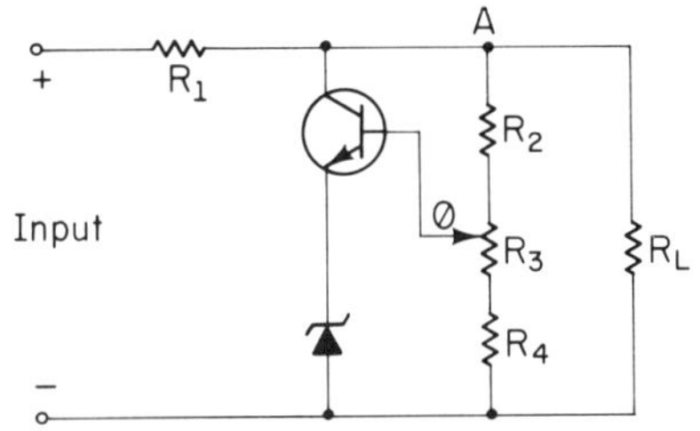

**Fig. 5-36.** Electronic Shunt Regulator.

The Zener diode controls the potential on the emitter of the transistor. Combined with this potential, the setting of $R_3$ determines the emitter-base bias, and thereby, controls the current through the transistor. The bleeder network composes of $R_2$, $R_3$, and $R_4$ is a high resistance to limit its shunting effect.

An increase in input voltage causes point *A* to go more positive. This increases the forward bias on the emitter-base junction by increasing the positive potential at the center arm of $R_3$. The transistor conducts more and causes an increased voltage drop across $R_1$. The increased drop in $R_1$ cancels the rise in potential at point *A*. This same action would result to compensate for an increase in load resistance.

A decrease in input voltage, or a decrease in load resistance, results in a smaller voltage at point *A*. This decreases the forward bias on the emitter-base junction and causes the transistor to conduct less. Less current through $R_1$ allows more voltage to point *A* which cancels the decrease.

The shunt regulators will waste a certain amount of power because the shunt current, of necessity, bypasses the load resistance. They are used primarily in areas where both load and shunt currents are of a small value. For larger currents the series regulators are more efficient.

### *principles of series regulation*

The series regulator places the variable component in series with the load resistance. This is a more efficient method of voltage regulation because the actual load current is used to regulate the load voltage. Again the simplest form is a variable resistor as illustrated in Fig. 5-37.

It should be easy to visualize the action in this circuit. The regulator resistor can be varied to compensate for any reasonable change of voltage. This includes changes in input level as well as changes brought about through variations in $R_L$.

An increase in input or an increase in $R_L$ will result in point *A* going more positive. An increase in the resistance of the regulator causes more voltage to drop across it. This cancels the increase at point *A*.

A decrease in input, or a decrease in load resistance, will cause point *A* to become less positive. A decrease in the regulator resistance will cause less voltage drop across it. This allows more voltage to point *A* which cancels the decrease.

To be effective the series resistance needs to be able to detect all changes and automatically vary its own resistance to counteract these changes. The electronic series regulator has such a resistor.

### *electronic series regulator*

The transistor is not only a sensitive variable resistor but an amplifier as well. It can be used very effectively as an automatic variable resistor as shown in Fig. 5-38.

The Zener diode is holding the transistor base at a relatively high positive potential. This provides forward bias for the base-emitter junction by keeping *B* more positive than *E*.

Any change in either input potential or load resistance will result in a change in potential at point *A*. Since this is also the emitter of the transistor, this change affects the base-emitter bias. An in-

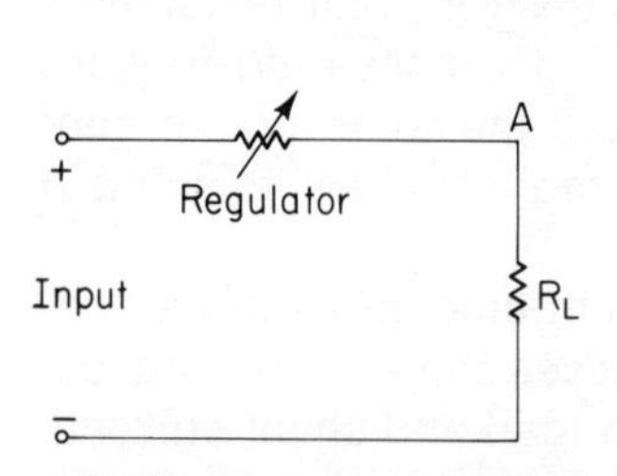


**Fig. 5-37.** Principles of Series Regulation.

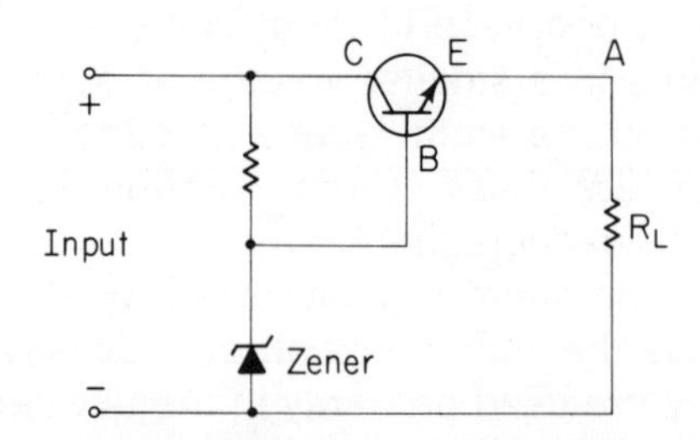


**Fig. 5-38.** Series Electronic Regulator.

crease in the input, or an increase in load resistance, will cause point *A* to become more positive. This decreases the forward bias and causes the transistor to conduct less. Less current through the transistor results in an increased internal resistance and a corresponding increase in voltage drop across the transistor. The increased voltage drop across the transistor cancels the increase at point *A*.

When the input voltage is too low, or the resistance of $R_L$ is decreased, the potential at point *A* becomes less positive. This increases the forward bias across the base-emitter junction and causes the transistor to conduct more. The increased conduction reduces the internal resistance of the transistor and results in a decreased voltage drop across it. The decrease in voltage across the transistor allows more voltage to point *A* which cancels the decrease.

The series regulator becomes a bit more elaborate when the adjustable feature is added.

### *adjustable series regulator*

One arrangement for an adjustable regulator is illustrated in Fig. 5-39.

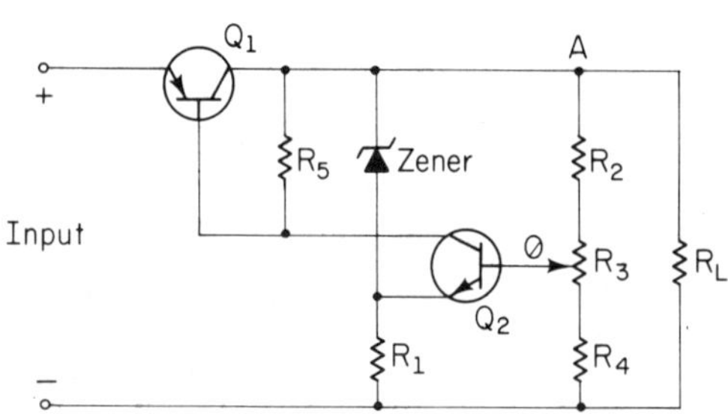


**Fig. 5-39.** Adjustable Series Regulator.

This configuration not only adds the adjustable feature, it is an improved regulator over all those we have seen thus far. It is stable over a wide range of voltage and current, and the output is constant within very close tolerances.

Point *A* is the spot where we want a regulated voltage. The potential at this point is determined by the input voltage and the voltage drop across $Q_1$. The voltage across $Q_1$ is determined by its level of conduction. The conduction of $Q_1$ is controlled by its bias which is dependent upon the voltage drop across $Q_2$. The voltage drop across $Q_2$ is regulated by its bias which is sensitive to any change in potential at point *A*.

Again, an increase of input voltage, or an increase of load resistance, causes point *A* to become more positive. The voltage drop across the Zener diode holds constant. Therefore, the entire increase appears across $R_1$. The increased voltage across $R_1$ makes the emitter of $Q_2$ more positive. The potential at the base of $Q_2$ also moves positive, but this change is less than that on the emitter. The result is a decrease in forward bias on the base-emitter junction of $Q_2$. $Q_2$ conducts less, and the voltage across it increases. This places a higher positive potential on the base of $Q_1$. $Q_1$ is a *pnp* transistor, and the increased positive on its base reduces its forward bias. $Q_1$ conducts less, its internal resistance rises, and the voltage drop across it increases. The increased drop across $Q_1$ cancels the increase at point *A*.

When the input voltage is too low, or the load resistance has been decreased, point *A* becomes less positive. The total drop is felt on the emitter of $Q_2$, and only part of it is on the base. Forward bias on $Q_2$ is increased which causes it to conduct more. The voltage at the collector of $Q_2$ decreases and places a less positive potential on the base of $Q_1$. This increases the forward bias on $Q_1$. $Q_1$ conducts more, its internal resistance decreases, and the voltage drop across it becomes less. The decreased voltage drop across $Q_1$ allows more voltage to point *A* which cancels the decrease.

The bias on $Q_2$ is adjusted by moving the center arm of $R_3$. Regulator adjustments of this type are usually made by turning the potentiometer with a screwdriver. Moving the arm of the potentiometer toward $R_2$ results in a reduced bias on $Q_2$. This in turn lowers the regulated voltage at point *A*. Moving the arm toward $R_4$ increases $Q_2$ bias and raises the regulated voltage level.

Some regulators are specified as current regulators while others are called voltage regulators. In our discussion, we have constantly referred to either a changing voltage input or a changing load resistance. Either of these conditions will change both current and voltage. Our regulators compensated for the voltage change. When you encounter a current regulator, you may analyze its actions in much the same fashion as we have done here.

## CHAPTER 5 REVIEW EXERCISES

1. What are the two most common sources of power?
2. Describe a situation which might require the use of a dynamotor.
3. Name the two functions of vibrators.

4. What is the phase relation of the two outputs from a two-phase generator?
5. What is the phase relation among the outputs of a three-phase generator?
6. Describe the difference between a step up and a step down transformer.
7. A power line is carrying 2500-V ac, and a transformer steps this down to 250-V ac. What is the primary to secondary turns ratio of the transformer?

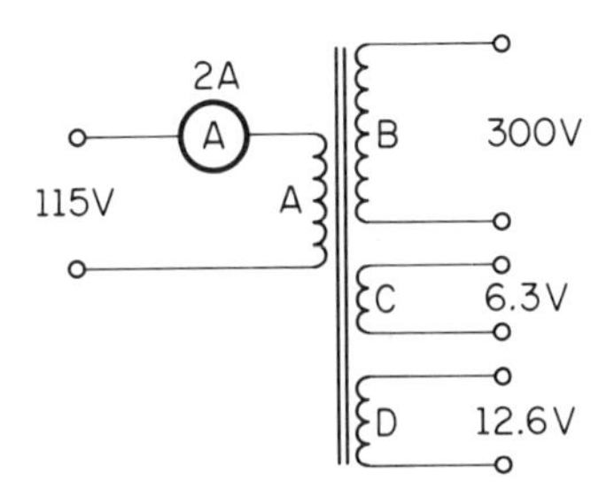


**Fig. 5-40.** Multiple Secondary Transformer.

Items 8 and 9 refer to Fig. 5-40.

8. What is the turns ratio from *A* to:
   (a) *B?*
   (b) *C?*
   (c) *D?*
9. How much current is available in secondary:
   (a) *B?*
   (b) *C?*
   (c) *D?*
10. A transformer has a 200-turn primary, a 500-turn secondary, and an input of 120 V. What is the amplitude of the output voltage?
11. You are constructing a transformer to reduce a 120-V source voltage to 24 V. You have a primary coil of 4000 turns. How many turns should you wind on the secondary?
12. Draw this transformer and indicate output of each secondary. It has a 10-turn primary with 10 V applied and the ratios of primary to secondary are 5:1 and 1:5.

13. A transformer raises an ac voltage of 120 to 120,000 V. What is the primary to secondary turns ratio?
14. A toy train operates on 12-V ac. Describe the requirements of a transformer that will enable the train to operate from a 120-V source.
15. Refer to Fig. 5-41. What is the value of:
    (a) $E_p$?
    (b) $I_p$?
    (c) $P_p$?

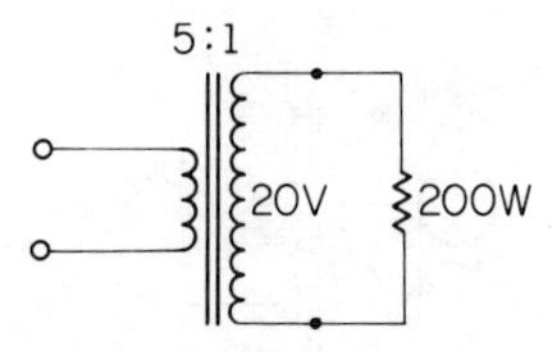


**Fig. 5-41.** Calculating Primary Values.

16. What is the maximum frequency that a power transformer can handle?
17. What type of installation would be expected to use 400-Hz transformers?
18. Draw a half-wave, solid-state diode rectifier that will change a sine wave to pulsating, negative dc. Show two cycles of the input and the corresponding output.
19. Draw a full-wave, solid-state diode rectifier that produces a positive output. Show two cycles of ac in and the corresponding output.
20. What is the advantage of a bridge rectifier over a conventional full-wave rectifier?
21. Name two jobs that a voltage multiplier performs simultaneously.
22. What type of circuits would be expected to use the output of a voltage multiplier?
23. A voltage quadrupler has 1000-V ac input. What is the output voltage?
24. What type of rectifier and filter would likely be found in a power supply designed to furnish a high and variable current?
25. List four actions which make voltage regulators necessary?

26. Draw a Zener diode shunt voltage regulator.
27. Draw a resistive circuit which will demonstrate the principles of series voltage regulation.
28. In a series electronic voltage regulator, which direction would the adjustment be moved in order to increase the regulated voltage?

# 6

# signal generation and control

Most practical electronic equipment is composed of a variety of circuit types. Each different type of circuit forms one or more stages, and a stage is designed to perform a specific task. One circuit generates a signal, another either multiplies or divides its frequency, and another increases the amplitude. One circuit fixes a dc reference for ac signals while another controls the maximum amplitude. It is our intent to examine the function of a few workhorse circuits that are common to most equipment.

## LIMITERS

A limiter is a circuit which controls the maximum amplitude of a signal. It is sometimes called a clipper because it eliminates any part of the signal which exceeds the specified limits. There are two general classes of limiters; series and parallel. The class is determined by the way the limiter is connected into the circuit.

### *series limiters*

A series limiter can be designed to eliminate all or any part of either the positive or negative alternation. It can also be used to eliminate a portion of both alternations. A series, positive limiter is shown in Fig. 6-1.

The sine wave input is applied to the cathode of the diode, and the output is taken across the resistor. Without a signal, there is no reason for the diode to conduct; there is no voltage drop across *R*, and the output is zero. The positive alternation of the input places reverse bias on the diode, and the diode remains cut off. The output remains at zero. This is time $T_1$ to $T_2$.

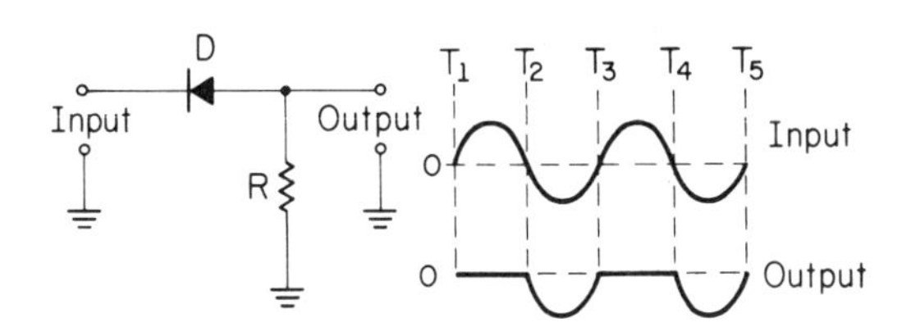


**Fig. 6-1.** Unbiased Series, Positive Limiter.

During the negative alternation of the input, the diode is forward biased. This allows current through the resistor to ground. The current variation will follow the input waveshape from $T_2$ to $T_3$. At $T_3$, another cycle begins with the diode cut off.

Since the diode has no dc bias, the entire positive alternation is eliminated from the output. That is why it is called a positive limiter. The series part of the name comes from the fact that the limiting device (the diode) is in series with the signal path.

Figure 6-2 illustrates the results of applying forward bias to this limiter.

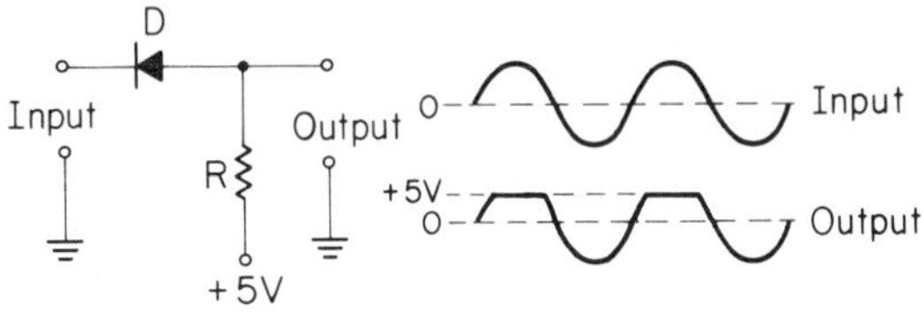

**Fig. 6-2.** Forward Biased Series, Positive Limiter.

The +5 V applied to the anode causes the diode to conduct under static conditions. The output, with no signal, is +5 V dc.

The positive alternations of the input signal oppose the forward bias, but the diode continues to conduct. When the input amplitude exceeds a positive five volts, the forward bias is cancelled and negative (signal) bias cuts the diode off. It remains cut off for the portion of this cycle that exceeds the dc bias.

The output is a reproduction of the input with limiting on the positive alternations. The limiter has removed the peaks from the positive alternations.

When more than 50 per cent of a signal needs to be removed, the limiter will have reverse bias. This is illustrated in Fig. 6-3.

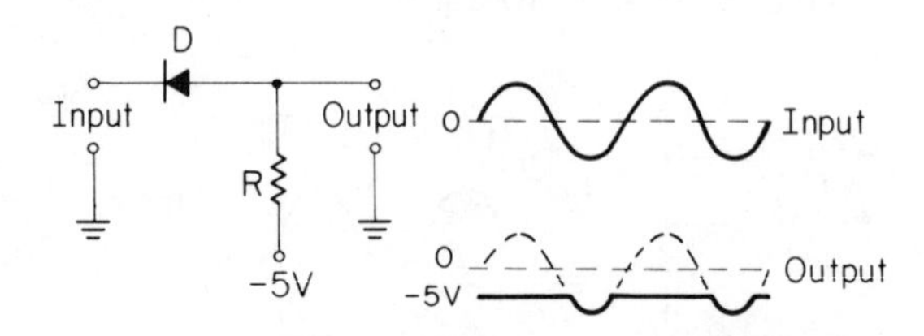


**Fig. 6-3.** Reverse Biased Series, Positive Limiter.

In a static condition this diode is cutoff, and the output is a negative 5 V dc. It remains in this state at all times except for the brief intervals when the input is more negative than the bias. The diode conducts during the negative peak of each alternation. The output consists of negative pulses which corresponds to the negative peaks of the input.

The negative portion of the signal may be limited by reversing the connections to the diode. In which case, we have a negative limiter.

***shunt limiters***

The shunt limiter has its limiting device parallel to the signal path. Whereas, a series limiter performs limiting during cutoff, the shunt limiter does its limiting during conduction. Figure 6-4 illustrates a shunt limiter circuit.

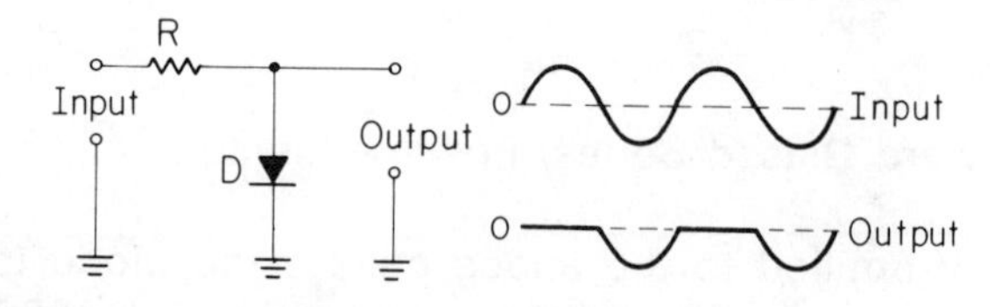


**Fig. 6-4.** Unbiased Shunt, Positive Limiter.

This circuit can develop an output only when the diode is cutoff. Without bias, the diode is conducting during the entire positive alternations of the input. During this time, the output is shorted out by the diode.

The negative alternations of the input cause the diode to cut off. So the negative alternations are passed on to the output circuit.

Less than 50 per cent limiting is accomplished by adding reverse bias. This is illustrated in Fig. 6-5.

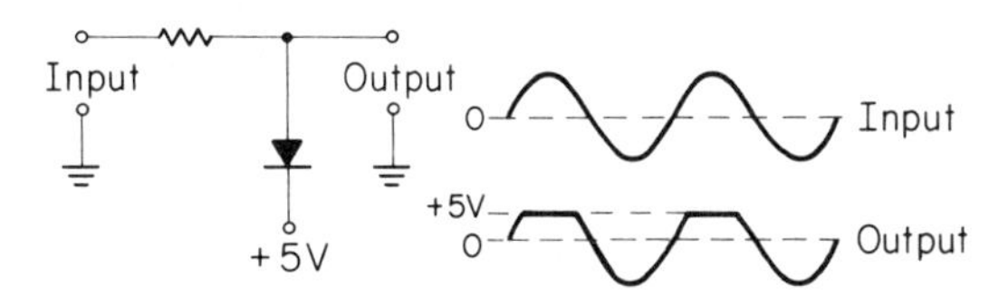


**Fig. 6-5.** Reverse Biased Shunt, Positive Limiter.

Here, the 5 V of reverse bias holds the diode cutoff until the incoming positive alternation exceeds that level. The diode conducts only during the peaks of the positive alternations of the input. These peaks are eliminated, and the remainder of the signal is passed on.

When it is necessary to eliminate more than 50 per cent of a signal forward bias is used. This is illustrated in Fig. 6-6.

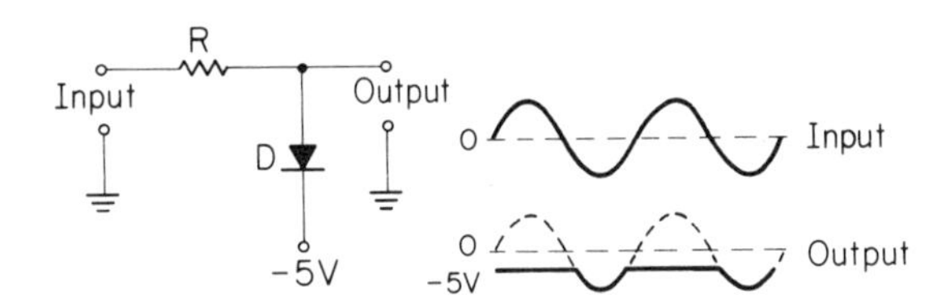


**Fig. 6-6.** Forward Biased Shunt, Positive Limiter.

The forward bias keeps this diode conducting most of the time. It cuts off only when the incoming signal is more negative than the dc bias. All of the signal is shorted out except the negative peaks.

Reversing the leads to the diode, transforms a positive limiter into a negative limiter.

Figure 6-7 shows a double diode shunt which limits a port: of each alternation.

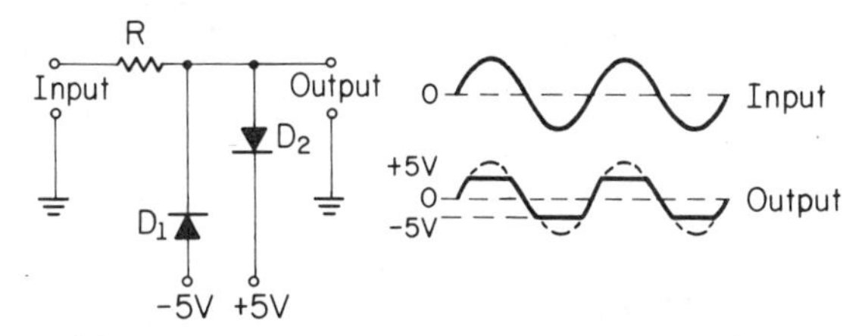


**Fig. 6-7.** Double Limiter.

This is actually a combination of a positive and a negative limiter, and both have reverse bias.

### *transistor limiters*

All the limiters that we have seen may be constructed with transistors instead of diodes. A transistor can perform either positive or negative limiting according to the bias arrangement. These circuits are generally reserved for situations which call for both amplification and limiting. This was previously mentioned with amplifiers as saturation and cutoff limiting. When the transistor is biased at cutoff, it is a negative limiter. If it is biased at saturation it is a positive limiter. Overdriving the transistor in both directions results in limiting portions of both alternations.

Another circuit which enjoys almost as much popularity as the limiter is the clamper.

## CLAMPERS

A clamper is a circuit which fixes the dc level for a signal. For this reason, it is sometimes called a dc restorer; meaning that it restores the signal to its proper dc reference.

### *need for clampers*

When a signal is passed from one stage to another, the dc level has a tendency to shift up or down. This is caused by the voltage charge on the coupling components. For instance, a signal with a zero reference acquires a 5-V dc reference when it passes through a capacitor with a 5-V charge. This is shown in Fig. 6-8.

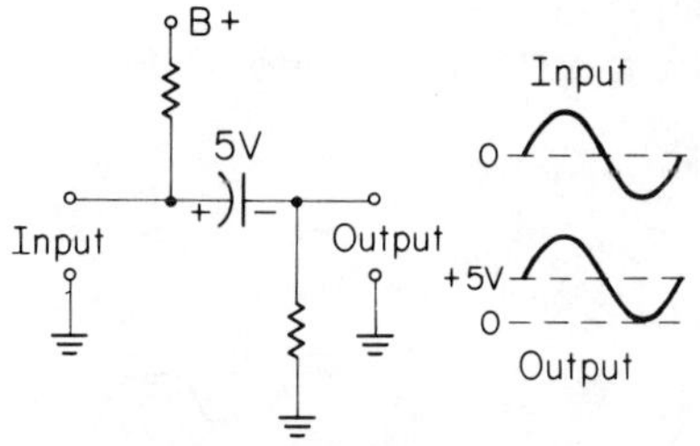

**Fig. 6-8.** Shifting of dc Reference.

If the capacitor had been charged in the opposite direction, the output would ride a —5 V.

This tendency to shift the dc reference makes the clamper a very essential circuit. Since the reference shifts in both directions, we need both positive and negative clampers.

### *positive clamper*

The positive clamper is a circuit which determines the dc starting point for a positive going signal. This starting point can be set at any desired level by setting the bias on the clamper. Figure 6-9 shows the result of zero bias.

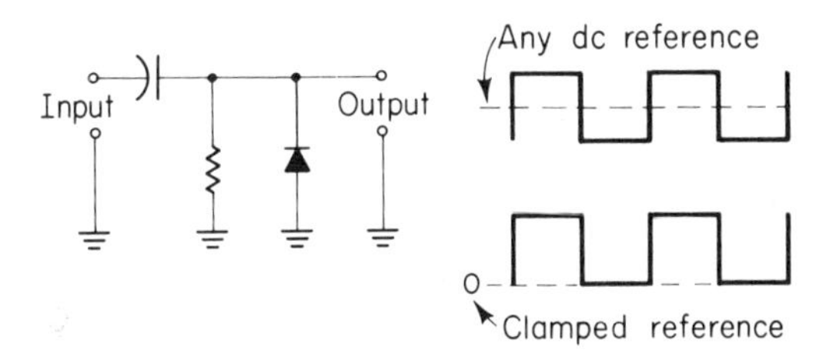


**Fig. 6-9.** Unbiased Positive Clamper.

If the incoming signal goes below zero, the diode conducts and charges the capacitor. This eliminates that output. The capacitor retains most of that charge. For subsequent negative inputs, the diode conducts only enough to replenish the charge on the capacitor. After one or two negative alternations, the entire input signal is producing a positive output across the resistor. So, a positive clamper with zero bias clamps the positive output to a zero reference line.

Placing forward bias on the diode will move the reference line in a positive direction. This is shown in Fig. 6-10.

In a static condition, the capacitor charges to the bias voltage

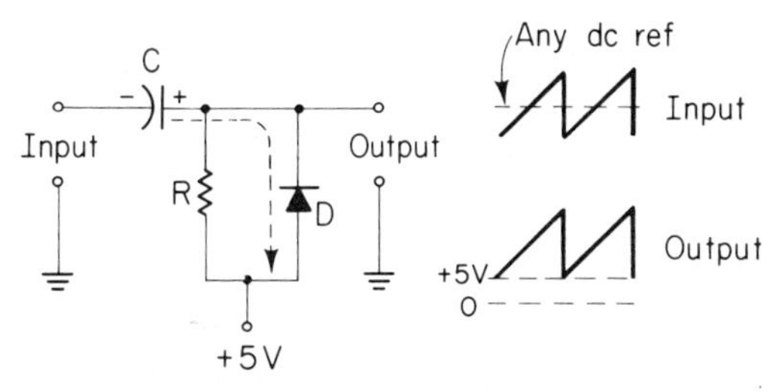


**Fig. 6-10.** Forward Biased Positive Clamper.

in the direction shown. The diode will conduct any time that the input drops below the +5-V level. The signal adjusts the charge on the capacitor to compensate for its input reference level. The result is a positive signal starting at +5 V.

The starting level can be set to a negative level by applying reverse bias. This arrangement is illustrated in Fig. 6-11.

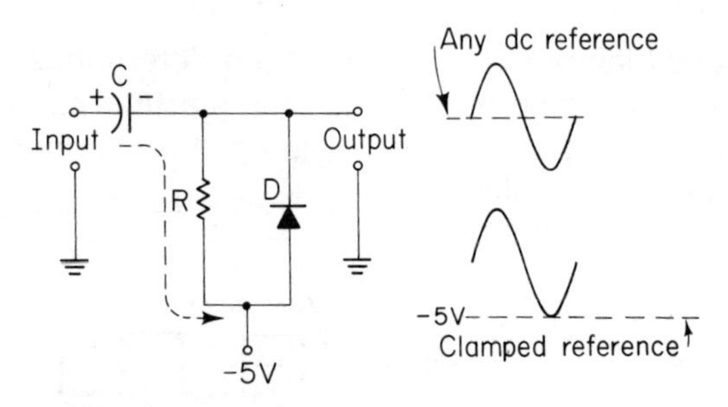


**Fig. 6-11.** Reverse Biased Positive Clamper.

In the static condition, current through the resistor charges the capacitor, as shown, to the bias level. The incoming signal will be entirely above this bias reference line.

### *negative clampers*

The negative clamper is no problem. Each of the positive clampers can be transformed into negative clampers by reversing the diode connections. The bias, in any case, sets the reference line. The positive clamper clamps signals above the reference line. The negative clamper clamps signals below the reference line.

## SWEEP GENERATORS

The sweep generator is in reality a generator of sawtooth wave shapes. The time base line on a CRT is a sweep. The term sweep generator has been attached because most sweep circuits operate with sawtooth wave shapes. For the same reason, it is also called a time base generator.

### *free running sweep generator*

This is our sweep generator in its simplest form. It consists of an *RC* network and an automatic switching device. This principle is illustrated in Fig. 6-12.

The switch is operated by the voltage level across the capacitor. When the switch is open, the capacitor charges slowly toward B+. When the charge on $C$ reaches a sufficient level, the switch closes. This provides a fast discharge path for the capacitor. When the capacitor discharges to a specified level, the switch opens and allows another charge cycle to begin.

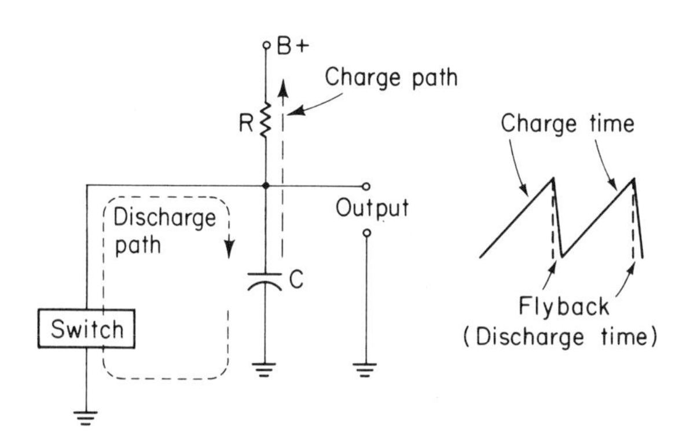


**Fig. 6-12.** Sweep Generator Principle.

The output waveform resembles the teeth on a saw. The slope of the rise time is determined by the $RC$ time constant. The flyback time (discharge) is dependent upon the reaction time of the switch. The peak value of the waveshape is the same as the potential required to open the switch.

A linear rise time is a primary consideration in sweep voltages. Since the capacitor charges at an exponential rate, the $RC$ time constant needs to be very large. Even with a large time constant, only the first 10 per cent of the charge has a near linear rise time. So, the B+ needs to be about 10 times the amplitude of the sawtooth output.

Practically speaking, the capacitor requires five time constants to reach a full charge, and a TC is $R \times C$

$$T = 5(R \times C)$$

where $T$ is total charge time in s, $R$ is resistance in $\Omega$, and $C$ is capacitance in farads (F).

What is one time constant for the circuit in Fig. 6-13?

$$\begin{aligned} \mathrm{TC} &= R \times C \\ &= (100 \times 10^{3})(10 \times 10^{-6}) \\ &= 1000 \times 10^{-3} = 1 \text{ s} \end{aligned}$$

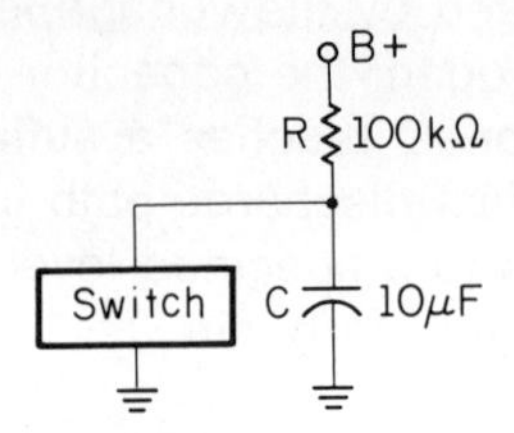


**Fig. 6-13.** Calculating One TC.

How long will it take this capacitor to reach full charge?

$$T = 5(R \times C)$$
$$= 5(1) = 5\text{ s}$$

Only the first 10 percent of the charge is linear enough for use, and the capacitor charges to 63 percent of maximum during the first time constant. What is the longest practical sweep time that can be obtained with this circuit?

In order to answer that question, we need to refer to the time constant graph in Fig. 6-14.

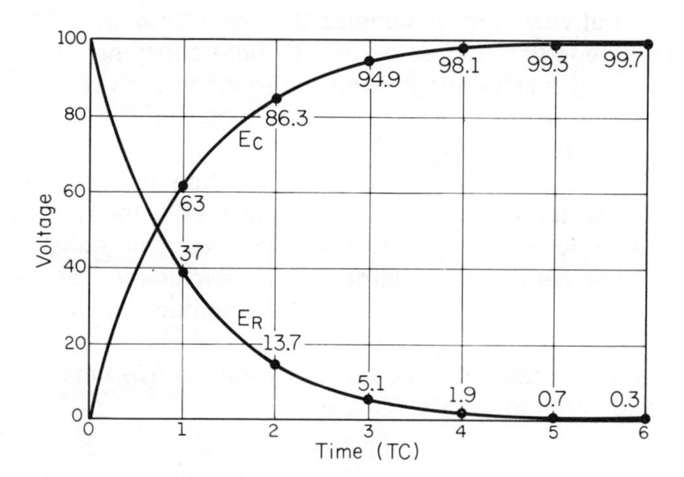


**Fig. 6-14.** Time Constant Graph.

Ten percent of the maximum charge is reached in about $\frac{8}{100}$ of a time constant. Therefore, our maximum sweep is $\frac{8}{100}$ of one second. This is 0.08 s or 80 ms.

Suppose that we need a sweep of 500 ms. We can obtain this by increasing the value of either *R* or *C* or both. Suppose that we have a capacitor of 30 $\mu$F. What size resistor would we use to obtain this sweep?

$$\text{Sweep time} = 0.08\ \text{TC}$$

$$0.08\ \text{TC} = 500\ \text{ms}$$

$$1\ \text{TC} = \frac{500 \times 10^{-3}}{80 \times 10^{-3}}$$

$$= 6.25\ \text{s}$$

$$6.25\ \text{s} = R \times C$$

$$6.25 = R \times 30 \times 10^{-6}$$

$$R = \frac{6.25}{30 \times 10^{-6}}$$

$$= 208.3\ \text{k}\Omega$$

### *synchronized sweep generator*

Any automatic switching device may be used as the control switch for the capacitor charge and discharge. The synchronizing signal may be a sine wave or a timing pulse. The four-layer transistor is a good switching device which lends itself to easy synchronization. This type of sweep generator is illustrated in Fig. 6-15.

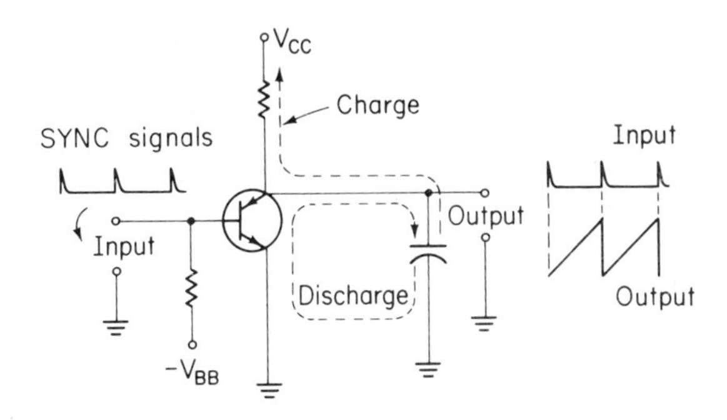


**Fig. 6-15.** Synchronized Sweep Generator.

The fixed reverse bias allows the generator to operate at a free run frequency. The synchronizing signals oppose the bias and tend to force conduction. As a result, one of the signals will trigger conduction at a time just slightly before it would occur naturally. This terminates the sawtooth a bit sooner than normal. Since another sawtooth begins immediately after one is terminated, the start time is controlled by triggering the termination action.

Remember, the generator is controlled by two individual actions. The rising sawtooth, due to capacitor charge, will overcome the bias and cause a natural termination if it is left to its own devices. The synchronizing signals cannot cause conduction until the capacitor charge has reached a certain level. The combining of these two actions is illustrated in Fig. 6-16.

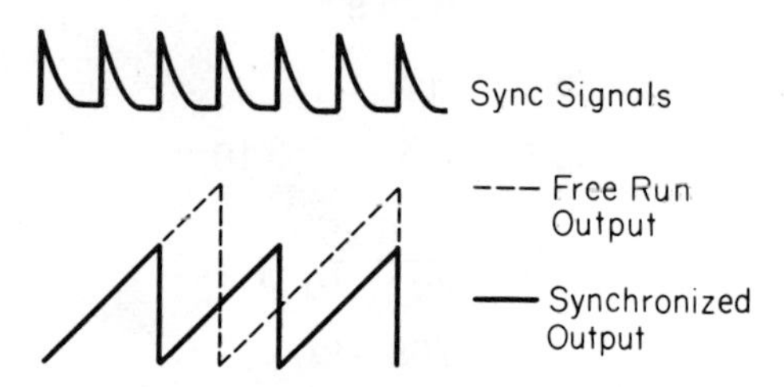


**Fig. 6-16.** Synchronizing Action.

The waveshape that is outlined with a broken line indicates what the output would be if the circuit was allowed to free run. With the synchronizing signals applied, every second signal triggers the transistor and terminates the sweep. Notice that this is just before the sweep would have terminated without the signal. Notice the slight difference in amplitude between the triggered output and the free run output.

For a sweep of a given time duration, linearity can be improved by increasing the time constant. This can be done by increasing the size of either resistor or capacitor or both. A longer time constant allows less (reactive) time for charge. This means that the sweep is terminated at a lower point on the charge curve. The sooner the sweep is terminated (in terms of percent of a time constant) the more linear the rise time.

There are situations where synchronizing the sweep voltage is not sufficient. It might be important that a sweep start at a specified time and terminate after a predetermined number of microseconds have elapsed. The gated sweep generator takes care of these special cases.

### *gated sweep generator*

The charge of a capacitor (or, in some cases, an inductor) still produces the sweep voltage, but the action is gated on and off. This type of circuit is illustrated in Fig. 6-17.

The transistor switch is gated open and closed by the square-wave input. When the input swings negative, it places reverse bias across the emitter-base junction. This cuts off the transistor and holds it cutoff for the duration of that alternation ($T_1$ to $T_2$). As the input swings positive, the input bias changes to a strong forward bias. This causes the transistor to conduct very hard from $T_2$ to $T_3$.

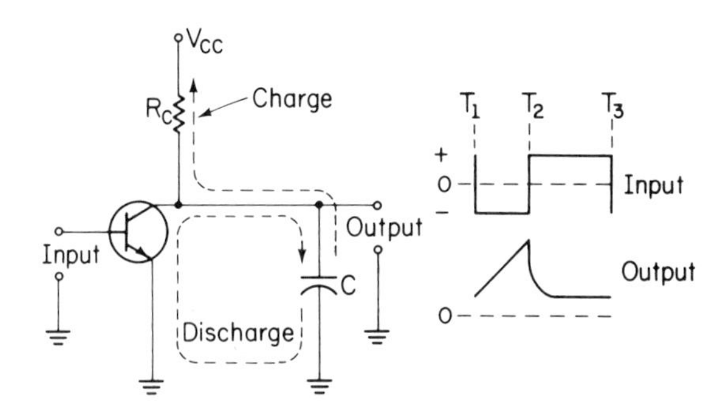


**Fig. 6-17.** Gated Sweep Generator.

During the cutoff time ($T_1$ to $T_2$), the capacitor is charging through the collector resistor ($R_c$) to $V_{CC}$. The rate of charge is determined by the size of $R$ and $C$. When the transistor is gated on, it provides a low-resistance discharge path for the capacitor. The charge on the capacitor does not return all the way to zero. The internal resistance of the transistor (emitter to collector) is in the discharge path. Therefore, each sweep starts from a slightly positive point. This internal resistance also slows the discharge (flyback) and accounts for the exponential curve on the trailing edge of the sawtooth. These are small problems, however. A clamper can return the base of the sawtooth to a zero reference, and the effects of a slow flyback are easily eliminated.

One of the most used circuits in electronics is the square-wave generator.

## SQUARE-WAVE GENERATORS

The square-wave generator finds a variety of uses. It comes in handy for gating and timing in many situations. When a square wave is needed for some purpose, the easiest way to produce it is generally by the use of a multivibrator. Multivibrators come in several styles, and we need to become familiar with a few of these.

### *free running multivibrator*

The free running multivibrator produces a series of square waves. It runs at a frequency determined by the circuit components and needs no input signal. Another name for this multivibrator is astable because it has no stable condition. An example of this type of square-wave generator circuit is shown in Fig. 6-18.

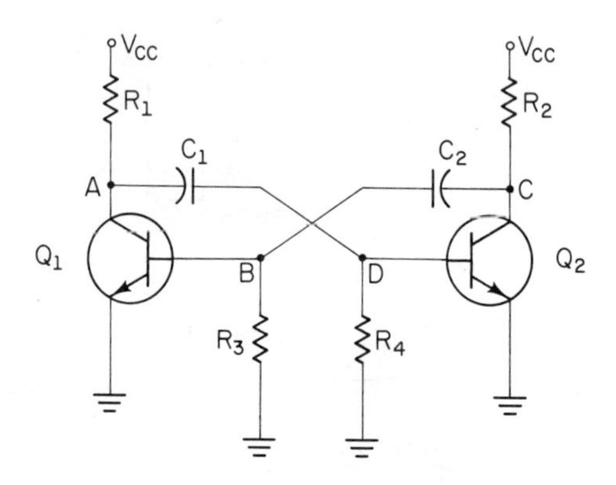


**Fig. 6-18.** Free Running Multivibrator.

When power is applied, both transistors start to conduct. Since a perfect balance never quite exists, one transistor will conduct more than the other. The voltage at the collector of the dominant transistor will take a sharp drop. The drop in potential causes the capacitor between that point (either *A* or *C*) and the base of the other transistor to start to discharge. The discharge path is through the base resistor to ground. This produces reverse bias on the transistor that is conducting less and cuts it off. The collector voltage of the cutoff transistor rises to $V_{CC}$, couples across the capacitor, and drives the conducting transistor to saturation.

Now, one transistor is saturated and the other is cutoff. This is not a stable state, however. In a short time the condition will reverse.

Let's assume a starting point, and follow the action through a complete cycle of events. In the process, we will describe and draw waveshapes of the voltages at points *A, B, C,* and *D.* Starting with $Q_2$ saturated and $Q_1$ cutoff the following conditions will exist. Point *A* will be $V_{CC}$ because there is no current through $R_1$. Point *B* will be far below ground because $C_2$ is discharging through $R_3$. Point *C* will be near zero because the saturation current of $Q_2$ is dropping most of $V_{CC}$ across $R_2$. Point *D* will be zero because there is no current through either $R_3$ or $R_4$. The waveshapes in Fig. 6-19 reflect all these conditions at $T_0$ and show all the changes for two complete cycles of events.

At $T_1$, $C_2$ has discharged enough to allow the voltage at point *B* to rise above the cutoff potential for $Q_1$. $Q_1$ comes on strong going all the way from cutoff to saturation. Its collector (*A*) drops from $V_{CC}$ to almost zero. $C_1$ has been charged to $V_{CC}$ and now starts to discharge through $R_4$. Point *D* is driven far below ground potential and cuts off $Q_2$. Point *C* would rise directly to $V_{CC}$, but the action is slowed a bit as $C_2$ charges through $Q_1$ and $R_3$. The levels at points *A, B,* and *C* hold steady until $T_2$. But the potential at point *D* is slowly rising as $C_1$ continues to discharge.

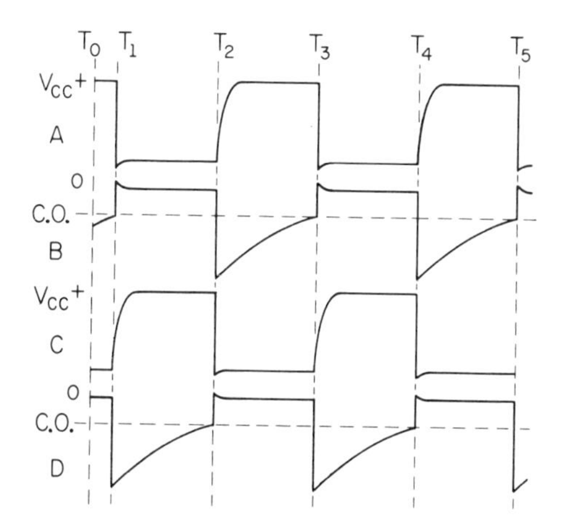


**Fig. 6-19.** Waveshapes for Fig. 6-18.

At $T_2$, $C_1$ has discharged enough to allow $Q_2$ to conduct. It instantly switches from cutoff to saturation. Point *C* drops sharply and forces $C_2$ to begin discharging through $R_3$. The resulting strong negative at point *B* drives $Q_1$ to cutoff. Point *A* would rise directly to $V_{CC}$ but can do so only as $C_1$ recharges through $Q_2$ and $R_4$. The levels at *A, C,* and *D* will now hold steady until $T_3$. But the potential at point *B* is slowly rising as $C_2$ continues to discharge.

At $T_3$, $C_2$ has discharged enough to allow $Q_1$ to conduct. This starts another cycle, and you should be able to trace it through without further assistance.

Outputs can be taken from both collectors. Waveshape *A* is the output from the collector of $Q_1$. Waveshape *C* is the output of $Q_2$. The only difference in these waveshapes is phase. They are 180° out of phase with one another. The amplitude of these outputs is determined by the value of $V_{CC}$. The time duration (and consequently, the frequency) is set by the TC or $C_2R_3$ and $C_1R_4$ which controls the cutoff time for the transistors.

Suppose that we need a circuit that will produce square waves 1 ms in duration. What capacitance must we use for $C_1$ and $C_2$ of the circuit in Fig. 6-20 to enable it to supply these square waves? (Assume 1.5 V reverse bias is cut off for both transistors.)

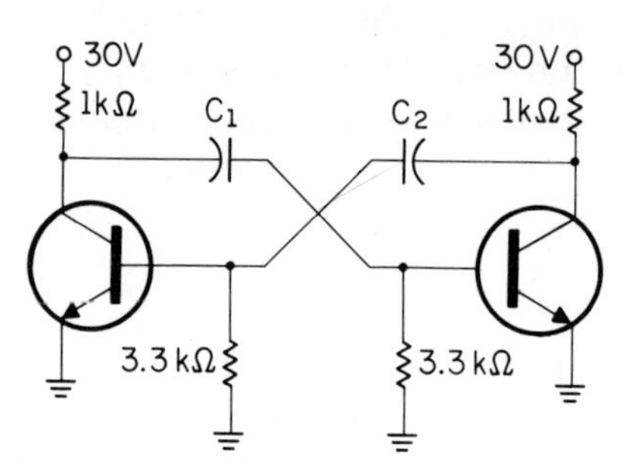


**Fig. 6-20.** Calculate Capacitance of $C_1$ and $C_2$.

$C_1$ and $C_2$ charge to 30 V and discharge through the base resistor to ground. Cutoff time is the same as square-wave duration which is the time required for the charge to drop from 30 to 1.5 V. This is a discharge of 94.9 percent of the 30-V charge. In other words, the transistors begin conducting when the total capacitor charge drops to 5.1 percent of the maximum charge.

Looking at the graph in Fig. 6-14, we determine that three time constants are required for a capacitor to lose this much of its charge. Now we need to determine what capacitance to use with a 10 kΩ resistor that will cause a time constant equivalent to $\frac{1}{3}$ of a ms.

The capacitances must be very small, probably in the picofarad range. Let's try 100 pF.

$$\begin{aligned} \text{TC} &= R \times C \\ &= 3.3\ \text{k}\Omega \times 100\ \text{pF} \\ &= (3.3 \times 10^{3})(100 \times 10^{-12}) \\ &= 330 \times 10^{-9} \\ &= 0.33\ \mu\text{s} \end{aligned}$$

Instead of allowing a multivibrator to free run, it is sometimes more appropriate to synchronize each change with an input timing signal. In this case, it becomes a synchronized multivibrator.

***synchronized multivibrator***

An input trigger of either positive or negative could control the action of our previous multivibrator. The controlling signal would need to be applied to both inputs if it is to control the switching time for both transistors. The synchronized circuit is shown in Fig. 6-21.

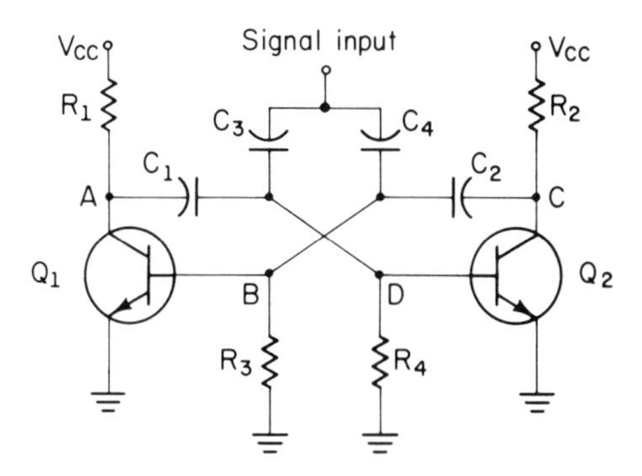


**Fig. 6-21.** Synchronized Multivibrator.

This circuit is identical to that in Fig. 6-16 with the exception of the coupling circuit for the input signal. $C_3$ and $C_4$ perform this coupling and serve no other function. Assuming the input signal to be a strong positive pulse, it will have no effect on the conducting transistor. But it will raise the potential on the base of the cutoff transistor and allow it to start conducting. This forces a switching action at that time which results in a reversal of conditions for both transistors. The next input signal operates on the other transistor, which is now cutoff, and forces another switching action.

The waveshapes and circuit actions are identical to those previously described with the exception of the forced switching. The frequency of the incoming signal would have to be slightly more than twice the free run frequency of the multivibrator. Of course, there is no necessity that the switching be accomplished with each input pulse. The circuit values can be arranged so that switching occurs with every second pulse, third pulse, or some other division of the input frequency.

The synchronized multivibrator is triggered by choice, but there is another type which must be triggered.

***bistable multivibrator***

This multivibrator has two stable states, and it requires a triggering action to change it from one state to the other. A sample circuit of this type is shown in Fig. 6-22.

Other names for this circuit are Eccles–Jordan multivibrator and flip-flop.

The function of the negative potential at the bottom of $R_5$ and $R_6$ is to improve stability. It asures that only one transistor will conduct at a given time and it speeds up the switching action.

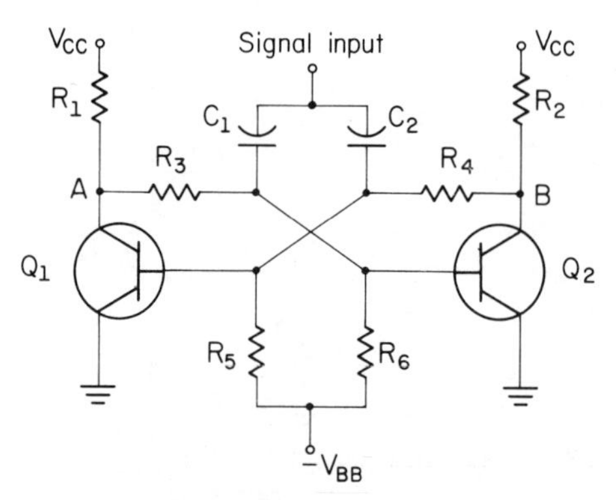


**Fig. 6-22.** Bistable Multivibrator.

When power is applied to the circuit, one transistor becomes dominant. This establishes a stable condition with one transistor saturated and the other cutoff. This condition will prevail until outside intervention takes place. The incoming signal will momentarily reduce the bias on the cutoff side and allow it to conduct. It jumps from cutoff to saturation, and its dropping collector voltage allows the fixed bias to cut off the other side. Now we have another stable state that will not change until another input initiates the action. Input and output waveshapes are shown in Fig. 6-23.

These waveshapes depict a condition where $Q_2$ was cutoff and $Q_1$ was saturated when the first input pulse arrived. The leading

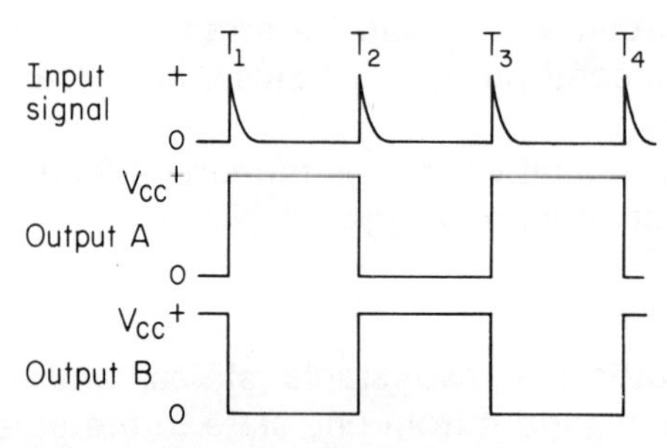


**Fig. 6-23.** Bistable Waveshapes.

edge of the pulse forced a switching action at $T_1$. Each subsequent pulse reverses the condition by initiating another switch.

Another multivibrator which has a variety of uses represents a compromise between the astable and the bistable multivibrator.

### *monostable multivibrator*

This circuit has one stable state, and a trigger pulse is necessary to force it out of this condition. The duration of the unstable state is determined by circuit values. At any rate, a trigger forces it to switch into an unstable state. After a predetermined period of time, it reverts to its original stable state and remains there until another trigger arrives. Such a circuit is illustrated in Fig. 6-24.

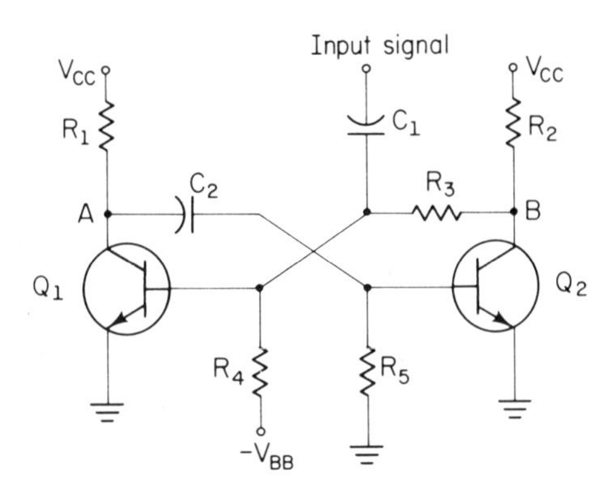


**Fig. 6-24.** Monostable Multivibrator.

This circuit has been dubbed a single-shot (one-shot) and at one time it was called a flip-flop. The term flip-flop has been transferred to the bistable multivibrator through popular usage.

In the static condition, $Q_1$ is biased below cutoff, and the bias on $Q_2$ keeps it conducting at or near saturation. This is a stable state that only a positive input can disturb. $C_2$ is charged to $V_{CC}$ and there is no current through $R_5$.

A positive input signal is coupled through $C_1$ to the base of $Q_1$. $Q_1$ conducts and its collector voltage drops. This forces $C_2$ to discharge and places reverse bias on $Q_2$. $Q_2$ cuts off, its collector voltage rises and places forward bias on $Q_1$. $Q_1$ goes to saturation and remains so as long as $Q_2$ is cut off. This is an unstable state. It will last as long as the discharge current from $C_2$ can develop cutoff bias for $Q_2$.

When $C_2$ discharges enough to allow $Q_2$ to conduct, its collector voltage will drop and return cutoff bias to the base of $Q_1$. $Q_1$ cuts off, its collector voltage goes to $V_{CC}$, and $C_2$ starts to charge. The charging current for $C_2$ places forward bias on $Q_2$ and drives it to saturation. Nothing more will happen until another positive signal arrives. The waveshapes for this circuit are shown in Fig. 6-25.

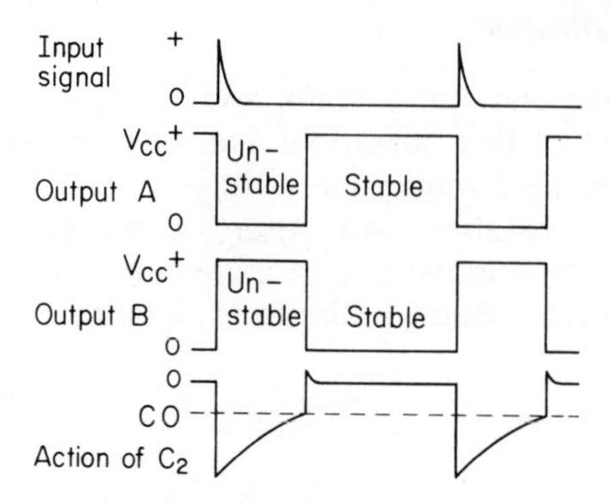


**Fig. 6-25.** Monostable Waveshapes.

The frequency of this circuit is controlled by the frequency of the input signal. There is a 1:1 frequency ratio. The duration of the unstable output is controlled by the *RC* time constant of $C_2$ and $R_5$.

Suppose that the unstable state remains in effect for exactly three TC. How much time would this be when $C_2$ is 10 $\mu$F and $R_5$ is 15 k$\Omega$?

$$1\,\text{TC} = R \times C$$
$$3\,\text{TC} = 3(R \times C)$$
$$= 3(15 \times 10^3)(10 \times 10^{-6})$$
$$= 450\text{ ms}$$

One common use of the monostable waveshape is delay in the application of a trigger pulse. When we differeniate wave shape *A* of Fig. 6-25, we have a positive trigger pulse each time the unstable state terminates. We can use these positive pulses to trigger a circuit. The effect would be equivalent to delaying the input signal for the duration of this unstable output. Using the same values of $R \times C$ as in our previous example, the delay is 450 ms.

With an input signal of a given frequency, we may use monostable multivibrators to produce signals any place between the input signals. Suppose that our input trigger frequency is 3 MHz and we need additional pulses exactly half way between the input signals. How much delay do we need?

$$T = \frac{1}{f}$$

where $T$ is time between pulses in s and $f$ is frequency in Hz.

$$T = \frac{1}{f}$$
$$= \frac{1}{3 \times 10^6}$$
$$= 0.33 \times 10^{-6} = 0.33\ \mu s$$

We need a delay equivalent to half the time between pulses. Therefore, our multivibrator needs to produce a square wave that is 16.5 $\mu$s in duration.

Using a 500-$\Omega$ resistor for $R_5$ and allowing three TCs for the unstable state, what value capacitor do we need for $C_2$?

$$3\ \text{TC} = 16.5\ \mu s$$
$$1\ \text{TC} = \frac{16.5}{3 = 5.5\ \mu s}$$
$$R \times C = 5.5\ \mu s$$
$$500\ C = 5.5 \times 10^{-6}$$
$$C = \frac{5.5 \times 10^{-6}}{500}$$
$$= 0.011\ \mu F$$

Some other waveshapes that are sometimes generated on the spot are timing pulses and sine waves. One method of producing either of these is to use an oscillator circuit. Let's examine a few of these.

## OSCILLATORS

An oscillator is defined as any nonrotating device which sets up and maintains fluctuations at a predetermined frequency. Oscillating current in a tank circuit is an example of such action. In some amplifier circuits, we cancelled feedback to prevent oscillations. Now we want a circuit that will create and maintain oscillations. There are many types of oscillators. Some produce radio frequency sine waves, some trigger pulses, and some square waves. In a sense, the free running multivibrator is an oscillator. We need to cover some preliminaries before studying oscillator circuits.

## *oscillating crystals*

Some crystals produce a voltage when they are subjected to a mechanical stress. This is known as the piezoelectric effect. A crystal will also produce a mechanical stress (vibrate) when a voltage is applied across it.

Crystals used in electronic equipment are thin sheets which have been cut from natural crystal. These sheets are ground to the proper thickness to produce the desired resonant frequency. Both quartz and rochelle salt crystals exhibit a marked degree of piezoelectric effect, and they have been widely used as a circuit components.

Crystals are cut in a special way to decrease their coefficient of temperature drift. The temperature coefficient is the relation between a temperature change and a frequency change. Some crystals have a positive temperature coefficient, some zero, and some negative.

The crystal, in conjunction with its holder, forms a circuit which has a predetermined resonant frequency. The holder provides external connectors and protects the crystal. It might even have an oven to maintain the crystal at a constant temperature. Figure 6-26 represents a tank circuit composed of a crystal and its holder.

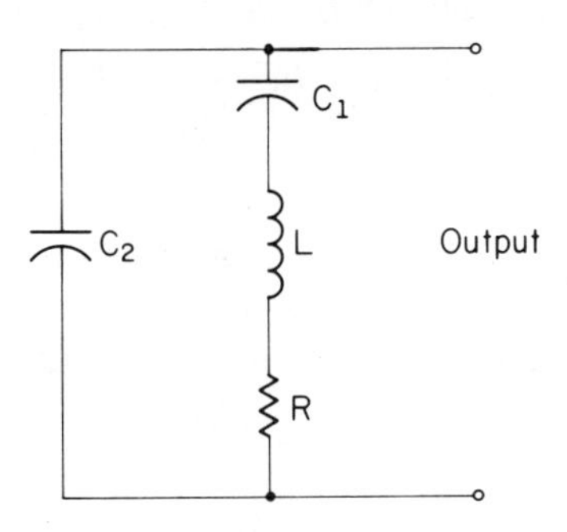


**Fig. 6-26.** Equivalent Crystal Tank.

The series combination of $L$, $C_1$, and $R$ represents the electric equivalent of the vibrating crystal characteristics. $C_2$ is the capacitance between the metal plates of the crystal holder. Together they form a parallel resonant tank circuit.

These crystal tanks are frequently used in oscillators to determine the operating frequency. The crystal is highly desirable because of its astounding frequency stability. This stability is a result of the very high Q of the crystal which may be as high as 400,000.

### *requirements for sustained oscillations*

There are four primary requirements that must be built into a circuit in order to sustain oscillations. These are:

1. A power supply.
2. A frequency determining device.
3. Amplification.
4. Regenerative feedback.

Of course, no circuit functions without power, but here the power supply gets special mention. Why? Because it replaces the circuit losses to keep the oscillations going.

The frequency determining device can be a common resonant circuit, an *RC* phase shifting circuit, or a crystal.

The other requirements are easy to obtain, also. Any amplifier will do the job of amplification, and there is no special trick to constructing a regenerative feedback circuit.

One widely used oscillator circuit has a crystal to control the frequency.

### *crystal-controlled oscillator*

In this circuit, the crystal holder is connected to the input of the amplifier. The other circuit components are selected to complement the work of the crystal. Let's examine the circuit in Fig. 6-27.

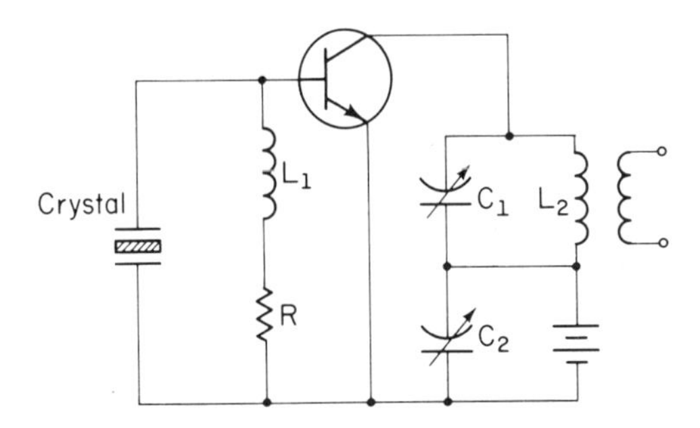


**Fig. 6-27.** Crystal-Controlled Oscillator.

First of all, the circuit frequency can be controlled by the crystal without being limited to the fundamental crystal frequency. It can be tuned to operate on the fundamental or on one of the harmonics. For this circuit, let's assume operation to be at the fundamental resonant frequency of the crystal.

The collector tank circuit is tuned to the resonant frequency of the crystal. The output will be taken across this tank circuit.

Regenerative feedback is supplied in two ways. Internally, energy is coupled back from the collector to the base and to the crystal. Externally, energy is coupled through $C_2$ and back to the crystal. $C_2$ also serves another vital purpose; it bypasses the ac oscillations around the dc power supply.

On the input to the transistor, $L_1$ and $R$ form a self-biasing network and a discharge path for the crystal holder capacitor.

When power is applied, a surge of current through the transistor sets up oscillations in the collector tank. The voltage from the tank alternations is fed back and applied across the crystal. The voltage alternately bends the crystal and allows it to spring back to its normal shape. The crystal, thus, generates a voltage of its own at its natural frequency. The voltage from the crystal is amplified through the transistor and reinforces the oscillations in the collector tank.

The oscillations from the collector tank are transformer coupled to another circuit. Generally, the next stage would be an amplifier tuned to provide maximum amplification of this particular frequency. What frequency is that? Fundamental frequencies range from 50 kHz to several MHz. After that, we have usable harmonics of the fundamental as high as the seventh harmonic. So, we select almost any frequency, then find a crystal that will produce that frequency.

The resonant frequency of the tank circuit is calculated by the formula:

$$f = \frac{1}{2\pi\sqrt{LC}}$$

where $f$ is frequency in Hz, $L$ is inductance of the coil in H, and $C$ is the capacitance in F.

We have a crystal frequency of 16.7 MHz and need an output equivalent to the third harmonic. $L_2$ is a 16-$\mu$H coil. We will obtain resonance on the third crystal harmonic when $C_1$ is set to what value? This can be answered by solving the resonant frequency formula for the value of $C$.

$$C = \frac{(0.159/f)^2}{L}$$

$$= \frac{(0.159/50 \times 10^6)^2}{16 \times 10^{-6}}$$

$$= \frac{(0.318 \times 10^{-8})^2}{16 \times 10^{-6}}$$

$$= \frac{0.101124 \times 10^{-16}}{16 \times 10^{-6}}$$
$$= 0.632 \text{ pF}$$

In order to produce resonance on the first harmonic, what value of capacitance would we set?

$$C = \frac{(0.159/f)^2}{L}$$
$$= \frac{(0.159/16.7 \times 10^6)^2}{16 \times 10^{-6}}$$
$$= \frac{(0.0095 \times 10^{-6})^2}{16 \times 10^{-6}}$$
$$= \frac{90.25 \times 10^{-18}}{16 \times 10^{-6}}$$
$$= 5.64 \times 10^{-12}$$
$$= 5.64 \text{ pF}$$

Changing our circuit, we insert an unknown crystal. With inductance of 70.4 $\mu$H and a capacitance of 250 pF, resonance is obtained on the third harmonic. What is the crystal frequency?

$$f_{\text{tank}} = \frac{1}{2\pi\sqrt{LC}}$$
$$f_{\text{xtal}} = \frac{1}{2\pi\sqrt{LC/3}}$$
$$= \frac{0.159/\sqrt{70.4 \times 10^{-6} \times 250 \times 10^{-12}}}{3}$$
$$= \frac{0.0012 \times 10^9}{3}$$
$$= 0.0004 \times 10^9$$
$$= 400 \times 10^3$$
$$= 400 \text{ kHz}$$

### *magnetostriction*

Magnetostriction effect is a term applied to iron, and it is somewhat similar to the piezoelectric effect of a crystal. When an iron rod is subjected to a rapidly changing magnetic field, it changes its dimensions as the field changes. The rod will alternately stretch and relax as the flux expands and decays. Other materials can be alloyed with iron to greatly enhance the magnetostriction effect.

As the rod changes its length, it causes a change in the magnetic flux, either increasing or decreasing the field. Any piece of material has a fundamental resonant frequency. This frequency is determined by the type of material and the size and shape of the object. So, with our iron alloy rod, if the changing magnetic field matches the resonant frequency of the rod, the rod will vibrate like a tuning fork. These oscillations will be of the same frequency as the resonant frequency of the rod.

The iron alloy type resonator is less stable than a crystal, but several things can be done to improve the stability. The alloying materials may be chosen to provide a negative temperature coefficient. Another method is to rigidly control the temperature of the rod. Either of these methods produces a resonator stable enough to be used as a frequency standard for low frequencies.

### *magnetostriction oscillator*

The magnetostriction resonator is often used as the frequency determining device in audio oscillators. Figure 6-28 illustrates this application of magnetostriction.

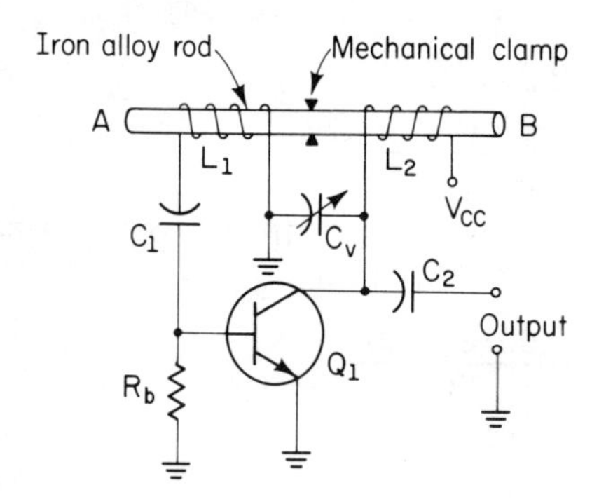


**Fig. 6-28.** Magnetostriction Oscillator.

When dc power is applied, $Q_1$ will conduct only slightly. If the dc would remain perfectly steady nothing else would happen, but the first fluctuation in collector current will trigger a chain of events. Suppose that a noise pulse causes a slight increase in collector current. This is all the trigger we need.

The surge of collector current will start a shock wave in the iron rod. The shock wave will travel to the left through the base coil $L_1$. This action is similar to a sound wave traveling through a solid.

The iron rod is compressed and elongated as the wave moves. This causes a change in flux of $L_1$ which places a positive signal on the base of $Q_1$.

When the original wave reaches the end of the rod (point *A*) it will be reflected back toward the opposite end (point *B*). The positive signal at the base causes $Q_1$ to conduct harder. Increased current in $L_2$ sets up a second shock wave to coincide with the first one which was reflected. Now a stronger shock wave starts down the rod toward point *A*.

This action is cumulative. The shock waves travel back and forth through the rod. They are reflected each time they reach an end of the rod, and the surges of current reinforce them to prevent them from dying out.

The rod, being mechanically clamped in the center, will vibrate at its resonant frequency. $C_r$ is a tuning capacitor for adjusting both coils to the resonant frequency of the rod. As far as collector current is concerned, these vibrations result in a slow rise and fall of current. These changes are coupled across $C_2$ as a low-frequency audio signal.

Numerous oscillators use some form of a tuned tank circuit as the frequency determining device. One of these is the audio oscillator.

## *audio oscillator*

An audio oscillator circuit is illustrated in Fig. 6-29.

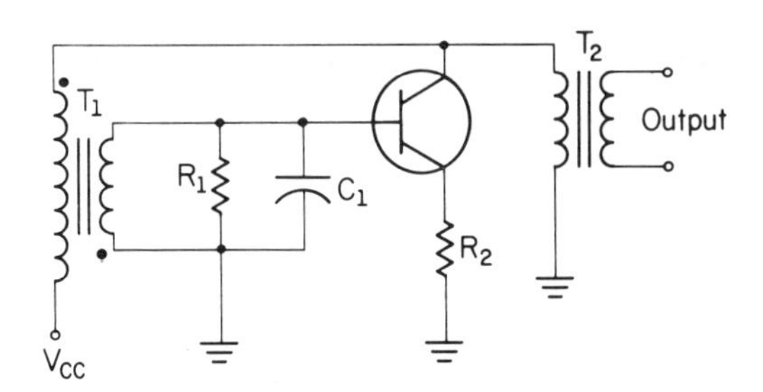


**Fig. 6-29.** Audio Oscillator.

The audio frequency allows the use of a tuned iron core transformer for frequency determination. Here, transformer $T_1$ has a tuned secondary to accomplish that. The frequency may be adjusted by varying $C_1$.

The regenerative feedback is furnished from the collector

across $T_1$ to the base. $R_1$ is a swamping resistor to control the amplitude of the feedback. Since $R_1$ is a variable resistor, the feedback amplitude can be adjusted.

$R_2$ provides a self-bias to prevent overdriving the transistor. It also produces a small amount of degeneration for temperature stability.

To be sure that we do not establish an impression that tank circuits are essential to oscillation, we will now examine an *RC* oscillator.

### *lag line oscillator*

Any amplifier that has sufficient regenerative feedback becomes a natural oscillator. An *RC* network can be constructed in such a manner that it will serve two purposes. First, it provides the required feedback. Second, it determines the frequency of oscillation. Figure 6-30 shows a sample lag line oscillator circuit.

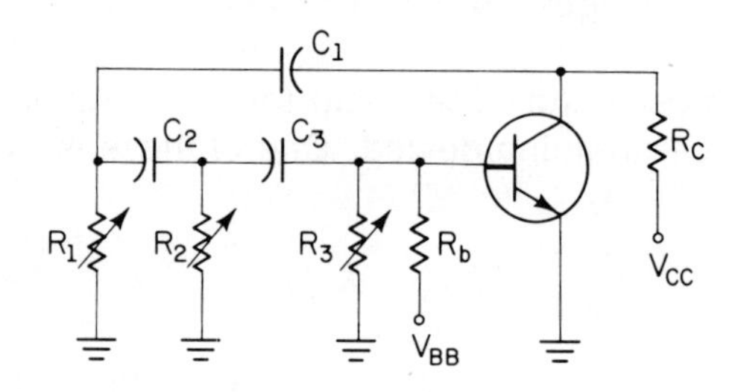


**Fig. 6-30.** *RC* Lag Line Oscillator.

The *RC* network, composed of $C_1R_1$, $C_2R_2$, and $C_3R_3$, determines the frequency of operation and provides regenerative feedback. This circuit is designed on the principle that the *RC* network will provide exactly 180° shift of phase for only one frequency. This particular frequency will be the frequency of oscillation.

Since this common emitter provides a 180° phase shift from base to collector, the *RC* lag line must furnish another 180° shift in order for feedback to be in phase with the base signal. Each *RC* combination needs to produce 60° of phase shift to accomplish the regeneration. Looking at just one *RC* combination, $X_C$ is a product of frequency. If $X_C$ and $R$ are equal, the phase shift is 45°. We need more than that. So, $X_C$ needs to be slightly more than 1.5 $R$. Since only one frequency can give this amount of $X_C$, only that frequency can obtain the proper phase shift. The frequency can be adjusted by varying the size of $R_1$, $R_2$, and $R_3$.

The proper amount of forward bias is furnished through $R_b$, and the output is taken across $R_c$.

How do the oscillations get started? Any electronic circuit produces noise. Noise frequencies range over the entire spectrum. When power is applied to this circuit, noise of many frequencies is generated. One of these frequencies will match the *RC* network, accomplish regenerative feedback, and be amplified over and over until it becomes the dominant frequency.

Another type of oscillator has wide usage because of its ability to produce a series of accurately timed pulses. This is the blocking oscillator.

### *blocking oscillator*

The name comes from the action of the circuit. The transistor amplifier is operated class *C*. Thus, it operates for a short time and is cutoff (blocked) for a long time. A sample of this circuit is shown in Fig. 6-31.

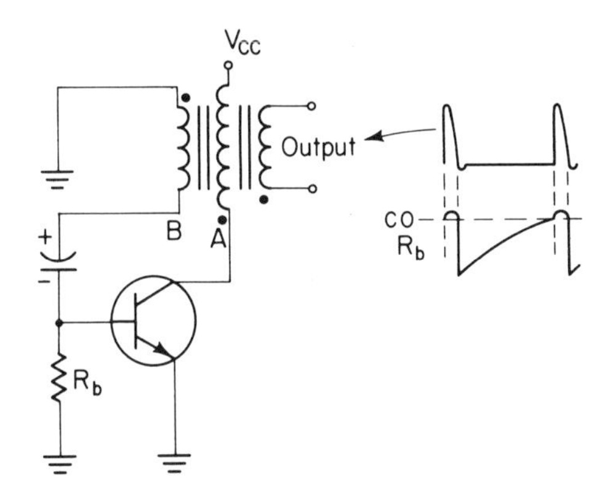


**Fig. 6-31.** Blocking Oscillator.

When power is applied, collector current causes the potential at point *A* to decrease. This places a positive potential at point *B*. The capacitor starts to charge to the indicated polarity. The resulting current from ground through $R_b$ places forward bias across the emitter-base junction. The transistor conducts harder, point *A* drops further, point *B* becomes more positive, and forward bias increases. This action is accumulative, and the transistor is rapidly driven to saturation.

The instant that point *A* ceases to change, there is no more feedback. The capacitor then starts to discharge. The resulting current (downward through $R_b$) places a strong reverse bias across the emitter-base junction. The transistor drops from saturation to cutoff. It remains in this cutoff state until the capacitor charge leaks off to almost zero.

When the capacitor has fully discharged, the transistor starts to conduct again. This begins another chain of events, and these actions repeat in a periodic fashion.

The *RC* time constant determines the time between the conduction periods. During each conduction period, a pulse of voltage can be obtained from the third winding on the transformer. This oscillator produces a chain of pulses at a free run frequency. They are used for timing purposes.

The free run frequency of the oscillator does not always exactly fit the need, in which case it may be synchronized by signals from some other source. Since the crystal oscillator is a very stable device, its output is frequently used as a synchronizing signal for the blocking oscillator. The circuit for a synchronized blocking oscillator is illustrated in Fig. 6-32.

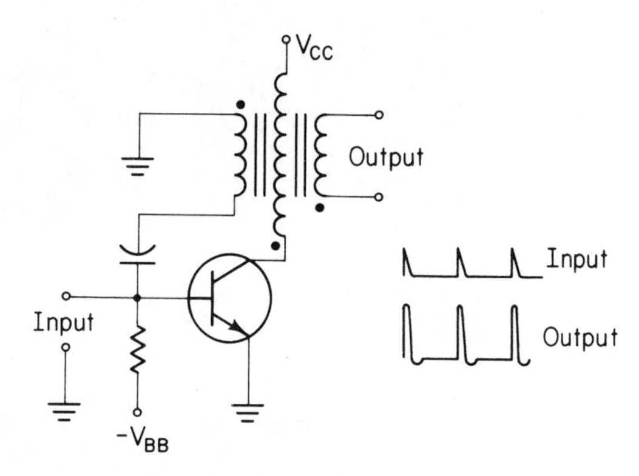


**Fig. 6-32.** Synchronized Blocking Oscillator.

This circuit is blocked by cutoff reverse bias across the emitter-base junction. Otherwise, the circuit is the same as that in Fig. 6-31. The input signals must be positive and strong enough to start conduction of the transistor. Once it is triggered into action, the cycle of events progresses exactly as before.

The reverse bias is not essential to the circuit operation. If the frequency of the incoming signal is slightly higher than the free run frequency of the oscillator, synchronization can be accomplished without fixed bias.

We have now seen circuits which produce a variety of waveshapes at a variety of frequencies. But suppose that the signal generator cannot be stabilized on the desired frequency. Suppose that we need a 100-kHz signal in one place and a 200-kHz signal in another. What shall we do, build two more multivibrators? We may, but we can build circuits that will perform some electronic arithmetic.

## ELECTRONIC ARITHMETIC

Situations are frequently encountered where it is necessary to either raise or lower the frequency of a signal. The circuits that perform these operations are multipliers and dividers.

### *frequency multiplier*

We have already seen circuits that could be used as frequency multipliers. When a crystal-controlled oscillator is tuned to a harmonic, the frequency is being multiplied by the number which represents the harmonic (3, 5, 7). In this case, the oscillator becomes a frequency multiplier. Figure 6-33 illustrates a basic multiplier.

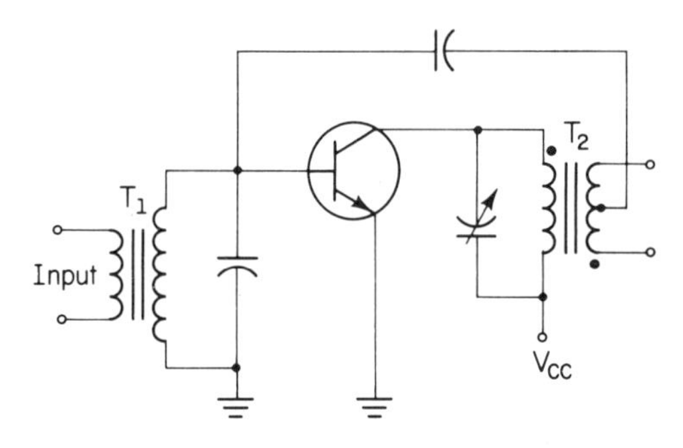


**Fig. 6-33.** Frequency Multiplier.

This is a basic oscillator with a tuned tank on the input and output. The secondary of $T_1$ is tuned to the input frequency. The primary of $T_2$ is tuned to some multiple of the input frequency. If we are to double the frequency, $T_2$ would be tuned to twice the frequency of $T_1$. Thus, every input would reinforce its second harmonic in the oscillations of the collector tank circuit. The regenerative feedback circuit is taken from the secondary of $T_2$, and every second output coincides with an input.

It is possible to build a frequency quadrupler in this manner. In which case, the collector tank would be tuned to the fourth harmonic of the input. It is not common practice to do so because of the increased circuit losses. Better performance can be obtained by using two frequency doublers.

### *frequency dividers*

We have already seen at least one circuit that can be used to divide frequencies. The bistable multivibrator is a natural. It requires two inputs to produce one output. Therefore, it divides its input frequency by two. A synchronized blocking oscillator becomes a divider when it is triggered in any manner except with every input pulse.

All types of counters are frequency dividers because they are designed to produce an output after receiving a specified number of inputs. A counter can be designed to divide the input frequency by any desired number. Figure 6-34 illustrates one method of frequency division.

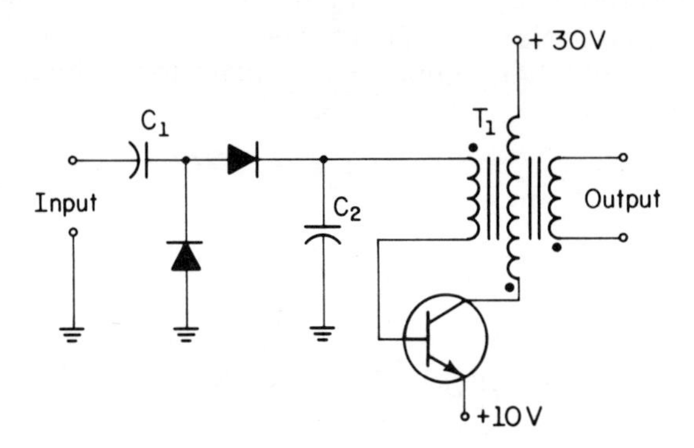


**Fig. 6-34.** Frequency Divider.

Here we have combined the step charging action of diodes and capacitors with a blocking oscillator. The 10-V reverse bias on the transistor keeps it cut off until the step charging circuit can produce a similar voltage on the base. The waveshapes for this circuit are illustrated in Fig. 6-35.

Capacitor $C_2$ is much larger than $C_1$, and they have the same charge path. Each positive input causes both capacitors to partially charge through $D_2$. $C_2$ has no discharge path. $C_1$ is forced to discharge through $D_1$ with each negative input. $C_2$ retains its previous charge and adds to it with each subsequent positive pulse.

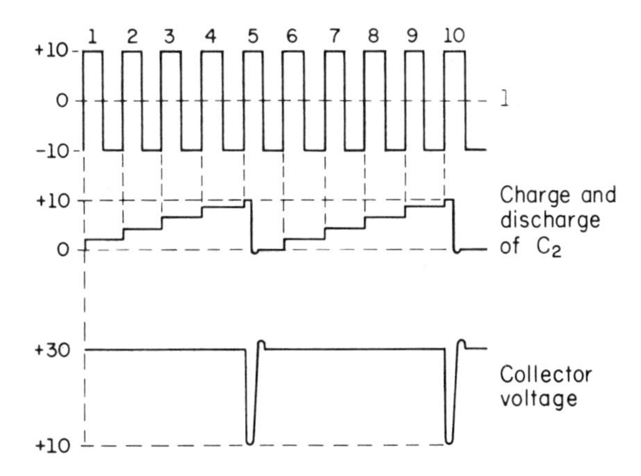


**Fig. 6-35.** Frequency Divider Waveshapes.

The charge on $C_2$ builds up at an exponential rate with positive pulses 1, 2, 3, and 4.

The 5th positive input raises the charge on $C_2$ to 10 V. This allows the transistor to conduct. The normal blocking oscillator action drives it to saturation. The resulting base current discharges $C_2$. The blocking oscillator produces one pulse and returns to cut-off. The 6th positive input starts another cycle of events.

The accuracy of frequency division hinges upon inputs of a constant amplitude and duration. Therefore, counters are usually preceded by waveshaping circuits.

## MODULATION

Modulation is the process of impressing an intelligence upon a carrier frequency. This was touched on in connection with RF amplifiers, but at that point it was incidental. Since this is a vital step in all electromagnetic communications it bears a closer examination. There are three types of modulation in popular use. These are continuous-wave, amplitude, and frequency.

### *continuous-wave modulation* (CW)

This is a process of keying a carrier wave on and off to form codes. Transmission of the International Morse Code is an example of the use of CW modulation. A code key can be connected into an RF amplifier to alternately turn the amplifier on and off. This is radiotelegraph transmission and it is a very reliable method of long range communication. Figure 6-36 illustrates one method of performing CW modulation.

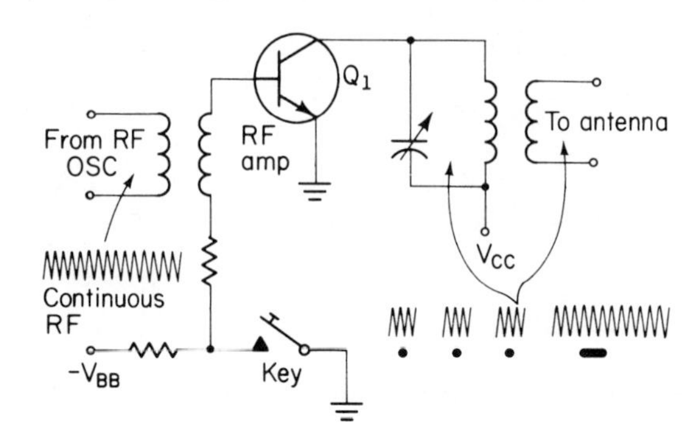


**Fig. 6-36.** Blocked Base Keying.

This time the RF amplifier is incidental and the modulation process is the important factor. $Q_1$ is biased to cutoff by the $-V_{BB}$. The continuous wave from the RF oscillator is on the base but can go no further.

When the key is closed, the $-V_{BB}$ is removed from the base and replaced by ground through the key. With 0 V on the base, the transistor will conduct and pass an amplified version of the RF oscillations. When the key is opened, $-V_{BB}$ returns, the transistor is cut off, and the output is interrupted. The output to the antenna illustrates the result of keying the code for the letter V; three dots and a dash.

### *amplitude modulation* (AM)

This is a process of causing the RF carrier wave to vary in amplitude at the frequency of the intelligence. In other words, it impresses an audio frequency on the amplitude of the RF carrier.

An audible sound wave into a microphone will be changed to an audio electric signal as illustrated in Fig. 6-37.

The sound waves cause the diaphragm to oscillate at the same frequency as the sound waves. This vibration is coupled through the driving rod to the armature. As the armature moves back and forth, the coil is moved through magnetic flux. An electric current is induced in the coil. The frequency of the current is the same as the frequency of the sound. Any variation of the sound frequency is

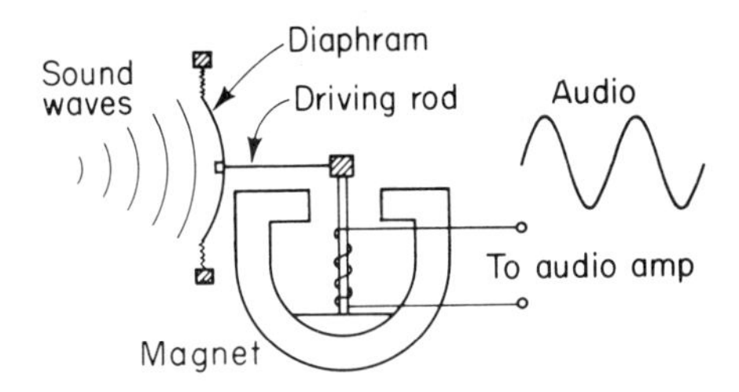


**Fig. 6-37.** Action of Magnetic Microphone.

reflected in the current frequency. This is a weak audio frequency signal, but after amplification it will be suitable for modulation. Since there is little change in the RF stage, the block diagram in Fig. 6-38 will illustrate the action.

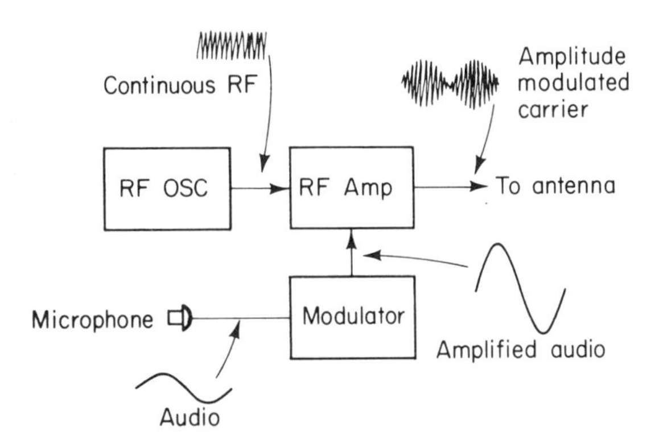


**Fig. 6-38.** Amplitude Modulation.

The modulator amplifies the weak audio signal from the microphone and uses it to vary the bias on the RF amplifier. The continuous wave from the oscillator goes through all the time, but its amplification is controlled by the audio signal. As a result, the output RF has the audio variations in its amplitude.

### *frequency modulation* (FM)

This is a process of impressing intelligence on a carrier wave by varying its frequency. This time we start a bit further back, and influence the action of the RF oscillator. This is illustrated in Fig. 6-39.

$Q_1$ is the RF oscillator and it oscillates continuously. One of the controlling factors on the frequency of oscillation is the capacitance across the primary of $T_1$. The capacitance is composed of $C_1$ and $C_2$. $C_2$ is the distributive capacitance between collector and emitter of $Q_2$.

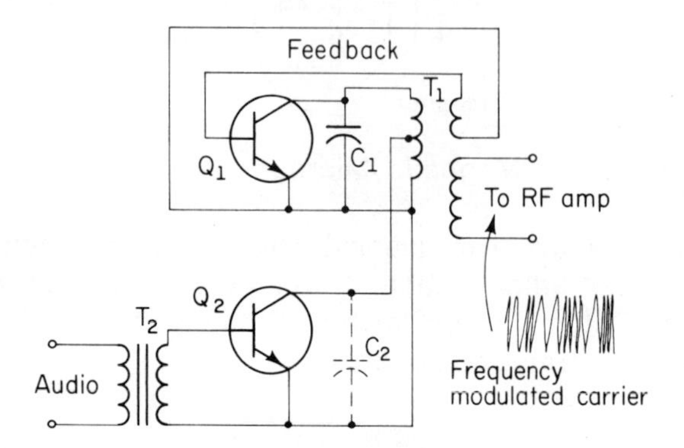


**Fig. 6-39.** Frequency Modulation.

The audio signal couples across $T_2$ to the base of $Q_2$. This varies the bias on $Q_2$ at the audio rate. The variation of the conduction of $Q_2$ varies the capacitance of $C_2$. The center frequency of the tuned tank (primary of $T_1$) varies with $C_2$ at the audio rate. The result is a raising and lowering of the oscillator frequency at an audio rate. This is frequency modulation. The audio has been impressed into the frequency of the RF oscillator. This varying frequency carrier (FM modulated) is sent to the RF amplifier for amplification and transmission.

### *optical modulation*

This type of modulation is not yet popular, but with the growing emphasis on laser communications, it is destined to become so. An optical modulator is shown in Fig. 6-40.

## DEMODULATION

Demodulation is the process of extracting the intelligence from a carrier frequency. This is primarily done through heterodyning and detection. Heterodyning is the mixing of two frequencies and extracting the difference frequency. Detection is rectification and filtering of the difference frequency to reproduce the audio. First we need an RF oscillator located in the receiver to provide a beat frequency; it is called a local oscillator. Since we have already studied several types of oscillators, we will just assume an output, and go on to the mixer.

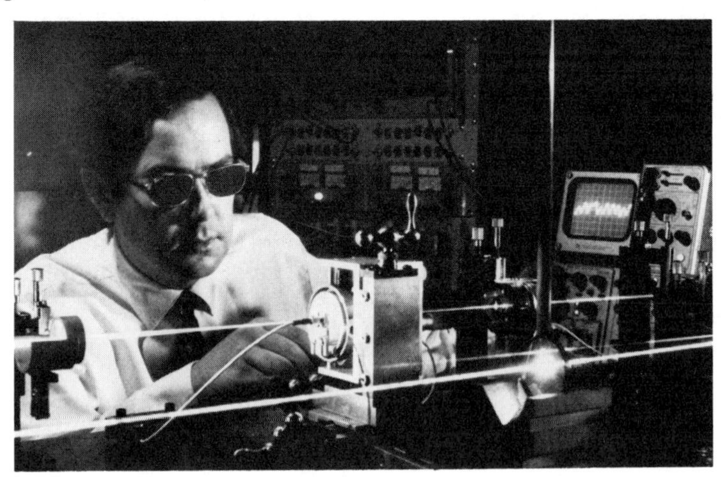

**Fig. 6-40.** An Optical Modulator in Use.

### *mixer*

The mixer accepts the incoming RF carrier from an RF amplifier and the locally produced RF from the oscillator, mixes them together, and extracts the difference frequency. This action is illustrated in Fig. 6-41.

The continuous RF from the local oscillator and the modulated RF carrier from a previous RF amplifier are both applied to the base of $Q_1$. The primary of $T_2$, which is the collector load for $Q_1$, is a tank circuit tuned to the difference between the two input frequencies.

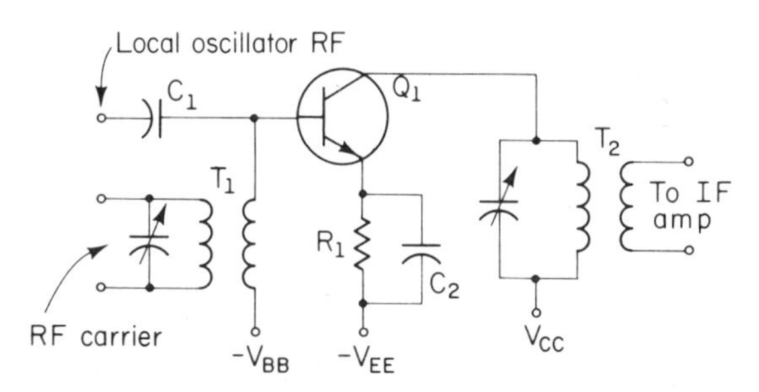


**Fig. 6-41.** Mixer.

Most communications receivers utilize an intermediate frequency of 455 kHz. Let's assume that to be our situation in order to have more definite figures to work with.

This means that the primary of $T_2$ is tuned to 455 kHz, which is the intermediate frequency. The IF is also the difference between the frequency of the RF carrier and the local oscillator. In most cases, the local oscillator is at the higher frequency. Let's assume that the RF carrier has a frequency of 1500 kHz. This gives us a local oscillator frequency of 1500 plus 455 which equals 1955 kHz.

The 1955 kHz beats with the 1500 kHz and produces all manners of sum and difference frequencies when we consider the bandwidth and the harmonics. The mixer must select from this maze the 455 kHz with a bandwidth of about 10 kHz. This is done primarily by the tuning of the collector tank as shown in Fig. 6-42.

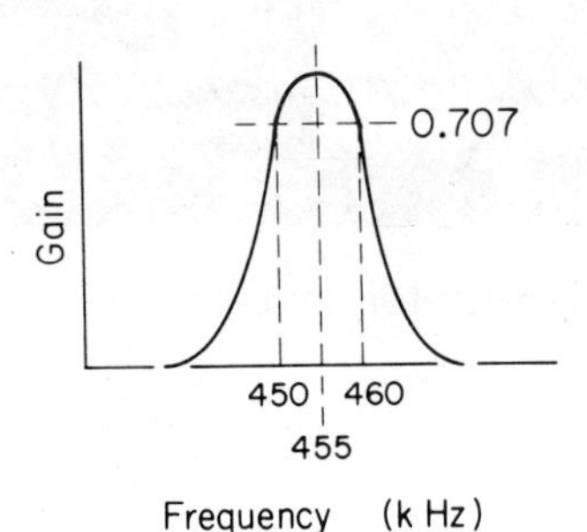


**Fig. 6-42.** Frequency Response of Mixer Tank.

The IF will now contain the same amplitude modulation that came in with the RF carrier.

### *converter*

A converter is a stage which combines the functions of the mixer and local oscillator. The converter is generally used with frequency modulation (FM) but may also be used with AM and CW. Figure 6-43 illustrates the converter circuit.

This is labeled as an FM converter because we are going to use it for FM, but it may also be used for AM and CW.

The FM carrier to the primary of $T_1$ is coupled across $T_1$ to the base of $Q_1$. $C_1$ provides regenerative feedback to the base of $Q_1$. The primary of $T_2$ is the oscillator tank circuit. The primary of $T_3$ is part of the collector load, and it is tuned to the difference frequency, 455 kHz. $T_3$ is also the output transformer to couple the signal to the IF amplifiers.

The signal now is a 455-kHz frequency varying at the audio rate. In other words, the intelligence has been transferred from the RF carrier to the intermediate frequency. It will now pass through several stages of IF amplification.

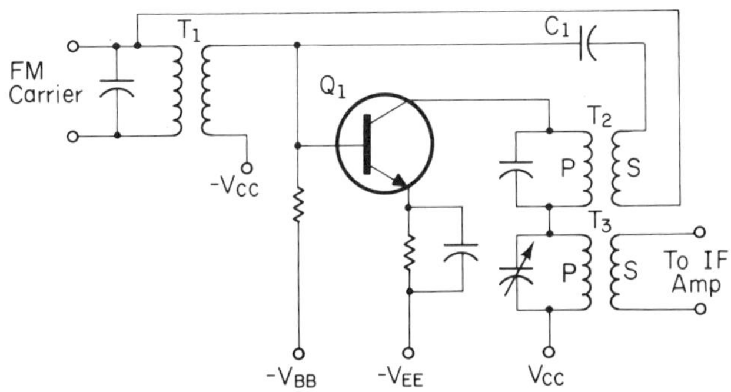


**Fig. 6-43.** FM Converter.

***the detector***

The next task is to eliminate the IF (which is now the carrier) and reproduce the audio intelligence. With AM, this is accomplished by detection and filtering. FM uses a frequency discriminator. The detector is shown in Fig. 6-44.

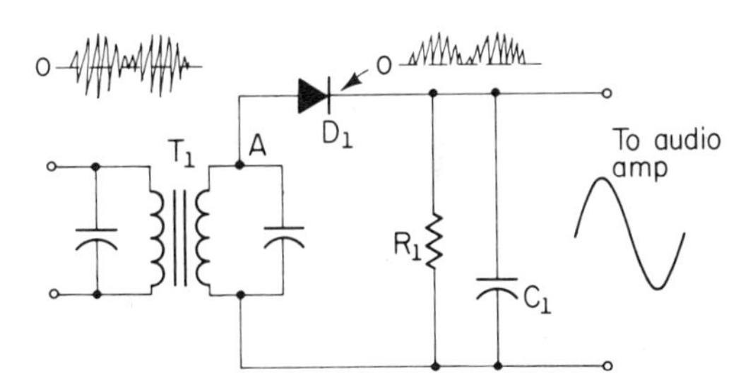


**Fig. 6-44.** Detector with Filter.

The incoming amplitude modulated IF is coupled across $T_1$. Each variation which causes point $A$ to be positive with respect to $B$ will allow the diode to conduct. These variations are developed across $R_1$. Without the filter capacitor ($C_1$), the rectified waveshape shown at the cathode of $D_1$ (the positive alternations of the input signals) would be the output.

The filtering action of $C_1$ and $R_1$ removes these variations and extracts the envelope. This envelope is the original intelligence which the microphone impressed on the carrier wave. It is now ready to go through an audio amplifier to a speaker. The speaker will change the electric audio signal to sound waves, and reproduce the sound which entered the microphone.

### *the frequency discriminator*

The discriminator is still a detector in the sense that it performs both rectification and filtering. The difference is that it must convert the frequency variations to amplitude variations. This is done by designing the circuit so that output amplitude is directly proportional to input frequency. The frequency discriminator is illustrated in Fig. 6-45.

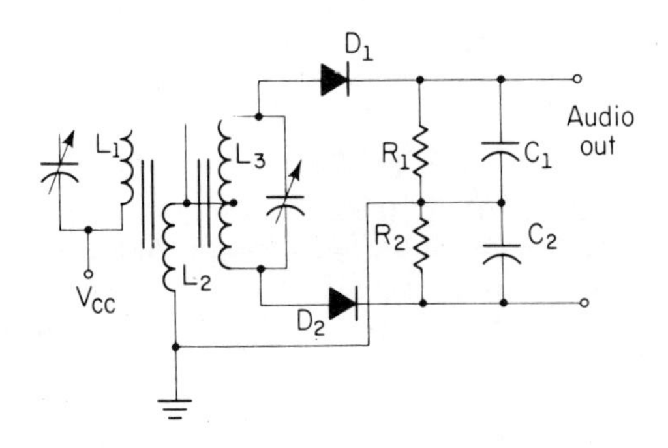


**Fig. 6-45.** Frequency Discriminator.

$L_1$ and $L_3$ of the three winding transformer are tuned to the center frequency of the incoming FM signal. $C_c$ and $L_2$ acts as impedance coupling for the same signal. Below the center frequency, this series circuit is capacitive; above the center frequency, it is inductive. So, the impedance of this coupling varies with the frequency coming in. The center of the impedance coupler is connected to the anode of both diodes through the center tap of $L_3$.

The incoming FM signal is also coupled inductively from $L_1$ and $L_2$ to the collector of both diodes. The diodes then have two voltages applied and their conduction is determined by the difference between these voltages.

$D_1$ conducts through $R_1$; $D_2$ through $R_2$, and the voltage across both resistors is positive with respect to ground. The output then represents the difference in conduction of the two diodes. $C_1$ and $C_2$ filter out the carrier and leave the audio sine wave.

## CHAPTER 6 REVIEW EXERCISES

1. Explain the difference between a limiter and a clamper.
2. What can be done to a limiter to set the maximum signal amplitude to a specific value?
3. Draw the output waveshape for the limiter in Fig. 6-46.

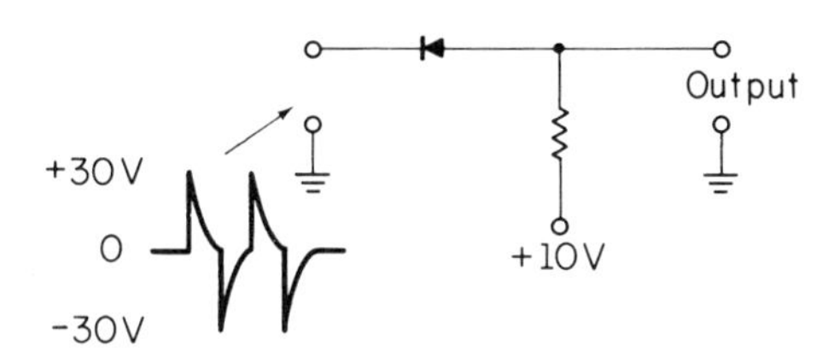


**Fig. 6-46.** Draw the Output.

4. Draw a diode shunt limiter which will change the waveshape of Fig. 6-47 from *A* to *B*.

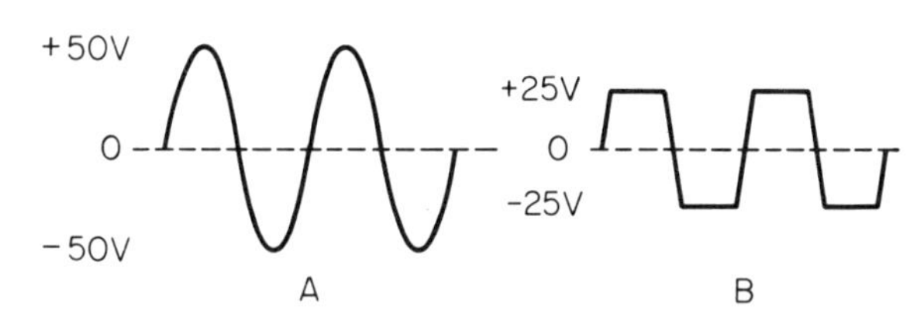


**Fig. 6-47.** Input and Output of Shunt Limiter.

5. Explain why clamper circuits are necessary.
6. A square wave of 6-V peak value is riding a dc level of −20 V. Draw a clamper circuit which will clamp the most negative portion of the signal to +10 V. Show the input and output waveshapes.
7. The first 10 percent of a capacitor's charge represents what portion of the total charge time?
8. A capacitor of 15 $\mu$F is in a sweep generator circuit with a 250-k$\Omega$ resistor. What is its total charge time?
9. The sweep generator of item 8 would produce a reasonably linear sweep for how many milliseconds?

10. What type of sweep generator is used when it is important for a sweep to start and stop at specific times?
11. Figure 6-48 is a schematic of a free running multivibrator.

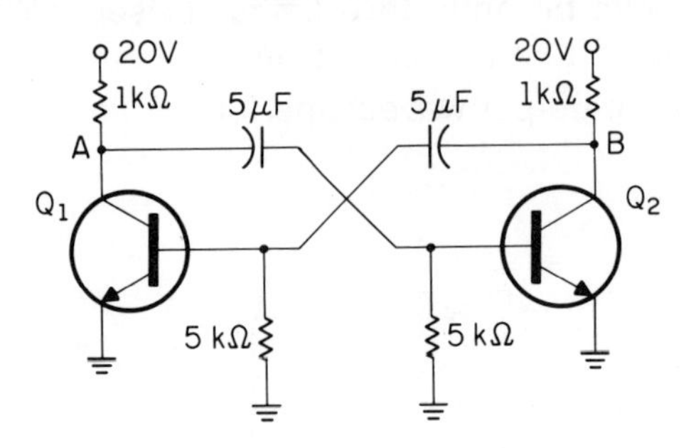


**Fig. 6-48.** Multivibrator with Circuit Values.

Assume that $Q_1$ is cut off at $T_0$ and draw the waveshapes for points *A* and *B*.

12. Assume that 1-V reverse bias is cutoff for each transistor in Fig. 6-48, what is the frequency of the output?

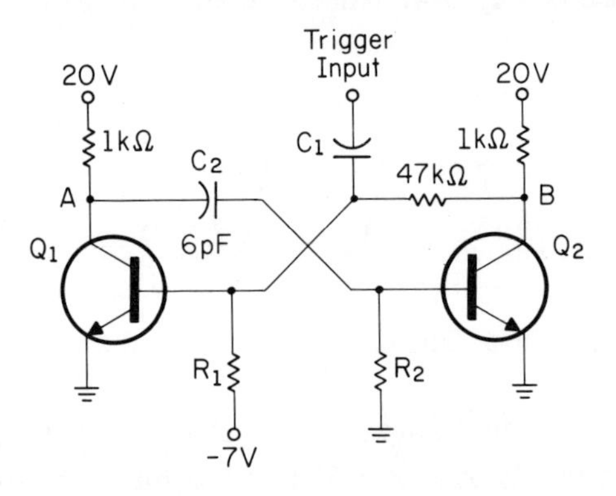


**Fig. 6-49.** Name the Circuit.

Items 13 through 16 refer to Fig. 6-49.

13. What type of circuit is this?
14. $Q_1$ has a fixed bias of 3 V. What is the value of $R_1$?
15. The output from point *B* is a positive square wave for 90 ns (nano $= 10^{-9}$) after the trigger occurs. Assuming that $Q_2$ starts to conduct with 2-V bias, what is the value of $R_2$?

16. Changing $C_2$ to 10 $\mu$F will change the positive output at point $B$ from 90 ns to what time duration?
17. When used as a frequency control for an oscillator, how many usable harmonics does a crystal produce?
18. A crystal has a fundamental frequency of 50 kHz, and it controls an oscillator tuned to the fourth harmonic. The tank has an inductance of 0.3 mH. What is the capacitance of the tank?
19. What starts the oscillations in a lag line oscillator?
20. What determines the frequency of a:
    (a) Free run blocking oscillator?
    (b) Synchronized blocking oscillator?
21. What type of modulation is being used when a continuous wave is keyed on and off?
22. Amplitude modulation is generally accomplished in which amplifier stage?
23. Briefly describe amplitude modulation.
24. Briefly describe frequency modulation.
25. Frequency modulation is normally accomplished in which transmitter stage?
26. What is demodulation?
27. What is heterodyning?
28. A converter does the job of what two stages?
29. What is the function of a detector?
30. The amplitude of the output of a frequency discriminator is directly proportional to the ______ of its input.

# 7

# transmission and reception

Most of the circuits encountered in transmitters and receivers have already been considered as separate items. The objective of this chapter is to bring them together and form practical units of electronic equipment. We will use block diagrams for this purpose while considering the structure of transmitters and receivers.

## TRANSMITTERS

A transmitter is that part of the electronic equipment which generates a carrier wave, modulates it with intelligence, amplifies it, and radiates it as electromagnetic waves from an antenna. There are many types of transmitters and numerous variations of each type. The type is named according to the modulation used; such as amplitude modulated, frequency modulated and continuous-wave. We will examine the function of some of the common types.

### AM *transmitter*

This transmitter impresses intelligence on a radio frequency carrier wave by amplitude modulating the carrier at an audio rate. The carrier can be produced by a crystal oscillator or by one of the other oscillators that we have studied. The audio is produced by a microphone which converts sound waves into electrical waves. Figure 7-1 is a block diagram of a practical AM transmitter.

The RF oscillator produces a continuous output at a predetermined radio frequency. Frequency stability is an important consideration in this oscillator. For that reason, a crystal-controlled oscillator is often used. The frequency is not of great importance. It can be doubled, tripled, and quadrupled as many times as necessary.

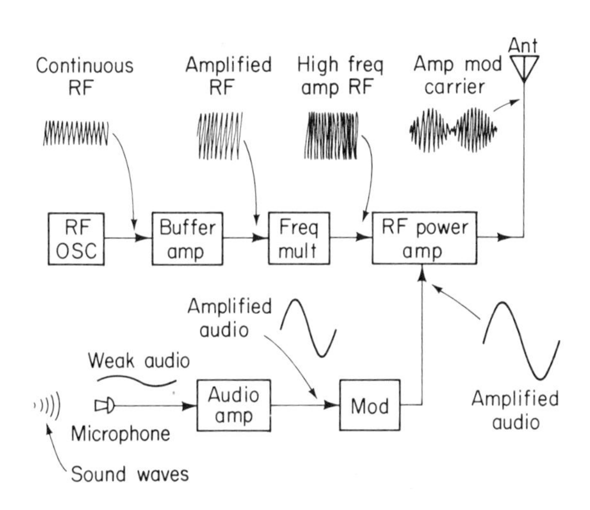


**Fig. 7-1.** Practical AM Transmitter.

This continuous RF is coupled into the buffer amplifier where it is slightly amplified. The main purpose of the buffer is isolation. It matches the impedance of the oscillator to the first stage of frequency multiplication. The object is to bridge this gap in a manner that will prevent interaction between the oscillator and the frequency multiplier.

The frequency multiplier may be one stage or several depending on the difference between the oscillator frequency and the final carrier frequency. It is safe to say that it consists of at least one frequency doubler. It is not rare to find a doubler followed by two triplers.

Here we have shown a direct link from the frequency multiplier to the final RF power amplifier. This is done in some cases, but if more amplification is needed, additional RF amplifiers may be inserted. With a high-amplitude radio frequency wave feeding into the power amplifier, we go back and pick up the audio channel.

The sound waves entering the microphone cause mechanical vibrations which the microphone converts to an electric audio signal. This is not a simple sine wave at a specified audio frequency; it is a complex signal with frequencies varying all over the audio band. This very weak, very complex audio signal is greatly amplified through one or more broad band audio amplifiers.

The modulator is a final stage of audio amplification with its circuitry interwoven with that of the RF power amplifier. The RF amplifier actually amplifies and passes the RF carrier frequency because that is the frequency it is tuned to handle. However, its gain is affected by the audio signal. As a result the audio is impressed on the amplitude of the RF carrier.

This complex waveshape is radiated from the antenna. Actually this wave is much more complex than we can adequately illustrate. The resultant carrier wave that we see at the antenna consists of a center frequency and two side bands. The center frequency is the same as the output from the final stage of frequency multiplication. The upper side band is the center frequency added to the audio frequency. The lower side band is the difference between the center frequency and the audio frequency. This is illustrated in Fig. 7-2.

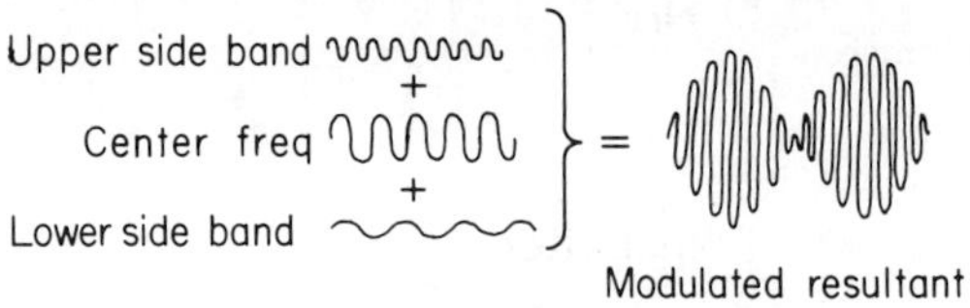

**Fig. 7.2.** Content of Modulated Wave.

Supose that the center frequency is 1200 kHz and the audio is 500 Hz. The upper side band is then 1200.5 kHz and the lower side band is 1199.5 kHz. Of course this could only be true for an instant; the audio signal frequency is constantly changing. A fixed audio signal could carry no intelligence more complex than a single tone. Therefore, the frequency of both side bands is shifting about with every change in the audio frequency.

The bandwidth is measured from the lowest to the highest side band frequency. This entire bandwidth must be passed to the antenna in order to transmit the full intelligence. The bandwidth at the frequencies just used is: $2 \times 500$ Hz $=$ 1 kHz. That is the bandwidth at that instant; the actual band width must include the variations. Therefore, the bandwidth is twice the highest frequency of the modulation signal. Suppose that our 500-Hz signal varies from

250 to 750 Hz. This makes the bandwidth 2 $\times$ 750 Hz which is 1.5 kHz.

The bandpass is the center frequency of the carrier plus and minus $\frac{1}{2}$ of the bandwidth. Our carrier frequency is 1200 kHz. This gives us a bandpass from 1200 kHz $-$ 750 Hz $=$ 1199.25 kHz to 1200 kHz $+$ 750 Hz $=$ 1200.75 kHz. This is illustrated in Fig. 7-3.

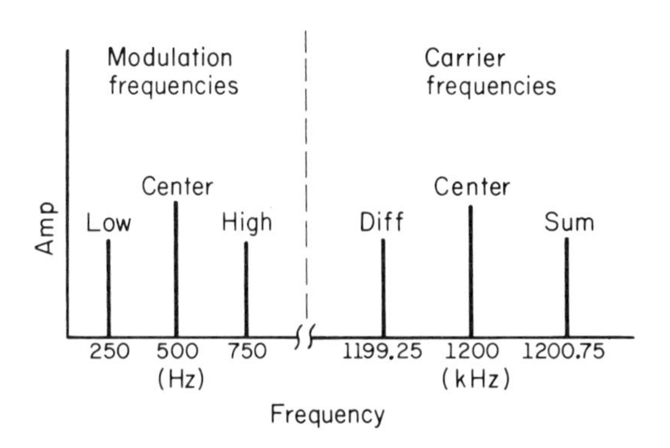


**Fig. 7-3.** Formation of Side Bands.

The tank circuit of the power amplifier output must be tuned to the carrier frequency (center frequency). The bandwidth of the tank must be broad enough to include both side bands as shown in Fig. 7-4.

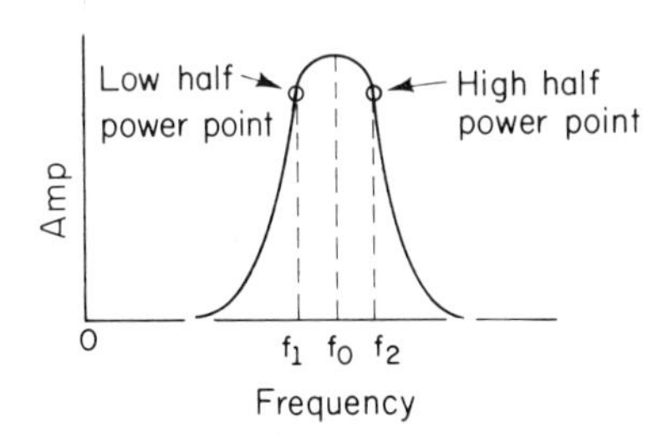


**Fig. 7-4.** Frequency Response of Tank Circuit.

In this drawing, $f_0$ is the resonant frequency of the tank circuit, and should match the center frequency of the carrier wave. $f_1$ is the frequency of the tank at its low half-power point. The frequency of $f_1$ must be, at least as low as the lowest side band frequency. $f_2$ is the frequency of the tank at its high half-power point. The frequency of $f_2$ must be, at least, as high as the highest side band frequency. Using our same frequencies from Fig. 7-3, $f_1$ would be 1199.25 kHz, $f_0$ would be 1200 kHz, and $f_2$ would be 1200.75 kHz.

The degree of modulation is a ratio between the amplitude of the modulation voltage and the carrier voltage. This may be expressed as $M = E_m/E_c$. When $E_m$ is volts of modulation voltage, $E_c$ is volts of carrier voltage, and $M$ is the degree of modulation.

The degree of modulation may be changed to percent of modulation by multiplying it by 100. So, percent of modulation $= M \times 100$. Since $M$ is always one or less, maximum percentage of modulation is 100 percent. It is desirable to operate near 100 percent because this provides maximum power for the intelligence carried by the side bands.

The oscilloscope or spectrum analyzer is generally used to check the percent of modulation. When this is done, the formula may be changed to

$$\%M = \frac{E_{max} - E_{min}}{E_{max} + E_{min}} \times 100$$

When striving for 100 percent modulation, care must be exercised not to overmodulate. Overmodulation will cause spurious, undesirable frequencies to be generated. These signals are best described as interference and distortion. Figure 7-5 shows the waveform for correct, over-, and undermodulation.

The maximum amplitude of the correctly modulated wave should be just twice the amplitude of the unmodulated carrier.

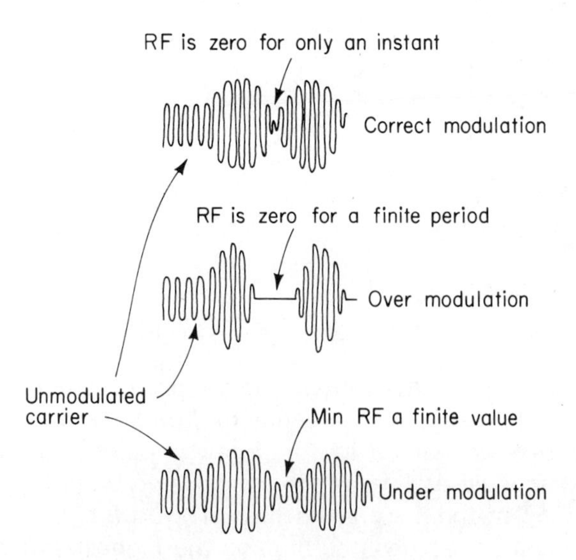


**Fig. 7-5.** Amplitude Modulated Waveforms.

Notice that the amplitude is greater than this with the over-modulated wave. With the undermodulated wave, the amplitude is less than twice.

## FM *transmitter*

This transmitter impresses intelligence on the carrier wave by varying its frequency. An RF oscillator produces a carrier. There is nothing different about this carrier except for the fact that it has both a critical amplitude and a critical frequency. In the modulation process, a complex audio signal causes this frequency to vary. The total amount of variation is determined by the amplitude of the modulating audio. The rate of frequency variation is determined by the frequency of the modulating audio, and of course, this audio is varying with every fluctuation of the sound waves. Figure 7-6 is a block diagram of a practical frequency modulated transmitter.

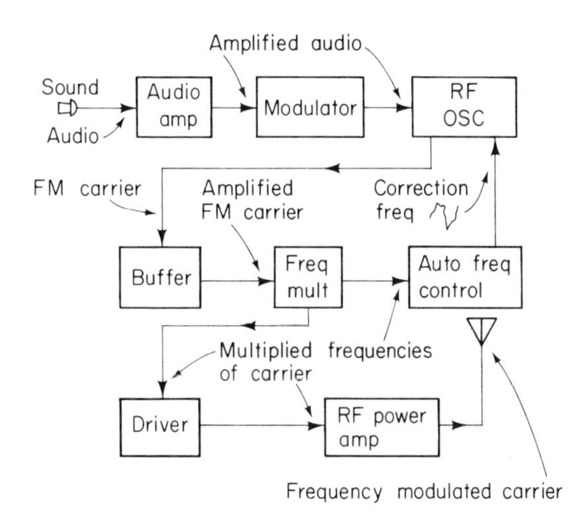


**Fig. 7-6.** FM Transmitter.

The RF oscillator is tuned to a fixed frequency, and without outside influence, it will produce this frequency continuously with a specified amplitude. But this oscillator has outside influence from two sources: the audio modulator and the automatic frequency control circuit.

The automatic frequency control circuit has a reference frequency which represents the desired center frequency of the transmitter. It constantly samples the actual center frequency from the final frequency multiplier stage. It then generates a correction signal which is proportional to the difference between actual center frequency and desired center frequency. This correction signal is coupled back to the RF oscillator, and it alters the characteristics of the oscillator enough to correct the error.

Starting with sound waves, these are converted to audio frequencies by the microphone. These audio frequencies are then processed through a network that does not appear on the block diagram. It is called a preemphasis network. Its function is to improve the signal to noise ratio of the high audio frequencies. This is done by increasing the amplitude of the high audio frequencies. This is a distortion that the receiver will have to remove, but it is necessary. Without this circuit, the high-frequency audio signals would be lost in the noise.

The complex audio wave is then amplified through one or more stages of broad band audio amplifiers. The modulator is the final stage of audio amplification, and its circuitry is interwoven with the circuitry of the RF oscillator.

The audio signals of the modulator alter the frequency characteristics of the oscillator. The result is a frequency variation to either side of the tuned frequency (center frequency) of the oscillator. This center frequency is also known as the rest frequency; it means the frequency of the oscillator when no modulation is present.

A low-frequency audio causes a slow change in oscillator frequency; a high-frequency audio causes a rapid change in oscillator frequency. Thus, the rate of change (times per second) is determined by the frequency of the audio signal. When the amplitude of the audio is low, it causes a small shift from the center frequency. A high-amplitude audio causes a wide swing from the center frequency. Therefore, the amplitude of the audio determines the amount (degree) of frequency variation to either side of the center (rest) frequency. This is illustrated in Fig. 7-7.

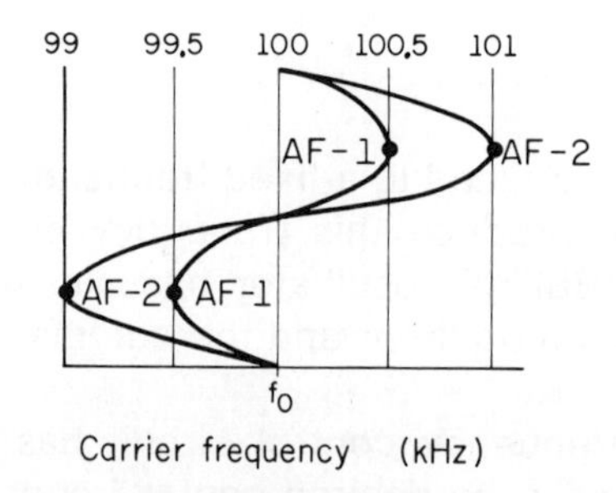


**Fig. 7-7.** Degree of Deviation.

Here we have superimposed two audio sine waves on a carrier frequency spectrum. They have the same frequency, but waveshape 2 is twice the amplitude of waveshape 1. AF-1 causes the oscillator frequency to deviate from the center frequency ($f_0$) by $\pm$ 500 Hz. AF-2 causes a deviation of $\pm$ 1000 Hz. Since both AF signals are the same frequency, the rate of deviation is the same. In other words, if both audio signals are 500 Hz, a complete audio cycle of deviation occurs at the rate of 500 times per second.

It must be understood that the actual audio signal is constantly varying in both frequency and amplitude. This means that both the rate and degree of carrier frequency variation are constantly changing. Figure 7-8 illustrates the modulating effect of two consecutive cycles of audio which are different in both frequency and amplitude.

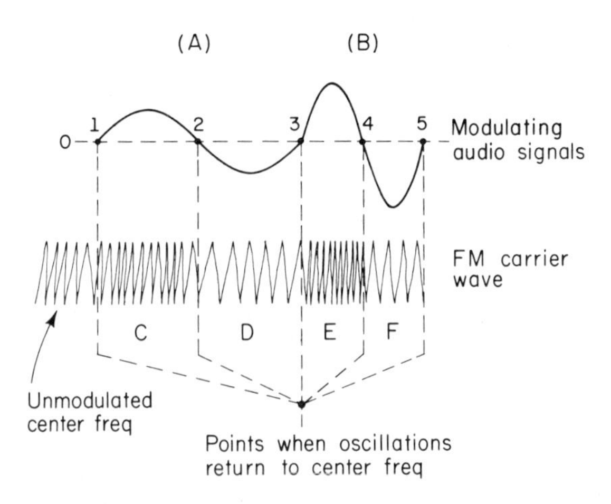


**Fig. 7-8.** Relation of Modulation Signals and Modulated Carrier.

Waveshape *B* is greater than waveshape *A* in both frequency and amplitude. Notice that the frequencies of the carrier are compressed during one alternation of the modulating signals. This is shown in areas *C* and *E.* The frequencies in area *E* are more compressed (higher frequencies) than those of area *C.* This is caused from the difference in amplitude of the audio signals.

Areas *D* and *F* are rarefaction areas, but the frequencies are lower in *F* than in *D.* Again this is a result of signal *B* having the higher amplitude.

Points 1, 2, 3, 4, and 5 represent points of modulation signal zero. The carrier returns to center frequency at these times.

The compression and rarefaction areas also contain the side bands, but the side bands are more complex than they appear. There are a great many side bands both above and below the carrier center frequency. Since these side bands contain all of the intelligence, it is desirable to retain as many of them as possible. For instance, with the values previously used, the carrier center frequency is 100 kHz, and we'll use the 500-Hz audio. The first high side band is 100 kHz + 0.5 kHz which is 100.5 kHz. The first low side band is 100 kHz − 0.5 kHz which is 99.5 kHz. The second side bands are obtained by adding the 0.5 kHz to the high side band and subtracting it from the low side band. So, our second side bands are 99 and 101 kHz.

The side bands develop on and on in this fashion in both directions as illustrated in Fig. 7-9.

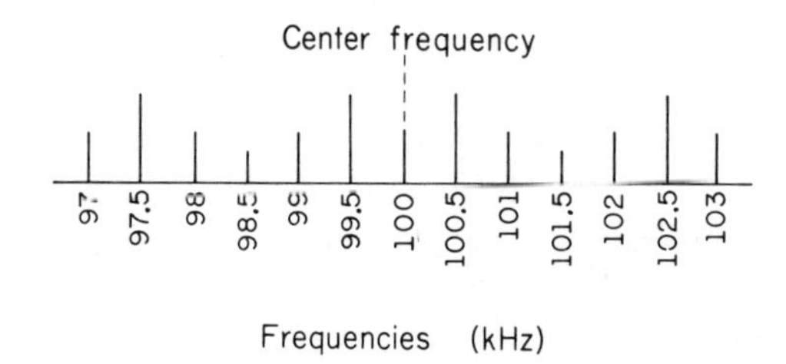


**Fig. 7-9.** FM Side Bands.

As the deviation from center frequency increases, the amplitude of the side bands decreases. When the amplitude of the side band signals is less than one percent of the carrier amplitude, the signals become insignificant. But if maximum intelligence is to be transmitted, all the significant side bands should be retained. The number of significant side bands can be determined by calculating a modulation index then referring to a table. $M = f_d/f_m$ where $M$ is the modulation index, $f_d$ is the frequency deviation of the carrier, and $f_m$ is the highest frequency of the modulating signal.

Going back to Fig. 7-7, we had a 500-Hz signal which caused a maximum carrier deviation of 1 kHz. The modulation index in this case is:

$$M = \frac{f_d}{f_m}$$

$$= \frac{1000}{500} = 2$$

The modulation index of 2 simply leads us to the table illustrated in Fig. 7-10.

We calculated the modulation index as 2. Reading across from the 2 in Fig. 7-10 we find 8 significant side bands and a bandwidth equivalent to 8 times the audio frequency. Since our audio signal is 500 Hz, this means that we have a bandwidth of 8 $\times$ 500 which is 4 kHz. This looks like a reasonable bandwidth, but what happens with a modulation index of 5 and an audio frequency of 2 kHz? We then have 16 significant side bands and a bandwidth of 16 $\times$ 2 kHz which is 32 kHz.

| Modulation index | Side bands | Band width |
|---|---|---|
| 1 | 6 | 6 x A |
| 2 | 8 | 8 x A |
| 3 | 12 | 12 x A |
| 4 | 14 | 14 x A |
| 5 | 16 | 16 x A |

**Fig. 7-10.** Modulation Index Table.

Obviously, this bandwidth can get out of hand if it is not controlled. This leads to two types of FM transmission; narrowband which limits the bandwidth and broadband which utilizes carrier frequencies in the range of several hundred MHz. The broadband gives higher fidelity, but the narrow band gives reasonable fidelity with more channels in a given spectrum of frequencies.

Referring back to our block diagram in Fig. 7-6, our complex FM carrier is now leaving the oscillator. It passes through a buffer stage which provides some amplification while isolating the oscillator from the frequency multipliers. The block called frequency multiplier usually contains three stages: a frequency doubler and two triplers. In this case, the frequencies going into the driver are 18 times the frequencies from the oscillator. So, the 100-kHz with 500-Hz variations is now an 1800 kHz with 9-kHz variations.

The center frequency of this waveform from the final frequency multiplier is constantly sampled in the automatic frequency control (AFC) circuit. This is compared with a frequency standard in the AFC circuit. Any deviation from this standard causes a correction signal to either raise or lower the frequency of the RF oscillator.

The driver is a final stage of amplification as well as a buffer between the final RF power amplifier and the last stage of frequency multiplication.

The RF power amplifier provides the required power to radiate the carrier from the antenna.

### *single side band transmitter*

This AM transmitter filters out all frequencies except those in one side band. Nothing is really lost by doing this since all of the intelligence is contained in each side band. The single side band transmitter can operate on a reduced bandwidth. This helps to relieve the crowded communications frequency spectrum, but perhaps more important, a lower power transmitter can produce a greater signal strength. Fig. 7-11 is a block diagram of a single side band (SSB) transmitter.

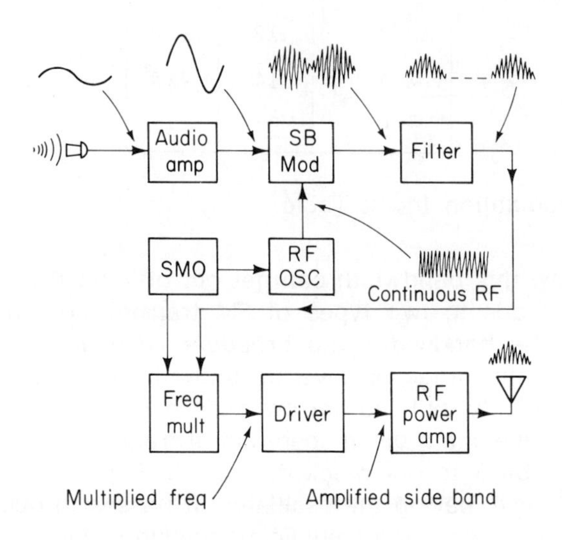


**Fig. 7-11.** Single Side Band Transmitter.

The audio signal from the microphone is amplified and coupled to a modulator. This is called a side band modulator because it has essentially two outputs. These are an upper side band and a lower side band. The second input to the side band modulator comes from the RF oscillator.

With either RF or AF input alone, the modulator has no output. Therefore, it is tuned to suppress both the center carrier frequency and the audio frequency. However, it does pass both the sum and difference frequencies. The output then consists of both the side bands. The two side bands are coupled into a filter circuit.

The filter is a high *Q* tank circuit; a crystal is often used for this purpose. It is tuned to pass a band of frequencies just wide enough to include the frequencies of one side band. The frequency response curve should be similar to that shown in Fig. 7-12.

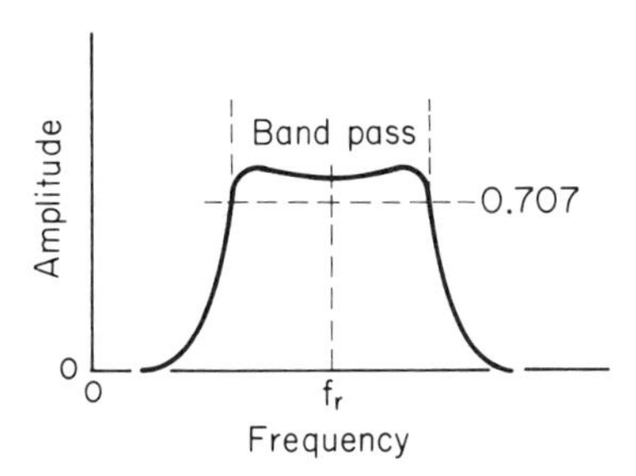


**Fig. 7-12.** Filter Response Curve.

The frequencies in the bandpass cannot accommodate the frequencies of both side bands. In fact, with the resonant frequency of the tank ($f_r$) being the same as the center frequency of the side band to be passed, all frequencies except that chosen side band will be rejected. Either side band may be used since both contain all the intelligence.

After filtering, the remaining amplitude modulated side band is coupled to the frequency multiplier. There are many variations in this portion of the transmitter, but a common method of frequency multiplication in SSB is heterodyning. This could be done by two stages of mixing. Since the object is to raise the frequency, the output would be the sum frequencies rather than the difference.

When the frequency has been increased to the desired transmission frequency, the signal is coupled into the driver. The driver amplifies it, and passes it to the RF power amplifier. After a final amplification the amplitude modulated, single side band signal is radiated from the antenna. With the RF power amplifier concentrating its energy into a single side band, a given transmitter can radiate a great deal more energy over the selected portion of the spectrum. In other words, a 500-W transmitter can concentrate the full 500 W into a relatively narrow side band rather than distributing that power over a full band. The end result is greater signal strength for any given frequency.

Single side band transmission requires a stable frequency. For instance, with a transmitted frequency of 4 MHz, a 50-Hz variation is sufficient to cause noticeable distortion of the voice. The SMO provides this high degree of frequency stability.

SMO is an abbreviation for stabilized master oscillator. The stabilizing device is usually a crystal which provides a frequency standard on the order of 100 kHz. This crystal is maintained at a constant temperature to enhance its natural stability.

The stabilized master oscillator works from this standard frequency to control the frequencies of the RF oscillator and frequency multipliers. It forces the standard frequency on these other stages by a synchronization process called phase locking. Corrections are made as soon as the signal at any point starts to differ in phase from the control frequency signal.

Now, let's change positions and look at this from the receiving end.

## RECEIVERS

A receiver is an electronic device which plucks electromagnetic signals from the air, amplifies them, extracts the intelligence, and reproduces the sound which entered the microphone of the transmitter. There is a type of receiver to match each type of transmitter. The type specifies the type of modulation the receiver is designed to handle; AM, FM, CW, etc. We will examine a few common types.

### AM *receiver*

This receiver is designed to accept the weak electromagnetic waves on its antenna, amplify them, and extract the intelligence. A good receiver will accept the desired band of frequencies and reject all others. The AM receiver works with the full band which means a carrier with upper and lower side bands. Figure 7-13 is a block diagram of an AM, superheterodyne receiver.

Many frequencies are present on the antenna. The setting of the channel selector control determines which frequencies will be allowed to enter. Turning the selector control tunes both the RF amplifier and the local oscillator. The RF amplifier (which may be more than one stage) is tuned to the center of the band for the desired frequency, and the local oscillator is tuned to 455 kHz above this frequency. For example, when receiving a broadcast at 1000 kHz, the receiver dial is set to 1000. Setting the dial tunes the RF amplifier to 1000 kHz and sets the local oscillator to 1455 kHz. The 455 kHz used here is meant as a sample only. While this is the most common intermediate frequency, many receivers are designed for other intermediate frequencies.

The bandwidth of the RF amplifier is such that it passes and

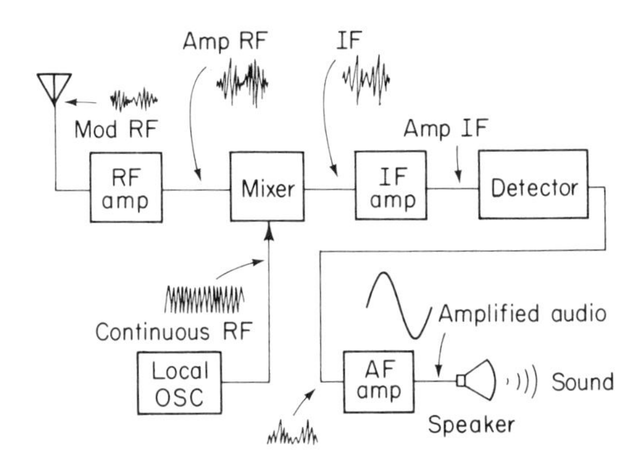


**Fig. 7-13.** AM Receiver.

amplifies both the upper and the lower side bands of the 1000-kHz carrier. The primary need for the RF amplifier is to increase the amplitude of the incoming signal before it can become lost in the noise. The very weak, amplitude modulated signal is amplified and passed to the mixer.

The local oscillator can be any RF oscillator that retains a fair degree of stability while being tuned over a wide range of frequencies. It will produce a continuous RF that is 455 kHz above the center frequency of the signal from the antenna. These oscillations are coupled into the mixer along with the incoming signal.

The mixer beats the signals together and produces many frequencies composed of sums and differences. The output of the mixer passes through a tank circuit which is tuned to 455 kHz. This tuned circuit generally has a bandwidth of about 10 kHz and passes frequencies from 450 to 460 kHz. All other frequencies will be eliminated. The local oscillator frequency is beating with all components of the modulated wave. All the difference frequencies should fit into this 10-kHz band with no problem.

Suppose that the received 1000-kHz signal is modulated by a 5-kHz audio. The upper band of 1005 kHz will beat with the 1455 kHz from the oscillator to produce a difference of 450 kHz. The lower side band of 995 kHz will beat with the same 1455 kHz and produce a difference of 460 kHz.

The output of the mixer is still a complex signal with frequencies varying around the 455-kHz center frequency. These signals are now amplitude modulated. The amplitude modulation on the incoming carrier wave has transferred through the mixer. The IF signal now carries the intelligence.

The block of IF amplification represents several stages of

amplifiers. This group of amplifiers is commonly referred to as the IF strip. These are relatively narrow band amplifiers and they generally employ tuned transformer coupling between stages. These are tuned to the 455-kHz center frequency. Since they always handle only this narrow band around 455 kHz, they can be designed and tuned for optimum gain and bandwidth.

The amplified IF is coupled from the final IF amplifier to either a detector or a second mixer. With a carrier frequency in the broadcast band similar to the sample used here, the double conversion is not necessary. If the carrier frequency is on the order of megahertz, the second mixer would probably be used.

It is the job of the detector to extract the intelligence and remove the IF component. This is done through a process of rectification and filtering. The rectifier removes half of each IF waveshape and a filter circuit filters out all IF components. Since the filter passes only audio signals, it preserves the envelope which corresponds to the original modulating signal.

The audio signals from the detector are passed to the audio amplifier. Here they are endowed with sufficient power to drive the speaker.

The audio signals vary the flux of the speaker coil and cause the diaphragm to vibrate in harmony with the audio. The vibrating diaphragm sets up sound waves. These sound waves are reproductions of the sound waves which entered the microphone of the transmitter a few microseconds before they emerge here. For all practical purposes, the sounds emerge from the speaker at the same instant they enter the microphone. Of course, there is a very short delay that can be noticed when a great distance is involved. For instance, radio waves travel from the earth to the moon in approximately 1.28 s. However, when a speaker is on a worldwide broadcast, his voice will emerge from the speaker of a receiving set anywhere on the globe before it reaches the back of the room in which he is speaking. In free space electromagnetic waves travel at approximately the speed of light; 186,000 mi/s.

One problem with receivers is the varying strength of the signal on the antenna. There are many reasons why signals fade in and out. Atmospheric conditions increase and decrease signal size. In addition to signals traveling in a straight line, they are reflected from numerous objects.

When a reflected signal reaches the receiver antenna in phase with the signal component which traveled a straight line, the result is a strong signal. When the reflected wave is out of phase, the result is a weak signal. If all receiver amplifiers provided a fixed gain for all signals, the sound would be constantly fading in and out.

The answer to this problem is to build in a sensing device which can control the gain of all amplifiers prior to the detector. This sensing device is called an automatic gain control (AGC) circuit. It senses the amplitude of the detector output and automatically adjusts the overall gain so that it is inversely proportional to signal strength. Figure 7-14 is a partial block diagram showing the AGC function.

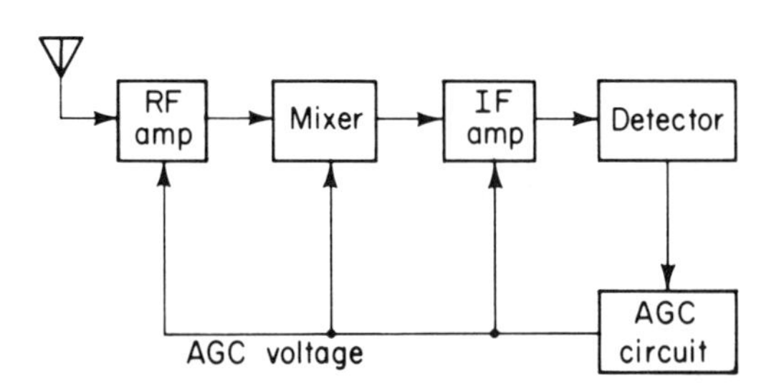


**Fig. 7-14.** AGC Circuit.

The output of the detector contains a dc component which is directly proportional to the average amplitude of the modulated IF. The AGC circuit filters undesirable components from this dc and uses it to control the bias on the previous stages.

Since AGC voltage is directly proportional to signal strength, it must be applied in such a manner that it will decrease the forward bias. When this is done, each stage that is controlled by AGC will have a gain inversely proportional to signal strength.

### *image frequency*

In our sample receiver, we had the local oscillator tracking above the selector frequency by 455 kHz. In this fashion, the desired signal mixes with the local oscillator frequency to provide the difference (IF) of 455 kHz. Suppose that we have selected a frequency of 1500 kHz; this places the local oscillator frequency at 1955 kHz. We have the proper IF, and all is going well. Now, suppose that a signal arrives at the antenna with a frequency of 2410 kHz. Chances are that some of this signal will get through to the mixer. If it does, it will beat with the local oscillator and produce a difference frequency of 455 kHz. This undesired signal will pass right through the receiver.

The signal just described is called an image frequency. If the local oscillator is tuned to track above the selector frequency, the image frequencies are equivalent to local oscillator plus the intermediate frequency. If the local oscillator tracks below the selector frequency, image frequencies are equivalent to local oscillator frequency minus the intermediate frequency.

Of course image frequencies are not desirable; they represent interference. One method of eliminating the effect of image frequencies is double conversion. One mixer stage near the antenna reduces the received frequency to a rather high IF. This places the image signals so far from the selected frequency that they are strongly attenuated by the RF stage. The second mixer reduces the high IF to a normal IF just before it reaches the detector.

Let's take a look at a different type of receiver.

## FM *receiver*

Only FM receivers are designed to receive and reproduce the intelligence carried by frequency modulated signals. Figure 7-15 is a block diagram of such a receiver.

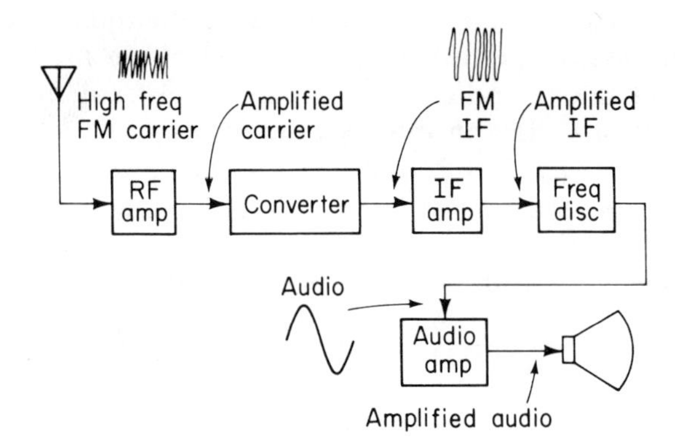


**Fig. 7-15.** FM Receiver.

The weak carrier with the frequency modulations is picked up by the antenna and coupled to the RF amplifier. The same waveshape, except with a greater amplitude, is passed on to the converter.

The converter could be a mixer. We selected a converter here because the mixer was discussed with the AM receiver. The converter is a combination mixer and local oscillator in one stage. The local oscillator portion of the converter is tuned by the selector control as previously discussed. The frequency is such that the center frequency of the carrier differs from it by an amount equal to the IF.

The side bands beat with this fixed frequency to produce frequencies above and below the IF.

The IF leaving the converter has been impressed with the same frequency modulation that came in from the antenna. This IF (with upper and lower side bands) is amplified through several IF amplifiers.

The frequency discriminator changes the frequency variations to amplitude variations. These amplitude variations represent the audio signals which the transmitter used for modulating the carrier.

The audio is processed through a deemphasis network as it leaves the frequency discriminator. The preemphasis network in the transmitter increased the amplitude of the higher frequencies. This provided a better signal to noise ratio, but it introduced signal distortion. The deemphasis network must remove this distortion. This job is accomplished by altering the bias on the audio amplifier. This bias is changed so that gain is inversely proportional to frequency. The low-frequency audio signals are then amplified more than the high-frequency audio signals.

The audio amplifier amplifies the signals and applies them to the speaker. The speaker reproduces the sound which entered the transmitter microphone.

The FM receiver is not bothered by signal fading and so may not use an automatic gain control circuit. However, it does have a closely related problem. It will receive very weak signals from one transmitter and very strong signals from another. An automatic gain control circuit will eliminate the need of resetting the volume when stations are changed.

The AGC circuit takes information from the frequency discriminator and produces a voltage which is proportional to signal strength. This voltage is used to control the gain on the RF amplifier. The RF amplifier will then have a gain inversely proportional to signal strength. This provides ample gain for weak signals while preventing strong signals from overdriving the converter. The AGC connections are shown in Fig. 7-16.

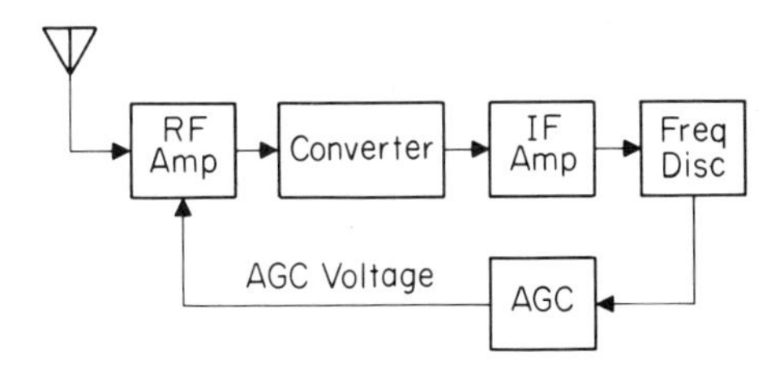


**Fig. 7-16.** FM Receiver with AGC.

FM receivers also use multiple conversion. Most FM receivers operate at very high frequencies, and the best performance cannot be obtained with a single mixer. Two or more stages of conversion are essential to optimum image rejection combined with optimum frequency selection. Figure 7-17 is a block diagram of an improved FM receiver.

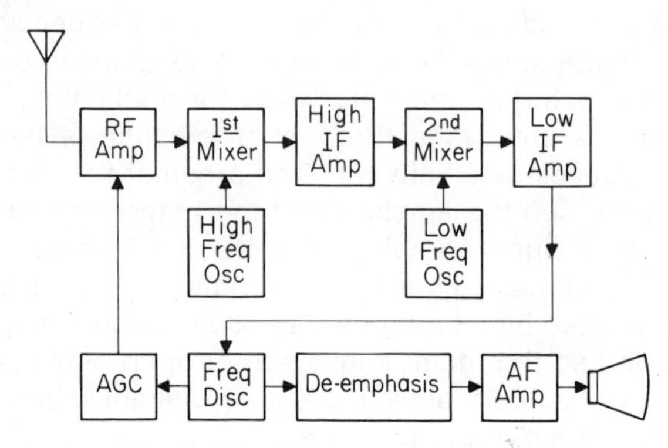


**Fig. 7-17.** Improved FM Receiver.

This receiver features double conversion and automatic gain control. In addition to better selectivity of frequencies and better rejection of images, it has the advantage of extra stages of IF amplification.

### *single side band receiver*

Since it operates on amplitude modulation, the single side band receiver has many features in common with the common AM receiver. It also has a few distinct differences as shown in Fig. 7-18.

The single side band amplitude modulated signal is picked up on the antenna, coupled to the RF amplifier, and passed to the mixer. In the mixer it is mixed with the RF from the local oscillator to produce an amplitude modulated IF. This IF is different; it has no center frequency as such. If the upper side band is being received, all the frequencies are above what would be the center frequency. With the lower side band being used, all the frequencies would be below the normal IF center frequency.

An IF filter now becomes necessary to prevent RF components from passing with the IF signal. After several stages of IF amplification, the signal goes to a detector.

The detector has another input from the CR oscillator. CR is an abbreviation for carrier reinsertion. This oscillator produces a continuous-wave signal which is equivalent to the missing IF center frequency. This is mixed with the side band frequencies to restore any missing component.

The detector then rectifies the signals and filters out everything except the audio envelope. This is amplified and used to drive the speaker.

Another popular receiver is the combination receiver.

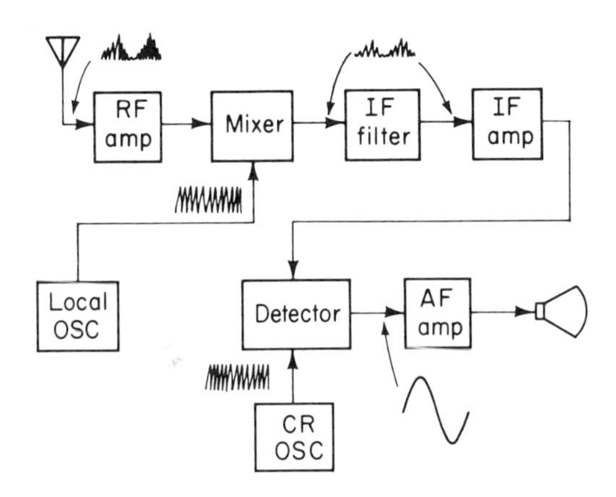


**Fig. 7-18.** Single Side Band Receiver.

## AM–FM *receiver*

This receiver can handle either amplitude or frequency modulation. It can be converted to an AM or an FM receiver by moving a switch. This does not indicate two complete receivers in the same case; many of the circuits are common to both AM and FM. The combination receiver has a selector switch which can be used to choose the type of modulation it will receive. In the AM position, this switch selects the circuits that are peculiar to AM and combines these with the common circuits. The result is an AM receiver. In the FM position, the switch selects the circuits that are peculiar to FM and combines these with the common circuits. This results in an FM receiver.

There are numerous possible arrangements of stages to form an AM–FM receiver. Figure 7-19 illustrates one arrangement. The common blocks are indicated by bold lines; the thin solid lines indicate AM only; and the broken lines are the FM only stages.

The RF amplifiers, mixer, IF amplifiers, AF amplifier, and speaker are common to both AM and FM. All the switches are

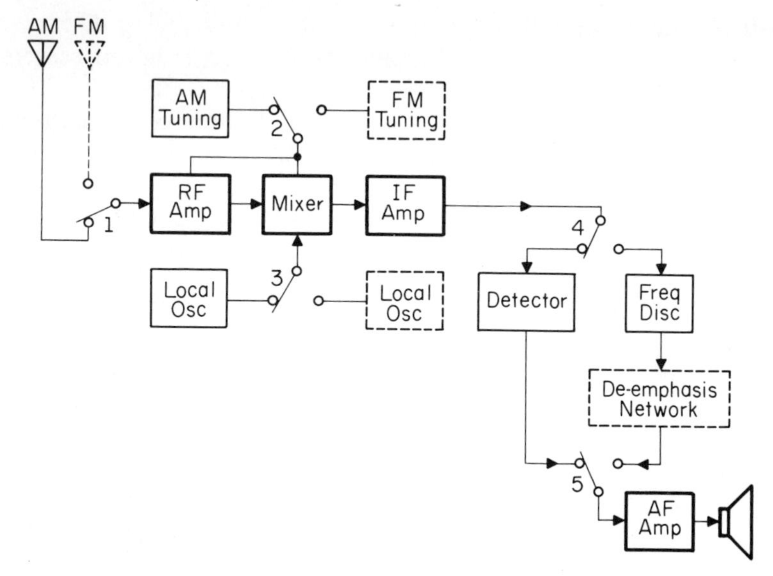


**Fig. 7-19.** AM–FM Receiver.

ganged to a common rotary shaft and change positions at the same time. The switches are presently in the AM position.

Switch 1 has selected the AM antenna and passes the AM signal to the RF amplifiers. Switch 2 couples the AM tuning controls to the RF amplifiers and mixer. Switch 3 couples the AM local oscillator output into the mixer. After several stages of IF amplification, the AM signal passes through switch 4 to the AM detector. The audio signal passes through switch 5 to the common AF amplifier before going to the speaker. With the switches as shown, we have a complete AM receiver.

When FM reception is desired, moving a single control changes the position of all five switches. Switch 1 changes antennas. The tuning circuits are changed by switch 2. Switch 3 replaces the AM local oscillator with an FM local oscillator. Switch 4 drops the detector and couples in the frequency discriminator along with the deemphasis network. The input for the AF amplifier is changed from the detector to the deemphasis network by switch 5. This composes a complete FM receiver.

## CHAPTER 7 REVIEW EXERCISES

1. What function does a microphone serve in the process of modulation?
2. In the AM transmitter, how is modulation accomplished?

3. What do the frequency multiplier stages accomplish?
4. Why do most designers prefer to use two frequency doublers rather than one frequency quadrupler?
5. What are the principle frequencies at the antenna of an AM transmitter?
6. An AM carrier of 1600 kHz is modulated with a 100-Hz audio. What is the frequency of the:
   (a) Upper side band?
   (b) Lower side band?
7. The audio signal of item 6 varies from 500 to 1500 Hz. What is the proper bandwidth?
8. What is the bandpass for the signals described in items 6 and 7?
9. When the modulation voltage is 50 V and the carrier voltage is 60 V, what is the:
   (a) Degree of modulation?
   (b) Percent of modulation?
10. Explain the action of the automatic frequency control in the FM transmitter.
11. What type of transmitter uses a preemphasis network?
12. What does the preemphasis network accomplish?
13. What characteristic of the audio signal establishes the:
    (a) Degree of frequency variation?
    (b) Rate of frequency change?
14. The center frequency of an FM carrier is 50 kHz and it is modulated by an audio of 1000 Hz. List the frequencies of these side bands:
    (a) First high.
    (b) First low.
    (c) Second high.
    (d) Second low.
15. When is an FM side band considered to be insignificant?
16. What is the point in calculating the modulation index?
17. We have a modulation signal of 1000 Hz which causes a deviation of 2000 Hz. What is the modulation index?
18. Refer to the table in Fig. 7-10 and determine the following information for the modulation index of item 17:
    (a) Number of significant side bands.
    (b) Bandwidth.
19. A frequency multiplier consists of a doubler and two triplers. What is the frequency ratio of the input to the output?

20. Name two advantages of single side band transmission.
21. The SB modulator produces a normal AM carrier. What portion of this signal is removed by the filter?
22. What is the most common IF in AM receivers?
23. What two stages are tuned by moving the selector control?
24. What receiver stage is used to reduce signal fading? How is this accomplished?
25. What type of receiver uses the deemphasis network?
26. What is accomplished by the deemphasis network?
27. An AM–FM receiver contains the following listed stages. Underline those that are common to AM and FM; place AM after those that are peculiar to AM; and place FM after those that are peculiar to FM:
    (a) RF amplifiers.
    (b) Mixer.
    (c) IF amplifiers.
    (d) Detector.
    (e) Frequency discriminator.
    (f) Deemphasis network.
    (g) Audio amplifier.
    (h) Speaker.

# transmission lines and antennas

In Chap. 7 we discussed transmission and reception, but we left a big gap between the transmitter and the receiver. We are now going to bridge that gap by utilizing transmission lines and antennas. One transmission line will carry the energy from the transmitter to the anntena and another will carry it from the receiving antenna to the receiver. So, when we understand the function of transmission lines and antennas, we have the big picture. In the process, we will take a brief look at how the waves traverse the distance between the antennas.

## CONSTRUCTION OF TRANSMISSION LINES

Any pair of conductors which couples energy from one point to another is a transmission line. The wires on power poles are transmission lines; the lamp cord is a transmission line. We are concerned with a more specialized transmission line; specifically, the lines which couple transmitters to antennas and antennas to receivers. Any two wires will perform a certain amount of coupling, but when transporting RF energy, care must be exercised that the line doesn't waste more energy than it delivers. At these critical points in our communications chain, conservation of energy is a primary concern. Several types of transmission lines have been designed in an effort to minimize the loss of energy. Let's examine the structure and electrical characteristics of a few of them.

### *open-wire line*

This line is composed of a pair of parallel conductors. The conductors are separated by a fixed distance, and this distance is maintained by insulating spacers placed at suitable intervals. This type of line may appear as either *A* or *B* in Fig. 8-1.

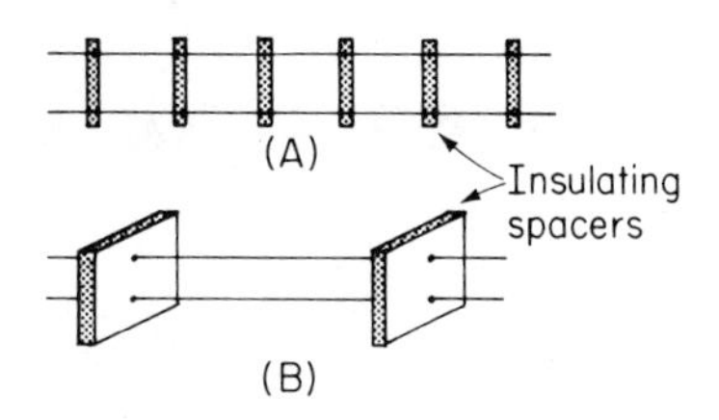


**Fig. 8-1.** Open-Wire Transmission Line.

This line has assets of ease of construction, economy, and efficiency. At radio frequencies, the spacing becomes critical. The separation in any case should be no more than 0.01 wavelength of the frequency being coupled. If excessive spacing is used, it will increase the radiation losses of the line. Since the wires are open, radiation loss is one of the characteristic weaknesses of this line, and proper spacing is one way to avoid aggravation of this weakness. For the same reason, the open-wire line should not be used in the vicinity of metallic objects. This radiation loss increases as frequency increases.

### *stranded-wire ribbon line*

This is an improvement over the open-wire line. Uniform spacing of the two wires has been assured by imbedding the wires in a solid insulation. This transmission line is very popular as a lead in from television antennas. Its general appearance is illustrated in Fig. 8-2.

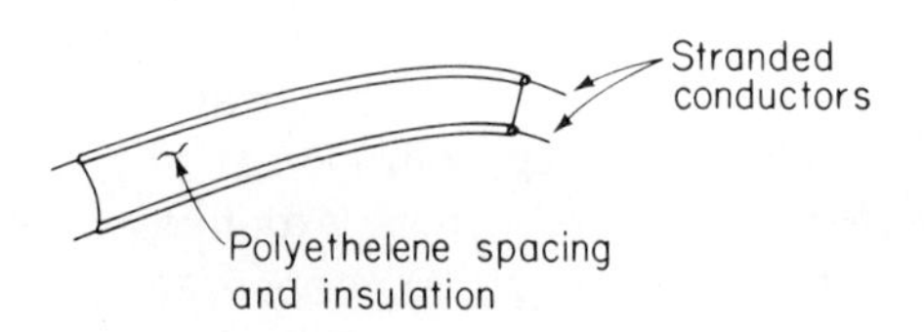


**Fig. 8-2.** Stranded-Wire Ribbon.

In addition to forming an insulation around each wire, a thin sheet of polyethylene separates the wires. This line is only good for relatively low power, but it is more immune to radiation loss than the open-wire line.

### *shielded line*

The open-wire line and the ribbon line have a common weakness. When in the vicinity of metallic objects, a capacitance is formed between the metal and each of the conductors. If one conductor is closer to the metal, it will have a higher capacitance than the other. This imbalance increases radiation losses. The shielded line was designed to overcome this weakness.

The shielded line is composed of two conductors imbedded in an insulating material. The insulating material is then covered with a metal braid shield. The shield is in turn wrapped by another coat of insulation. The metal shield is grounded, and this provides a ground which completely encloses both conductors. The conductors are not only uniformly spaced from each other; they are also uniformly spaced from ground. This line is shown in Fig. 8-3.

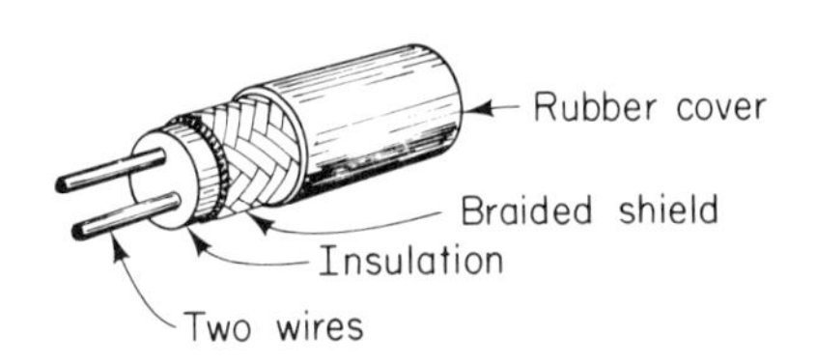


**Fig. 8-3.** Shielded Line.

The metal braid shield is a uniform distance from both conductors. When this shield is grounded a uniform capacitance between each conductor and ground is established throughout the length of the line. The shield and extra layer of insulation also effectively isolate both conductors from outside interference. This line may be laid directly on metal objects without causing a noticeable loss.

The shielded line can carry relatively high power and handle frequencies up to 30 MHz. The coaxial line is closely related to the shielded line.

### *coaxial line*

In this transmission line, the metal shield becomes one of the conductors. The inner conductor is insulated from, and is held in the exact center of, the outer conductor. This line is illustrated in Fig. 8-4.

The line represented here is a popular construction. It is flexible and features quick connectors on both ends. A line of this type $\frac{1}{2}$ in. in diameter can carry power of 1 kW at frequencies up to 30 MHz. The better grade connectors provide moisture proofing to prevent moisture from seeping in and altering the characteristics of the line.

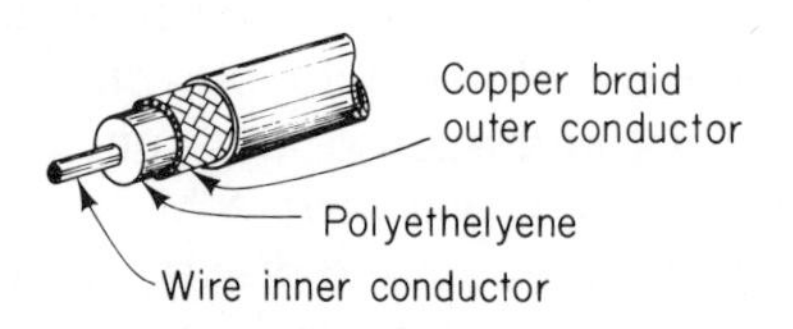


**Fig. 8-4.** Coaxial Line.

Another method of constructing the coaxial line is to use two solid conductors. The outer conductor would be hollow, and the insulated inner conductor would be in its center. In this case, the cavity is usually filled with an inert gas. Depending on the intended use, additional layers of insulation and armor braid may be added to the outside.

The coaxial line is a very efficient and widely used transmission line. The chief advantage lies in its ability to keep down radiation regardless of its environment. It may be strung on poles, buried in the ground, submerged in water, and strung along steel bulkheads. It can be laid between walls, through air ducts, and in elevator shafts. All of this is accomplished with a minimum radiation loss because very little energy penetrates to the surface of the cable.

The RF resistance in Ω/ft of a coaxial line is calculated by:

$$r = \sqrt{f}\left(\frac{1}{d_1} + \frac{1}{d_2}\right) \times 10^{-3}$$

where $f$ is frequency in MHz. $d_1$ is the inside diameter of the outer conductor in in., and $d_2$ is the outside diameter of the inner conductor in in.

A given line has $d_1$ of 0.75 in. and $d_2$ of 0.1 in. What is the resistance of 20 ft of this line at a frequency of 30 MHz?

$$\begin{aligned} r &= [\sqrt{30}\left(\frac{1}{0.75} + \frac{1}{0.1}\right) \times 10^{-3}]20 \\ &= [5.48 \times 11.33 \times 10^{-3}]20 \\ &= 1.24\ \Omega \end{aligned}$$

What is the resistance of the same line for 300 MHz?

$$r = [17.32 \times 11.33 \times 10^{-3}]20$$
$$= 3.85\ \Omega$$

The coaxial line serves very well for all common communication frequencies, but some of the uncommon frequencies are growing in popularity. Microwave communications are not rare anymore. Whereas the coaxial line serves very well for 30 MHz, it is less efficient at 3000 MHz, and still less efficient for frequencies of 30,000 MHz. A different type of transmission line is required at these high frequencies.

### *waveguides*

Some would argue that a waveguide is not a true transmission line, but it does couple energy from transmitter to antenna and from antenna to receiver. That makes it a transmission line despite its differences from other types of lines.

At microwave frequencies, energy can be piped through a tube similar to water through a pipe. Tubes used for this purpose are called waveguides. They come in two popular styles; round and rectangular. These shapes are illustrated in Fig. 8-5.

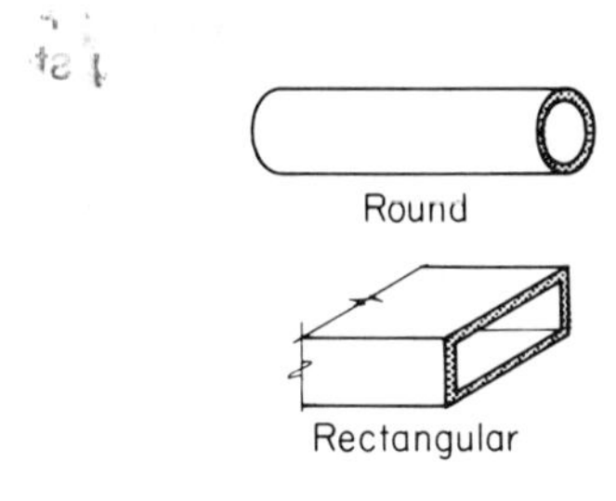


**Fig. 8-5.** Shapes of Waveguides.

## LINE CHARACTERISTICS

All transmission lines have certain common characteristics. These are the characteristics which we need to consider at this time. The physical characteristics of a line determine how it will react electrically. Of course, even a perfectly constructed line has its own limitations of power and frequency due to its characteristic losses.

### *composition*

All conductors have two components which hinder current; resistance and inductance. When two conductors are placed close together, the third component shows up: capacitance. In a straight line of low-resistance material the dc resistance is negligible. This leaves the ac opposition of inductance and capacitance. These two properties are distributed over the entire length of the line. This is illustrated in Fig. 8-6.

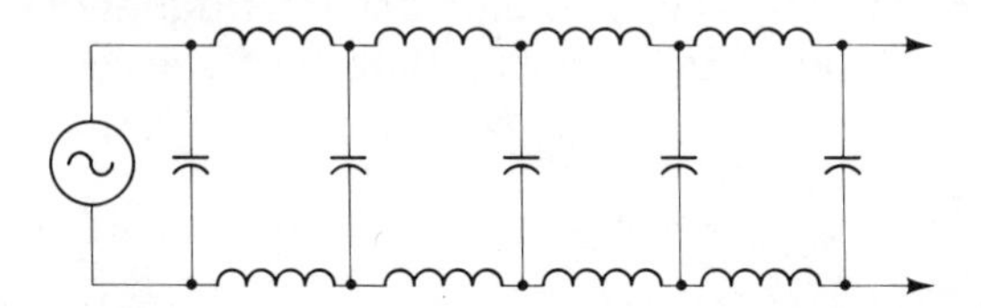

**Fig. 8-6.** Electrical Composition of a Transmission Line.

The generator shown in this drawing represents a radio frequency signal source. It is the job of the transmission line to transport the energy from the signal source to its destination. When transmitting, the signal source is the RF power amplifier, the final stage in the transmitter, and the destination is the antenna. When receiving, the antenna becomes the signal source, and the destination is the RF amplifier, the first stage in the receiver. Along the way, its path is composed of continuous inductance and capacitance. The physical composition of the line (such as size and material of the conductor, distance separating the two conductors, and dielectric constant of the insulating material) determines the amount of inductance and capcitance per unit length of the line.

### *characteristic impedance*

Each transmission line has its own particular impedance. This characteristic impedance is based on the assumption that the line is infinitely long: that it has no termination point. If this was true, energy could be fed into one end of the line, and it would propagate down the line continuously never to return. The entire input would be absorbed by the line. Under these conditions, the impedance of the line is equal to $E/I$. In this infinitely long line, the voltage and current are in phase. This means that the characteristic impedance is pure resistance.

This characteristic impedance is a result of physical construction and remains a fixed value for a particular transmission line. If the quantity of inductance and capacitance is known, the characteristic impedance can be calculated by $Z_0 = \sqrt{L/C}$: Where $Z_0$ is the characteristic impedance in $\Omega$, $L$ is inductance in $H$, and $C$ is capacitance in F. A section of line which has 30 pF of capacitance and 0.15 $\mu$H of inductance would exhibit a characteristic impedance of 70.7 $\Omega$.

$$Z_0 = \sqrt{\frac{L}{C}}$$

$$= \sqrt{\frac{0.15 \times 10^{-6}}{30 \times 10^{-12}}}$$

$$= \sqrt{5000} = 70.7\ \Omega$$

Each line is considered to be composed of an infinite number of such sections, and each section exhibits the characteristic impedance. Keep in mind that this impedance is different with different types of lines. This characteristic impedance uses no energy.

### *terminating the line*

In the transmitter, the transmission line is terminated with the antenna; at the receiver it terminates at the input to the first RF stage. Therefore, the practical line has a definite length of some finite value. These lines can be made to exhibit all the characteristics of the infinitely long line by a proper termination. The proper termination is a resistance equal to $Z_0$ placed as a load across the end of the line.

As far as the signal source is concerned, this is an infinitely long line which exhibits an impedance of $Z_0$. Nearly all of the energy sent down the line will be delivered to the load. The load will absorb all this energy so that none returns to the source. Of course, there are some losses in the line. The slight resistance of the wires and the dielectric leakage causes some signal attenuation, but we have an impedance match which causes maximum power transfer to the load.

### *electromagnetic energy*

The application of a potential difference (voltage) between the two wires of a transmision line sets up an electric field between the two wires. This field is similar to that between the plates of a capacitor. Since we are dealing with RF voltage, the direction of the elec-

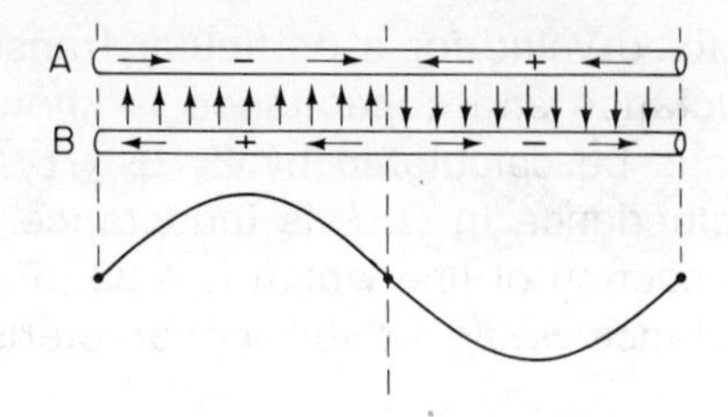


**Fig. 8-7.** Direction of Electric Field.

tric field will change with each alternation of the waveshape. This is illustrated in Fig. 8-7.

The direction of the field is described as the direction a positive charge would move if placed in the field. The waveshape as shown, then, is the potential on conductor *B* taken in respect to conductor *A*. The arrows represent the current direction for each half wavelength section of the line. Notice that the direction of current and the direction of the field reverse when the polarity of the signal changes.

In addition to the electric field, there is a magnetic field surrounding each conductor. Figure 8-8 illustrates the direction of the magnetic field.

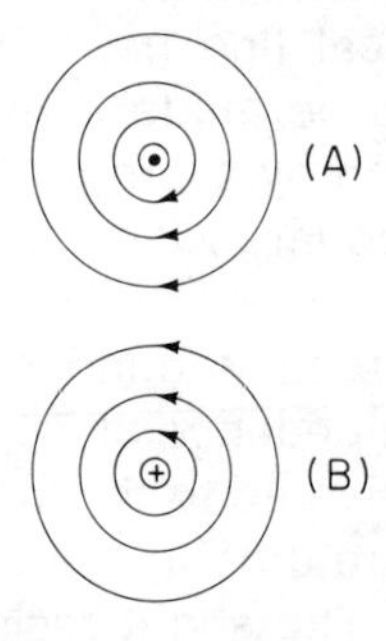


**Fig 8-8.** Direction of Magnetic Field.

This is an end view of the conductors at an instant when electrons are coming up out of the page through conductor *A* and going into the page through conductor *B*. Notice that in the area between the conductors all lines of force are in the same direction: right to left. On the next alternation of the waveshape, current will reverse in both wires with a resulting reversal of field direction. Figure 8-9 illustrates the relationship of the electric (*E*) lines and the magnetic (*H*) lines between the conductors.

Parts A and B show the same section of line on different alternations of the waveshape. Notice that both fields change directions by 180° when polarity reverses. This means that the *E* lines and *H* lines are always perpendicular to each other.

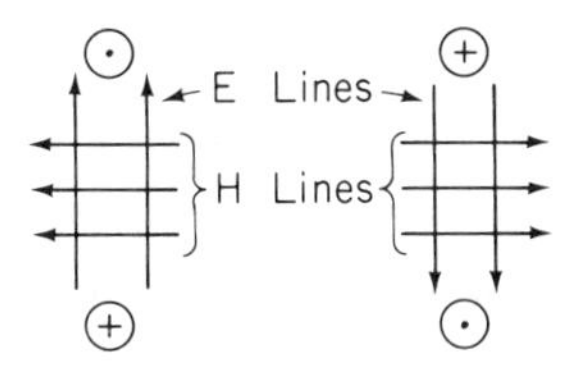


**Fig. 8-9.** Relation of Fields Between Conductors.

These fields advance from the source toward the load at a rate determined by the conducting medium and the frequency of the applied voltage. The first lines appear at the input to the line with the first increment of voltage. These lines move away from the source and are followed by the lines resulting from subsequent voltage values. Even through voltage reverses polarity, current reverses direction, and fields reverse with current, the direction of energy propagation is always away from the source.

Each period of the input voltage produces the identical pattern of the preceding period.

### *wavelength*

Wavelength is velocity divided by frequency. In free space this is $\lambda = v/f$, where $\lambda$ is wavelength in m, $v$ is the velocity of light in m/s, and $f$ is frequency in Hz. This is usually seen as $\lambda = 300/f$ because the speed of light is approximately 300,000,000 m/s. When the formula is $\lambda = 300/f$, the $f$ is in MHz, and part of the calculation has been completed. A wavelength of a 60-Hz wave is $300 \times 10^6/60 = 5 \times 10^6$ m. This is equivalent to slightly more than 3100 statute miles. For 30 MHz it becomes:

$$\lambda = \frac{300}{30} = 10\text{ m} = 10 \times 39.4\text{ in.}$$

$$= 394\text{ in. or nearly 33 ft}$$

The speed of light (300,000,000 m/s, 186,000 mi/s, or 984 ft/$\mu$s) is the reference velocity for all electromagnetic propagation in free space. This means that a given frequency will propagate a given distance during each period of the voltage. This distance is equivalent to the wavelength.

The velocity of propagation is less in a transmission line than it is in space. Just how much less is determined by the velocity of propagation (VP) constant of the line. The VP constant is expressed as a percentage with the velocity of light being 100 percent. A transmission line with a VP constant of 95 percent will propagate energy at the rate of $984 \times 0.95$ which is 943.8 ft/$\mu$s.

The length of a transmission line is measured in wavelengths. Since the wave is slightly shorter in a line than it is in space, wavelength of a line takes into consideration the VP constant. Wavelength in a line, then, is equivalent to the distance the energy will travel down the line during the time for one period of voltage. A 20-MHz signal in free space has a wavelength of 49.25 ft. The same signal in a line with a VP of 80 percent has a wavelength of 39.4 ft. Using this VP and this frequency, a line of 10 wavelengths is 394 ft in length.

### *energy distribution*

When a transmission line is properly terminated, the energy travels down the line as a wavefront. All parts of the line are subjected to all values of current and voltage. That is, any selected point along the line will experience all values of current and voltage once with each period of energy that is propagated past that point. There are no points of null nor peak; the line reacts as if a constant value of voltage was applied. If values could be taken at a specified instant, there would be null and peak points. Under these circumstances all values of current and voltage could be found in a distance of one wavelength. One wavelength in either direction, the value of voltage and current repeats. A distance of a half wave in either direction, the amplitude repeats, but the polarity is inverted. This is illustrated in Fig. 8-10.

Select any point on the drawing and notice the value of voltage and current at that point. Now, move an even number of half wavelengths in either direction, and you will find the same values of voltage and current. Move an odd number of half waves in either direction, and you will find the amplitudes the same but polarities are reversed.

The conditions described for the infinitely long line or a line terminated with its characteristic impedance is a nonresonant condition. The load is resistive, current and voltage are in phase and there are practically no losses in the line. When the terminating load is not equal to $Z_0$, we have a different ball game.

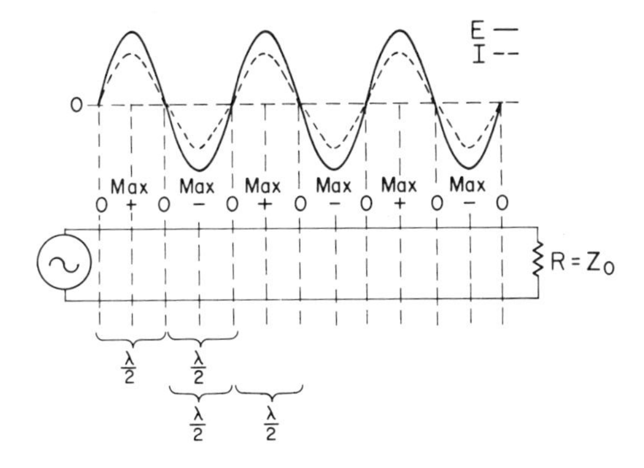


**Fig. 8-10.** Repeating Instantaneous Values.

## RESONANT LINES

Any live line which terminates in any manner except a resistor equal to $Z_0$ is a resonant line.

### *mismatched impedance*

When a transmission line is improperly terminated it results in a loss of power. The line is improperly terminated when the load is of any value except the characteristic impedance. This is a mismatch of impedance, and the greater the degree of mismatch the greater the power loss. In most cases, a mismatch of less than 25 percent can be tolerated. Over 25 percent mismatch results in a substantial loss which is caused from energy being reflected from the load.

The energy is still propagated down the line to the load, but the mismatched condition causes voltage and current to arrive out of phase. As a result the load only absorbs a portion of the energy. The remainder of the energy is reflected from the load back toward the source.

The line is now a two way street with electromagnetic energy traveling in both directions. Direct energy travels from the source to the load; reflected energy travels from the load to the source. These waves combine and result in standing waves all along the line. Forward and reflected waves will be in phase at one point and out of phase at another. This results in points of different potential along the line. There will now be nodes and peaks of voltage instead of the steadily advancing wavefront. The energy absorbed by the load is the difference between the forward energy and the reflected energy. This absorbed energy is the useful energy. The remainder is wasted because of the mismatched impedance.

### *open-end line*

Any transmission line which has standing waves is a resonant line. Terminating the line with an open circuit is another way to create a resonant line. Resonance implies a particular frequency and the resonant frequency is the frequency the line was designed for. Therefore, reference to wavelength means wavelength of the resonant frequency.

An infinite impedance exists across the open end of a line. This is equivalent to a parallel resonant circuit at that point. At the open end, then, current is minimum and voltage is maximum.

A quarter wave back from the open is a condition of very low (almost zero) resistance. This is equivalent to a series resonant circuit. At this point, current is maximum, and voltage is minimum.

One half wavelength back from the open, we have parallel resonance again with the same conditions that exist at the open end.

All lengths of the open-end line can be summarized in terms of odd and even quarter wavelengths. At the open end, we have parallel resonance. At all points that are an even number of quarter wavelengths from the end, the parallel resonant condition is repeated. At all points that are an odd number of quarter wavelengths from the end, the open end conditions are reversed; changed from parallel resonance to series resonance. Since the parallel resonant impedance approaches an open and the series resonant state is practically a short, these points may be described in terms of opens and shorts. If this is done, we have an open at the end of the line and at all points that are even quarter wavelengths from the end. All points that are an odd number of quarter wavelengths from the open will exhibit a shorted condition.

Many uses are made of the open end line, and the characteristics just described are vital to the understanding of their behavior. These conditions are illustrated in Fig. 8-11.

In this drawing, we have used a line that is three wavelengths. The standing waves of current and voltage indicate amplitudes at all points along the line. The impedance is indicated at the quarter wavelength points from the open end.

When the total length of the line is not an exact multiple of the quarter wavelength, the open-end line acts as either a capacitor or an inductor. For lengths less than a quarter wavelength, it is a capacitor. For lengths more than one quarter wavelength but less than two, it is an inductor.

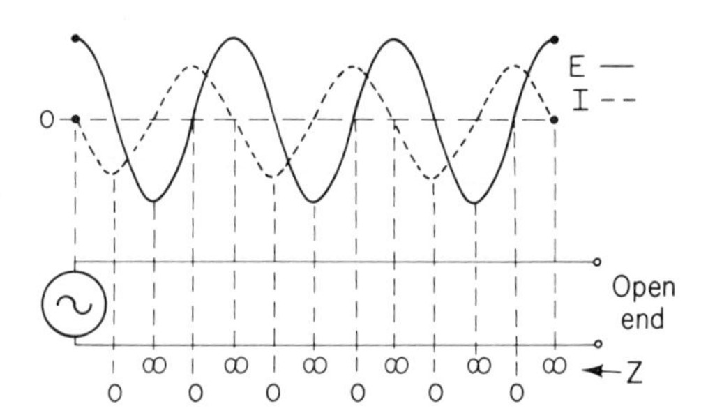


**Fig. 8-11.** *Z, E,* and *I* at Quarter Wavelength Points from Open End.

### *shorted-end line*

Terminating a transmision line in a short also forms a resonant line. At the shorted end, the impedance is, of course, zero. This is equivalent to the series resonant condition. The shorted end, then, will have zero voltage and maximum current. Moving back down the line from the shorted end, the condition is repeated at all points that are even quarter wavelengths from the short. The shorted condition is inverted (open) at all points that are odd quarter wavelengths from the short. These conditions are further illustrated in Fig. 8-12.

For lengths less than a quarter wavelength, the shorted line

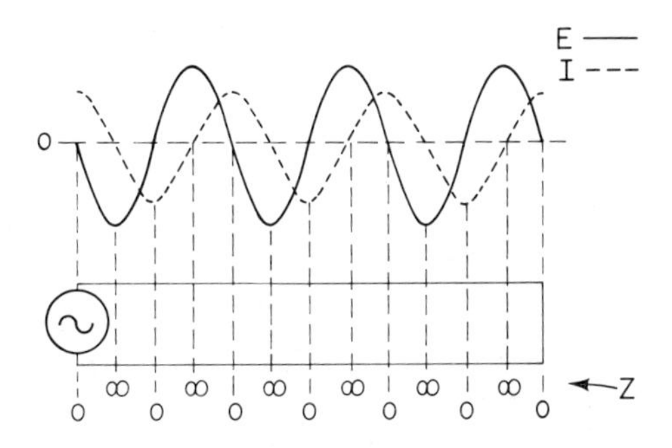


**Fig. 8-12.** *Z, E,* and *I* at Quarter Wavelength Points from Shorted End.

acts as an inductor. For lengths that are more than a quarter wavelength but less than a half, it acts as a capacitor.

### summary of resonant lines

Figure 8-13 summarizes all the conditions of open and shorted end resonant lines from 0° to 360° (all parts of a full wavelength).

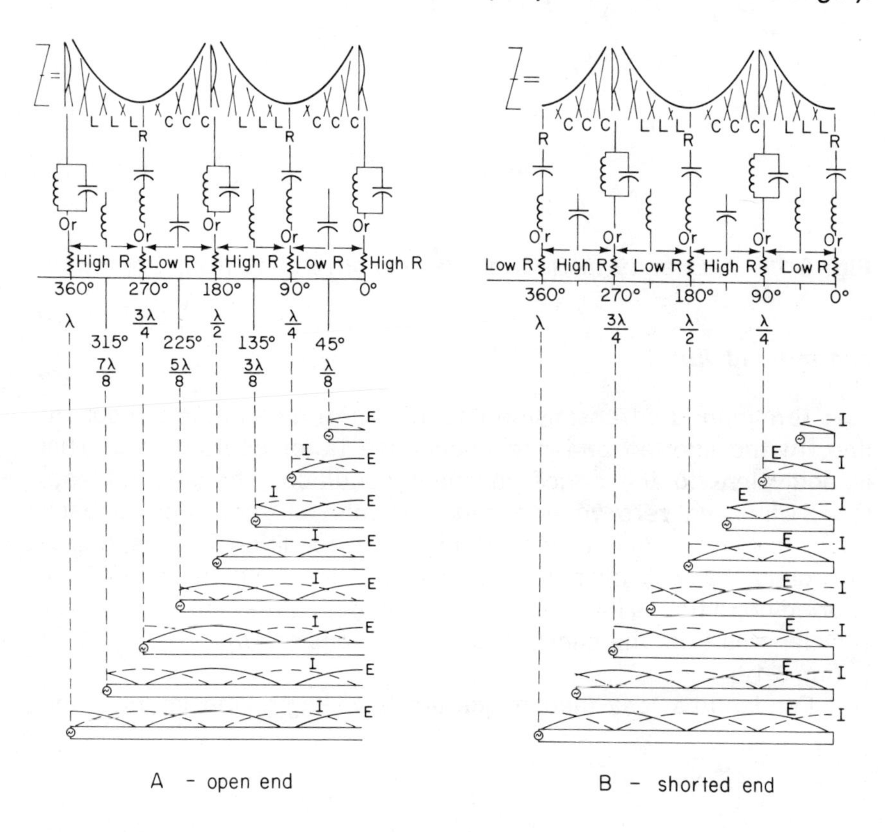


**Fig. 8-13.** *Z, E,* and *I* at Various Points.

### uses of resonant sections

As you might have guessed by this time, resonant sections of transmission lines are used as transformers, resonant circuits, filters, shorting devices, impedance matching devices, and insulators.

The half-wave open section is a parallel resonant circuit; it can be used as a transformer. The quarter-wave open section is a series resonant circuit; it can be used as a filter. The quarter-wave short is a parallel resonant circuit; it can be used as an insulator.

If your receiver antenna is picking up an undesired frequency, a quarter-wave open section can eliminate that frequency. The section would be connected as shown in Fig. 8-14.

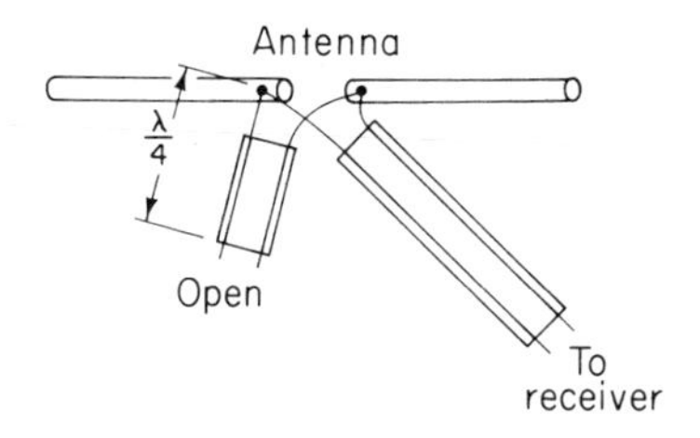


**Fig. 8-14.** Quarter-Wave Section as a Reject Filter.

The quarter-wave section would be cut to the frequency of the interference signal. It will then reflect a short for this frequency. The reflected short will appear across the entrance of the transmission line going to the receiver. This shorts out the interference. Since the quarter-wave section is a high *Q* resonant circuit, it will have little effect on other frequencies. When used in this fashion, it is a band rejection filter.

If the quarter-wave open section is connected in series with one wire of the transmission line, it becomes a bandpass filter. This is illustrated in Fig. 8-15.

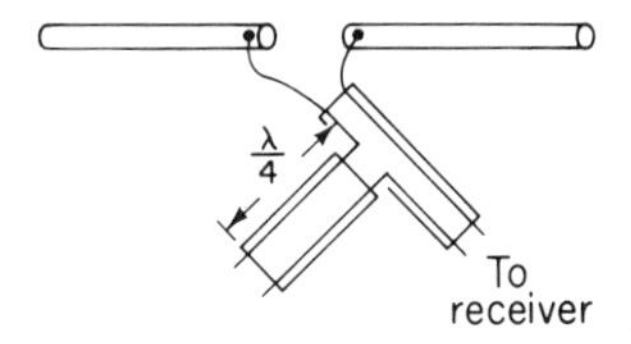


**Fig. 8-15.** Quarter-Wave Section as a Pass Filter.

This section is cut to match the desired frequency. For this frequency (and a narrow band) the break in the conductor is shorted out. All frequencies outside of this band will find an open conductor.

The quarter-wave shorted section can be used as a filter and as an insulator at the same time. This is illustrated in Fig. 8-16.

The quarter-wave, shorted-end sections are cut to the center frequency being transmitted. To this frequency, the shorted end reflects an open at the entrance. The shorted end may be welded to metal to give support to the transmission line. Even with ground at the shorted end, the transmission line is electrically insulated from ground.

For harmonics of the fundamental frequency, it is a different story. The second harmonic (the first harmonic is the fundamental) finds these to be half-wave shorted sections. They find practically zero resistance to ground. The fourth harmonic finds a full-wave line with the same result. All even harmonics can be eliminated by such simple filters.

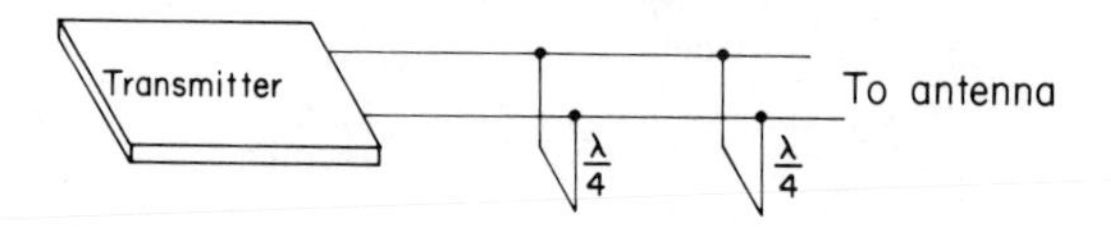


**Fig. 8-16.** Harmonic Filters and Insulators.

The impedance of a quarter-wave, shorted-end section ranges from zero at the shorted end to almost infinity at the open end. Two widely different impedances may be matched by inserting such a section between them. This is illustrated in Fig. 8-17.

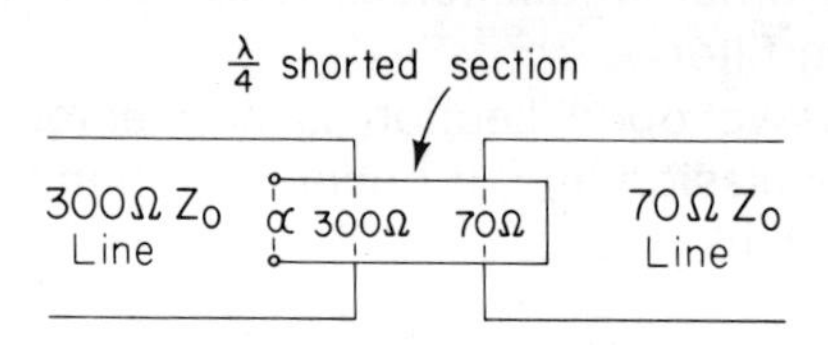


**Fig. 8-17.** Impedance Matching.

Making a direct connection between these two lines would result in standing waves on both lines. The 300-Ω line is matched by connecting it to the 300-Ω point on the quarter-wave section. The 70-Ω line is matched by connecting it to the 70-Ω point on the shorted section. The impedance of both lines is matched; there are no standing waves on either line; maximum power can be transferred in either direction.

Remember that the wavelength of a signal in a transmission line is less than it is in free space. The velocity of propagation is a key factor. It is necessary to know the velocity of propagation before calculating the physical length of the sections. Since wavelength in meters (300 × VP)/*f*, it is easy enough to convert to in.

for any portion of a wavelength. A quarter wavelength in in. is $[(2955 \times VP)/f]$ (MHz). For half, three-quarters, and any other length it is a matter of multiplication.

**Problem:** A transmission line has a propagation velocity of 68 percent. How many in. of this line is a quarter-wave section to a frequency of 100 MHz?

Solution:

$$\frac{\lambda}{4} = \frac{(2955 \times VP)}{f} = \frac{(2955 \times 0.68)}{100} = 20.08 \text{ in.}$$

The figure was rounded off to obtain 20.08. The answer could be rounded off a bit more with no ill effects. The .08 in. could be dropped leaving 20 in. as the quarter-wave section.

## LOSSES IN TRANSMISSION LINES

There are basically three types of losses in a transmission line. Let's take a brief look at each of these.

### *copper loss*

Copper loss is due to the resistance of the conductor. The zero resistance conductor does not exist. The loss is $I^2R$. For a short line, this may be negligible; for a long line, it might be considerable.

At high frequencies a transmission line suffers a greater copper loss because of skin effect. Each moving electron builds up its own magnetic field. The electrons change direction with each change in voltage polarity with a consequential reversal of the individual magnetic fields. At a frequency of 100 MHz these changes are taking place at the rate of 200,000,000 times per second.

The interaction of these fields retards the progress of electrons at the center of a conductor. Nearly all movement of electrons is confined to a thin layer at the surface of the conductor. This effect is so strong that the entire center of a conductor can be removed with no noticeable effect on current.

Since all the current is concentrated on the surface (skin) of the conductor, the cross-sectional area is effectively reduced. Reducing the cross-sectional area increases the resistance and results in a greater energy loss. This loss increases with frequency.

The depth that RF energy penetrates a copper conductor can be calculated by:

$$\delta = \frac{6.62}{\sqrt{f}}$$

where delta ($\delta$) is depth of penetration in cm and $f$ is frequency in Hz.

What is the depth of penetration for a frequency of 300 MHz?

$$\begin{aligned}\delta &= \frac{6.62}{\sqrt{f}} \\ &= \frac{6.62}{\sqrt{300 \times 10^6}} \\ &= \frac{6.62}{17.32 \times 10^3} \\ &= 0.38 \times 10^{-3} \\ &= 0.00038 \text{ cm}\end{aligned}$$

### *dielectric loss*

This is a loss due to the heating effect on the insulating material between the two conductors. The electric field distorts the electron paths in the atoms which form the dielectric. The changing field causes a constant agitation of these electrons. The agitation uses a certain amount of energy which is dissipated in the form of heat.

### *radiation loss*

This loss is caused by the fact that some lines of force do not return to the line when the signal alternation changes. These lines are radiated into the surrounding medium and represent a loss of energy.

Another loss closely associated with radiation loss is induction loss. When a line is in the vicinity of a metallic object, the magnetic lines induce a current into the neighboring metal. This is a transformer action which saps energy from the line.

All these losses become more pronounced as frequency increases. The open- and ribbon wire lines can't be used above 3000 MHz. Shielded and coaxial cables are still used, but their losses are becoming considerable. If frequencies were to go higher a different type of transmission line had to be constructed.

## WAVEGUIDES

This is a transmission line for microwave frequencies. It is not normally included in a basic electronics course, but microwave communications are becoming commonplace. Therefore, an introduction to waveguides has become essential to basic electronics.

### *structure*

It is a small step from the coaxial line to the waveguide. Electromagnetic waves can transfer energy through a line that has no center conductor. When the center conductor and accompanying insulation is removed from a coaxial cable, a hollow tube remains. This hollow tube is closely related to a waveguide. The waveguide does not have to be round. They are constructed to fit practical situations which means round, square, rectangular, or eliptical as the need arises. Figure 8-18 illustrates some sections from a rectangular waveguide.

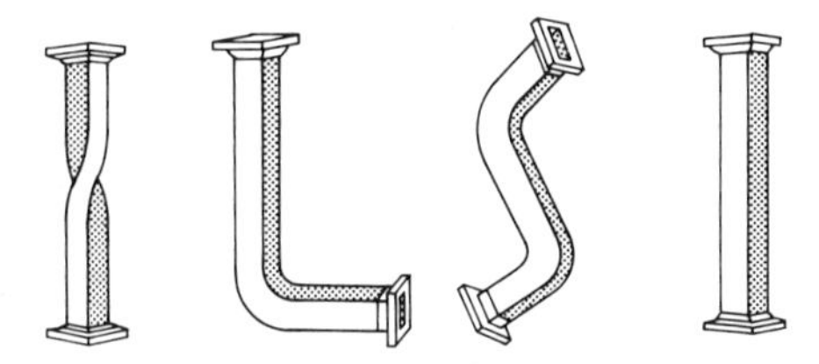

**Fig. 8-18.** Waveguide Sections.

Waveguides are constructed on the premise that an infinite number of shorted quarter-wave sections can be added to a two wire transmission line. As more and more sections are added to both sides of the line, the line takes on the appearance of a box. The conductors are then part of the walls. This is illustrated in Fig. 8-19.

### *wave propagation*

Current and voltage are present in the center of the wide walls much the same as they appeared in the two-wire line. However, in waveguides, the concern is related directly to wave propagation. Energy is radiated into the cavity and propagates through the waveguide somewhat similar to free space propagation. The wavefront

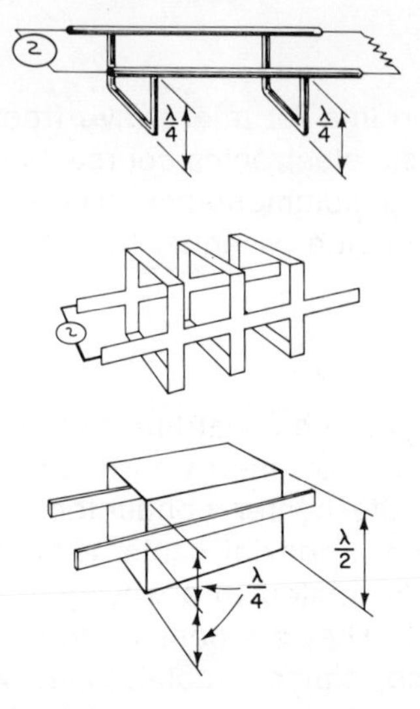


Fig. 8-19. Constructing a Waveguide.

moves diagonally through the cavity reflecting from first one narrow wall then the other. Figure 8-20 shows one method of injecting energy into a waveguide.

Both positive and negative alternations are radiated from the probe. This produces a wavefront which would resemble a widen-

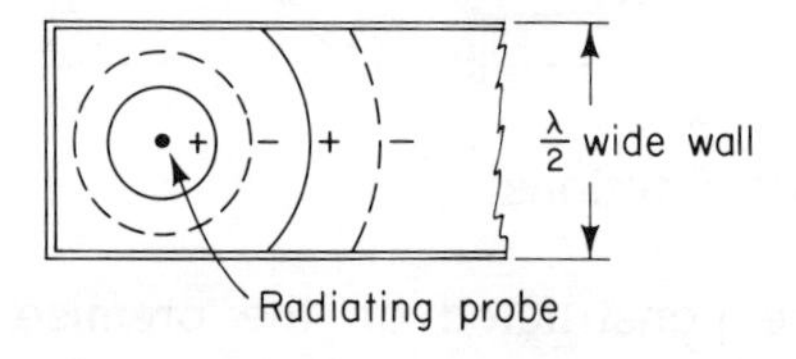


Fig. 8-20. Energy Entering Waveguide.

ing circle except for the confining walls. But the walls are there, and as a wave approaches the wall, it is reflected as light from a mirror. The phase reverses with each reflection. The progress is illustrated in Fig. 8-21.

In order for a wavefront to exist, both electric and magnetic fields must be present. The *E* and *H* lines form complex patterns, which repeat at points separated by a distance of a half wavelength. The entire fields are confined inside the cavity and are perpendicular to one another. The *E* lines are parallel to the wide walls and tangent to the narrow walls. The *H* lines are perpendicular to the wide walls. This is illustrated in Fig. 8-22.

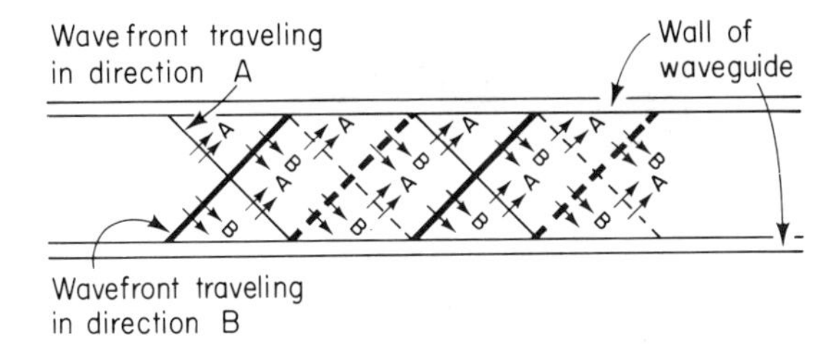


**Fig. 8-21.** Wave Movement.

This drawing represents the fields at one point at a particular instant. They are in constant motion and change direction by 180° when the polarity changes.

Energy may be coupled from a waveguide in several ways. A flared feed horn on the end will radiate energy into space. An opening in the wall will allow energy to escape. A probe can be inserted to remove energy by the same method that we used for insertion.

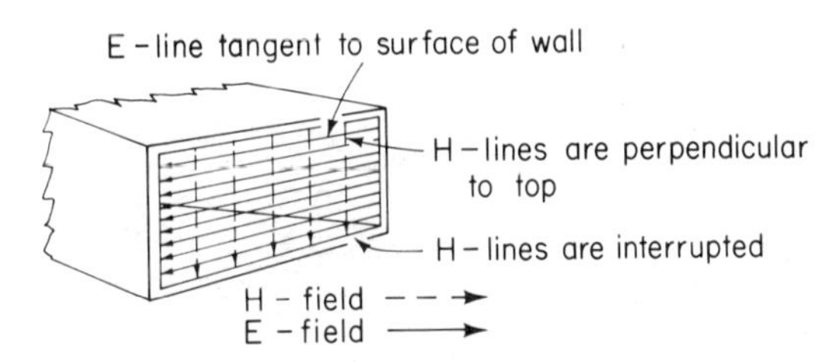


**Fig. 8-22.** Fields in a Waveguide.

## LIGHT GUIDES

None of our present transmission lines are capable of carrying light frequencies. But the need for a light frequency transmission line is present. These will undoubtedly be called light guides and experimentation is underway. Figures 8-23 and 8-24 indicate that some progress is being made.

The transmission line transports our electromagnetic energy from the final stage of the transmitter to the transmitting antenna. It also carries the energy from receiving antennas to the first stage of the receiver. If proper care is exercised in selecting and using a transmission line, it does its job with a very small waste of energy. Now we need a device that will release this electromagnetic energy into space and another that will recover that energy after it has completed its space journey. Both of these devices are called antennas.

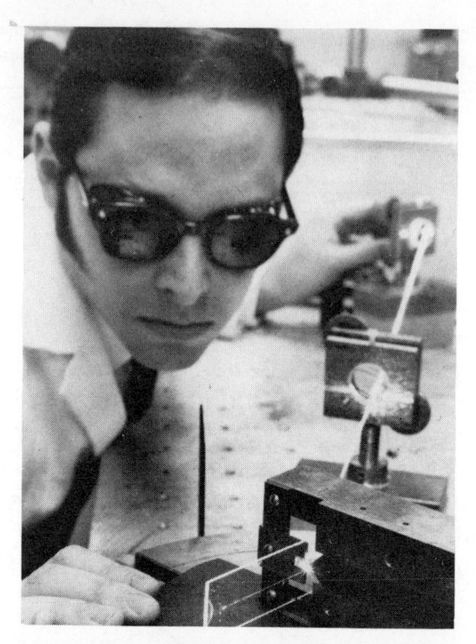

**Fig. 8-23.** A Transmission Line for Light.

**Fig. 8-24.** Teaching Light to Turn Corners.

## ANTENNAS

The transmitting antenna is an active device which couples electromagnetic energy to free space. It is driven by relatively high-power levels to cause a strong energy radiation.

Any exposed piece of conducting material can be a receiving antenna. Electromagnetic waves induce a current in any conductor they encounter. This constitutes reception of the radiated signal.

Of course, just any piece of conductor will not make a good antenna. For efficient reception, the antenna is refined to extract the weak signals from space and couple them to the receiver with a minimum of loss.

Any antenna that is used for transmission can also be used for reception. The characteristics are essentially the same for both sending and receiving. Let's examine the mechanics of coupling a signal into space.

### *electromagnetic radiation*

If a quarter wavelength of a two-wire transmission line is folded back as shown in Fig. 8-25, it becomes a very effective antenna.

An antenna formed in this fashion is called a half-wave dipole. The fold, being a quarter wavelength from the open end of the line, is made at a point of reflected zero resistance. At the fold, voltage is zero, and current is maximum. At both ends of the folded section, current is zero, and voltage is maximum.

At any given point in a transmission line, current is always in opposite directions in the two conductors. This is also true of the quarter-wave section that we folded back. Current was toward the end on one side and away from the end on the other side. This direction does not change when the fold is made, but now the currents are additive. Current and voltage at the peak of one voltage alternation are shown in Fig. 8-26.

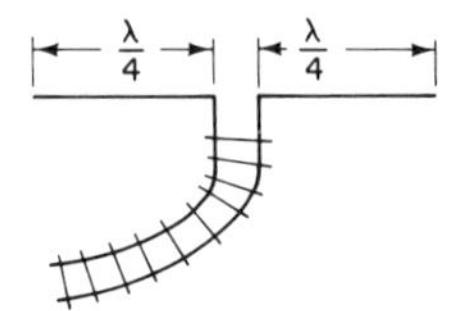


**Fig. 8-25.** Forming an Antenna.

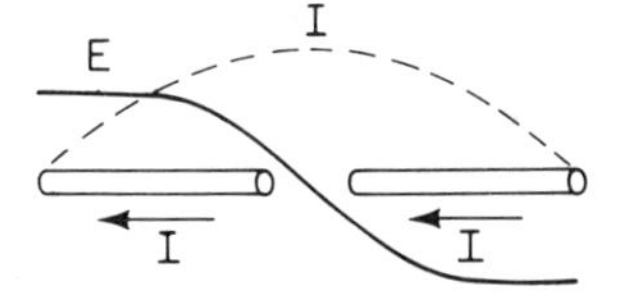


**Fig. 8-26.** Voltage and Current Waves on a Dipole.

Electrons, which constitute antenna current, flow in the direction indicated as long as the voltage retains this polarity, but the voltage is constantly changing. On the next alternation, voltage polarities will be reversed, and electron flow will be in the opposite direction. The current is 90° out of phase with the voltage, and both are rising and falling at a very rapid sinusoidal rate.

The magnetic lines encircle the antenna at all points. This field builds up, collapses, and builds up in the opposite direction in synchronization with the changing current. Figure 8-27 illustrates the magnetic field when current and voltage are the same as depicted in Fig. 8-26.

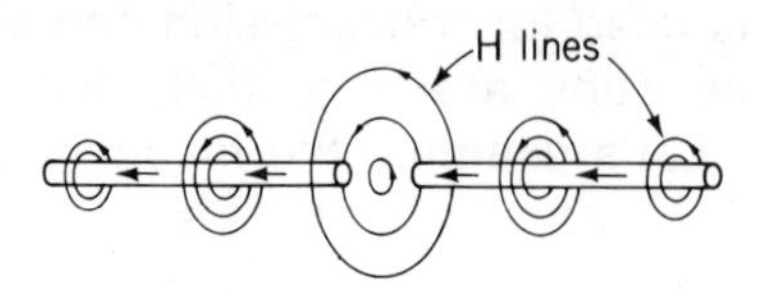


**Fig. 8-27.** Magnetic Field About a Dipole.

A magnetic field must be accompanied by an electric field, and the two fields must be perpendicular to each other. The magnetic field in Fig. 8-27 is accompanied by the electric field in Fig. 8-28.

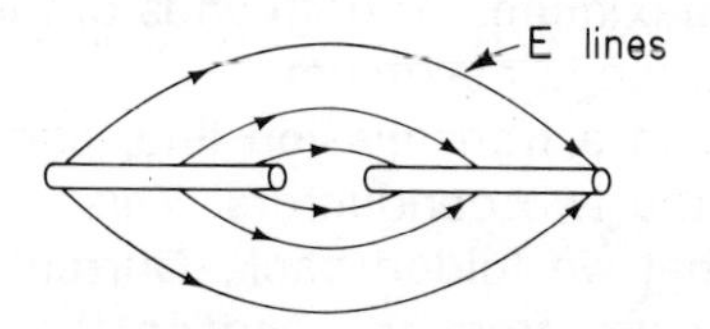


**Fig. 8-28.** Electric Field About a Dipole.

The power contained in the respective fields about a transmission line may be calculated as follows:

$$P_m = \frac{LI^2}{2}$$

where $P_m$ is power in the magnetic field in W, $L$ is H of inductance per unit length, and $I$ is current in A.

$$P_e = \frac{CE^2}{2}$$

where $P_e$ is power in the electric field in W, $C$ is capacitance in F per unit of length, and $E$ is voltage in V.

Since the antenna is basically an extension of a transmission line, we may apply these formulas to our dipole. Suppose that we have a dipole which contains 30 mH of inductance and 0.48 $\mu$F of capacitance. When 1000 V is applied, there are 4 A of current. What is the power in the magnetic field?

$$P_m = \frac{LI^2}{2}$$

$$= \frac{(30 \times 10^{-3})(16)}{2}$$

$$= 240 \text{ mW}$$

What is the power in the electric field?

$$P_e = \frac{CE^2}{2}$$

$$= \frac{(0.48 \times 10^{-6})(1 \times 10^6)}{2}$$

$$= 0.24 \text{ W} = 240 \text{ mW}$$

Radiation of energy is made possible by the buildup and collapse of the electric and magnetic fields. At radio frequencies, the changes take place so rapidly that portions of the fields become trapped in space. As the fields expand in the opposite direction, the trapped fields are repelled into space at the speed of light. Figure 8-29 illustrates trapped $E$ lines being detached from an antenna.

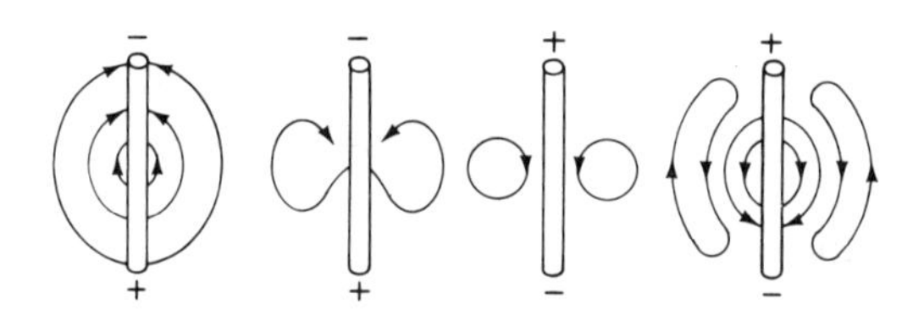

**Fig. 8-29.** Trapping and Repelling $E$ Lines.

In part A, the $E$ field is fully expanded. In part B, the current is rapidly decreasing to zero, and the collapsing field is starting to form loops. In C, the polarity has changed and blocked further collapse of the $E$ field. The lines that are still outside the antenna are trapped. They detach themselves from the antenna. In D, the $E$ field is expanding in the opposite direction and driving the trapped loops into space.

### *electromagnetic fields in space*

The $E$ lines that are detached from the antenna are accompanied by $H$ lines because portions of the magnetic field are trapped in the same way.

Now the electromagnetic field travels through space at the rate of 186,000 mi/s, and expands in all directions as it moves. The directions of the lines remain the same as they were when they left the antenna. The *E* lines are roughly parallel to the antenna, and the *H* lines are perpendicular to the *E* lines. At points along the propagation path separated by a distance of a half wavelength, both *E* and *H* lines are in opposite directions. This is the same pattern that we discussed in connection with transmission lines.

If the antenna is mounted parallel to the surface of the earth, the *E* lines are also parallel to the surface of the earth while the *H* lines are perpendicular to this plane. In this case, the antenna is horizontally polarized. That is to say, the *E* lines and the antenna are horizontal with respect to the earth's surface.

When the antenna is perpendicular to the earth's surface, the *E* lines are perpendicular, and the *H* lines are horizontal. This antenna is vertically polarized.

For maximum reception of electromagnetic energy, the receiving antenna should be polarized the same as the transmitting antenna.

### *receiving electromagnetic energy*

The wavefront approaches the receiving antenna as a huge sheet of energy. This sheet is composed of a gridwork of *E* and *H* lines. As this front, with its following energy, rushes past the antenna, a few of the *H* lines cut across the conducting material of the antenna. The magnetic lines cutting the antenna induce a voltage on the antenna which causes current. This very weak signal, frequently on the order of a few microvolts, is coupled through the transmission line from the antenna to the first RF amplifier.

Such a small percentage of the total energy reaches any given point that a very high transmitter power produces only very weak received signals. If the transmitting and receiving antennas are polarized in different planes, the *H* lines are parallel to the receiving antenna. In this case, practically no *H* lines cut the antenna, and practically no signal is induced.

The transmitting antenna is equivalent to the primary of a transformer. The receiving antenna is the secondary of this transformer. This is a high-loss transformer, but it couples usable energy across many miles of space which separates the primary from the secondary. The transmitting antenna converts the electric energy to electromagnetic energy. The receiving antenna converts the electromagnetic energy back to electric energy.

## TYPES OF ANTENNAS

Antennas are constructed in many shapes and in sizes to match the frequency they use.

### *Marconi antenna*

Marconi discovered that he could obtain the effect of a half wave antenna with a quarter wavelength of conducting material. He found that a quarter wave section mounted perpendicular to a conducting surface exhibited all the characteristics of a half-wave antenna. This creates a mirror effect which reflects the missing half of the antenna.

The quarter wave section of conductor is the physical half of the antenna, and the reflected image of this quarter wave section forms the other half. This is illustrated in Fig. 8-30.

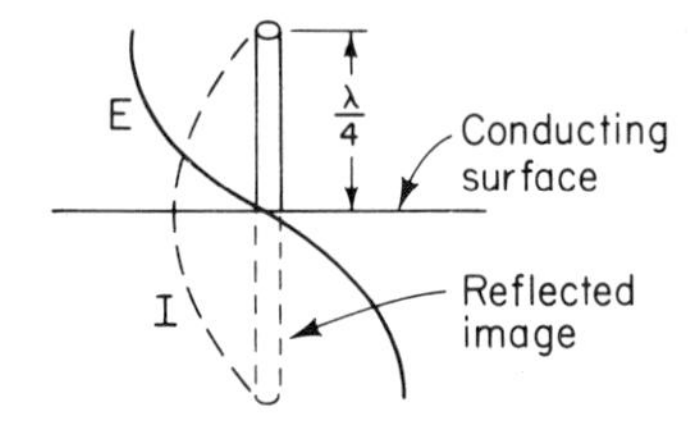


**Fig. 8-30.** Marconi Antenna.

The quarter wave section is shorted to the conducting surface. Most any surface will do; water, earth, or metal, but the better the conductor, the better the results. This is a good antenna for frequencies up through the radio broadcast band. Most radio towers are quarter wave sections of a Marconi antenna; the surface they stand on reflects the other quarter wave section.

### *whip antenna*

A whip antenna is a flexible conductor usually secured on one end and mounted vertically. They are effective for frequencies between 1.8 MHz and 30 MHz. The average whip is a quarter wavelength to a frequency of 6 MHz. When it is used for other frequencies, inductors or capacitors are connected in series with the antenna to change its electric length.

### *Hertz antenna*

Any half-wave antenna is a Hertz antenna. It is similar to the dipole previously discussed and may be fed at the center or at one end. Hertz antennas are effective up to 400 MHz.

### *folded dipole*

This is two half-wave elements connected at the ends as shown in Fig. 8-31.

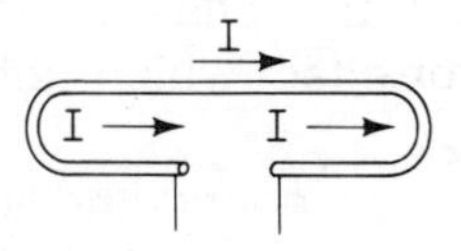


**Fig. 8-31.** Folded Dipole.

The input impedance of some antennas is so low that it is difficult to match with the high impedance of the transmission lines. The plain dipole may have an input impedance as low as 15 Ω. Connecting another half-wave section to each end of the dipole changes it to a folded dipole. This results in a great increase in impedance; it increases the impedance about four fold. This higher impedance is closer to the transmission line impedance and is much easier to match.

The folded dipole also radiates more power. One of the elements is fed direct and the other is excited by radiation from the first. The current in both elements is in the same direction. When the current is the same magnitude in both elements, it doubles the field density in space.

### *directional antennas*

Many uses of antennas require that they radiate energy in all directions. All directions constitutes a nondirectional antenna. For instance, a public broadcasting radio station uses, for the most part, nondirectional antennas. There are many other cases where directional antennas are essential. To name a few, we have navigational equipment, direction finders, homing devices, radar, and satellite trackers.

The antennas that we have discussed thus far have been of the nondirectional type. Some of these same antennas can be made directional by adding reflectors and directors. Electromagnetic energy can be reflected in much the same manner as a beam of light is reflected.

Most ordinary television receiver antennas are directional. This is the parasitic array. The parasitic array is sometimes called a yagi, and it can be used for either transmitting or receiving. This antenna is usually composed of from four to seven elements. One element is active (the actual antenna), and the others are parasitic in nature. That is they are excited by radiation from the driven element.

A parasitic element 5 percent longer than the antenna and placed 0.2 wavelength away becomes a reflector. When the parasitic element is 5 percent shorter than the antenna and is placed 0.1 wavelength away, it becomes a director. A four element yagi is illustrated in Fig. 8-32.

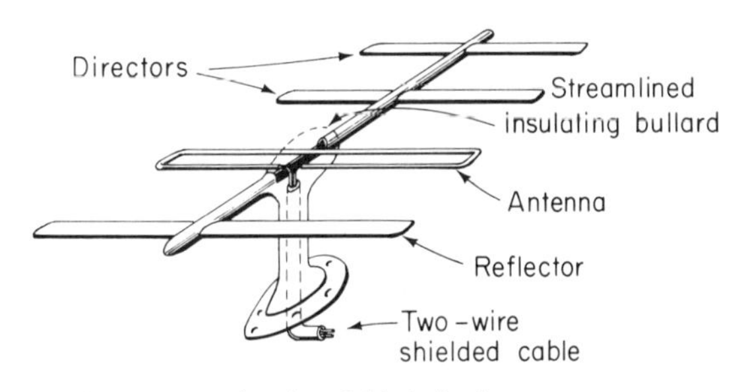


**Fig. 8-32.** Yagi Array.

The shortest element on a parasitic array is the front. All elements forward of the antenna are directors. The longest element is the reflector, and it is mounted on the back of the array. The electromagnetic energy is beamed forward and may attain a pattern as narrow as 50°. This is not highly directional, but the energy is, at least, concentrated in one general direction.

We have now examined the mechanics of radiating and receiving electromagnetic energy. Let's travel with a wavefront, and observe some of the objects it encounters on its trip between antennas.

## FACTORS AFFECTING WAVE PROPAGATION

The moving wavefront encounters many conditions that either aid or hinder its progress. To start with, a particular segment of a wavefront has its destiny pretty well laid out when it leaves the antenna. The type of antenna, the surrounding terrain, the height of the antenna, and the direction of radiation all have a bearing on wave behavior. After leaving the antenna, the progress of the wave will be affected by weather conditions, the time of day, obstructions, sun spots, and type of terrain.

### *types of waves*

As the wave of energy fans out from the transmitting antenna it resembles a huge, invisible doughnut with the antenna threaded through the hole. The curved surface of this doughnut represents the expanding wavefront. The names for radiation waves are derived from the direction of movement of certain small segments of this wavefront. The three most important waves are ground waves, direct waves, and sky waves. These are illustrated in Fig. 8-33.

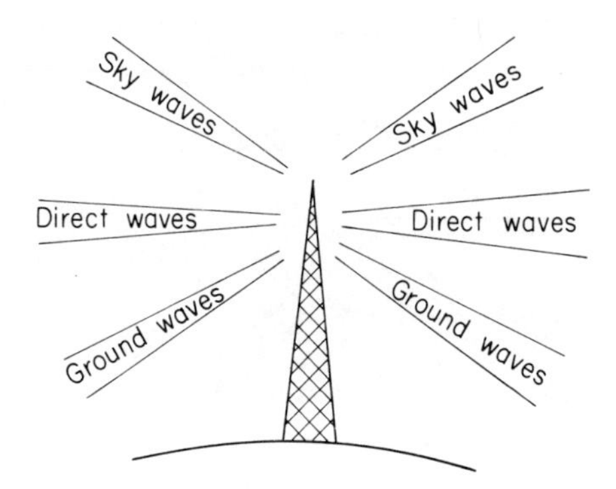


**Fig. 8-33.** Principal Waves from Antenna.

The portions of the wavefront which are radiated parallel to the surface of the earth are called direct waves. They travel in a relatively straight line from the antenna to the horizon. These waves are blocked by the curvature of the earth after traveling distances from 30 to 75 mi.

The ground waves leave the antenna at a downward angle and travel along the earth's surface. The distance that can be spanned by these waves is dependent on frequency and type of terrain. Frequencies below 250 kHz may reach a distance of 1000 mi over land; more over water. For frequencies of 30 MHz, the distance is cut to about 75 mi.

We may calculate the line of sight distance to the horizon by this formula:

$$d = \sqrt{\frac{3h}{2}}$$

where $d$ is distance to the horizon in mi and $h$ is antenna height in ft.

What is the line line of sight distance when the antenna is 200 ft high?

$$\begin{aligned} d &= \sqrt{\frac{3h}{2}} \\ &= \sqrt{\frac{600}{2}} \\ &= \sqrt{300} \\ &= 17.32 \text{ mi} \end{aligned}$$

The portions of the wavefront which leave the antenna at elevated angles are called sky waves. Some texts group all waves into two categories; sky waves and ground waves. According to such a grouping, the previously mentioned direct waves would be included with the ground waves. Sky waves are primarily responsible for most of the reliable, long distance radio communications.

### *the ionosphere*

In the early days of radio, two scientists advanced a theory in an effort to explain why certain radio signals could be received over a great distance. They speculated that the upper portion of the atmosphere consisted of an electrified layer of gaseous particles that reflected the radio waves back to earth. This layer was called the ionosphere because it was assumed that atoms in that region were charged in some fashion. The ionosphere is also called, after the two scientists who discovered it, the Kennelly-Heaviside layer.

This region of our atmosphere is better understood today. We know, for instance, that the ionosphere is composed of four rarefied layers. They have been labeled *D, E, F,* and $F_2$ as illustrated in Fig. 8-34.

The rarefaction begins between 40 and 50 mi above the earth. It

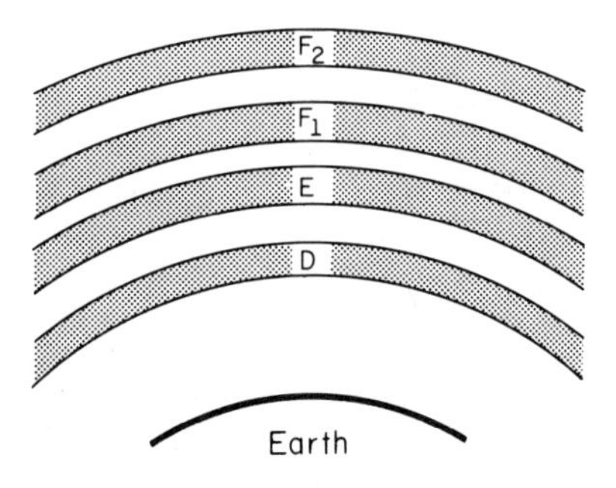


**Fig. 8-34.** Layers of the Ionosphere.

is marked by a greater concentration of ionized particles. The negative ions are believed to be atoms with a higher than normal energy level. This is attributed to solar radiation and ultraviolet bombardment. The rotation of the earth, our orbit around the sun, solar storms, time of day, and season of the year are some of the factors that affect the concentration of ions. The ionosphere is constantly changing as the various factors raise and lower the energy levels of the ions.

The varying density of ions at various heights produces the layer effect. The *D* layer lies roughly between 40 and 50 mi altitude. It is only lightly ionized and disappears at night. It has little effect on radio waves except for absorbing more energy than the lower atmosphere.

The *E* layer is much stronger and better defined. It begins at an altitude of about 50 mi and extends upward to 90 mi. The greatest concentration of ions appears at noon at an altitude of 70 mi. The *E* layer weakens at night but is still very much in evidence. This layer refracts electromagnetic energy. The degree of refraction is dependent upon the concentration of ions and the frequency of the wavefront. Frequencies up to 20 MHz may, at certain times, be refracted enough to return to the earth. Higher frequencies penetrate the *E* layer but lose energy doing so.

The *F* layer begins at an altitude of 90 mi and extends to the top of the atmosphere. At night, $F_1$ and $F_2$ combine into a single layer. During daylight hours, they separate into two distinct regions. Normally, $F_2$'s greatest intensity appears in the early afternoon, but this pattern has many variations.

In addition to the regular ionization patterns, stray patches appear in a random fashion at *E* layer heights. These are called sporadic *E* ionization. Their effect frequently enables signal reception over abnormally long distances.

### *refraction of sky waves*

The ionosphere generally acts as a conductor and absorbs energy from the electromagnetic waves. However, frequencies up to about 30 MHz are bent back toward the earth. The actual frequency, the ion density, and the angle of incidence determine the exact amount of refraction. This is not a reflection as light from a mirror. It is rather a bending effect as light waves passing through water. This effect is illustrated in Fig. 8-35.

The angle for refraction back to earth is different for different frequencies. The higher the frequency, the lower the angle with respect to the earth's surface. In Fig. 8-33, sky wave 1 leaves the transmitting site at an angle. It penetrates to the $F_2$ layer, refracts, and strikes the earth again at point *A.* From point *A,* it rebounds to the ionosphere and is refracted again to point *B.*

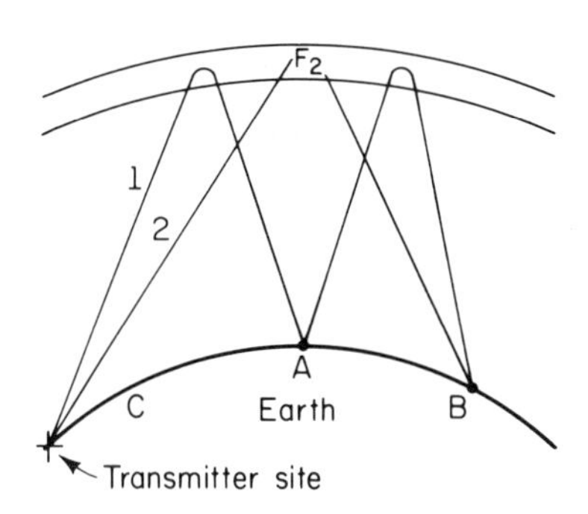


**Fig. 8-35.** Refraction of Sky Waves.

Waveshape 2 is radiated at a lower angle and is refracted at a larger angle. It covers the distance in one bounce. Signals radiated at an angle too close to vertical, or at a too high frequency, will penetrate the ionosphere. Some refraction always occurs, but in these cases, it is not enough to return the beam to the earth.

### *skip zones*

The distance along the surface of the earth between the maximum range of the ground waves and the return of the sky wave is a dead zone as far as the signal is concerned. It is beyond the reach of the ground waves but is too close to receive the sky waves. This is known as the skip zone. In Fig. 8-33, the skip zone for signal 1 begins at point *C* and extends to point *A.* The signal can be received again in the vicinity of point *A,* then it skips the distance to point *B.* The skip zone for signal 2 begins at point *C* and extends all the way to point *B.*

### *weather conditions*

Most of us are familiar with radio interference caused from local electrical storms. This is especially severe with amplitude modulation where the noise is demodulated along with the intelligence. There are many other effects of weather.

Frontal systems are especially noted for their manipulation of electromagnetic waves. These systems frequently create a temperature inversion which is a layer of warm, dry air above a layer of cool, moist air. The warm, dry air conducts waves at a higher velocity than the cool, moist air. Waves become trapped in these layers of air and may be bent upward never to return or curved downward to cover phenomenally long distances. This effect is illustrated in Fig. 8-36.

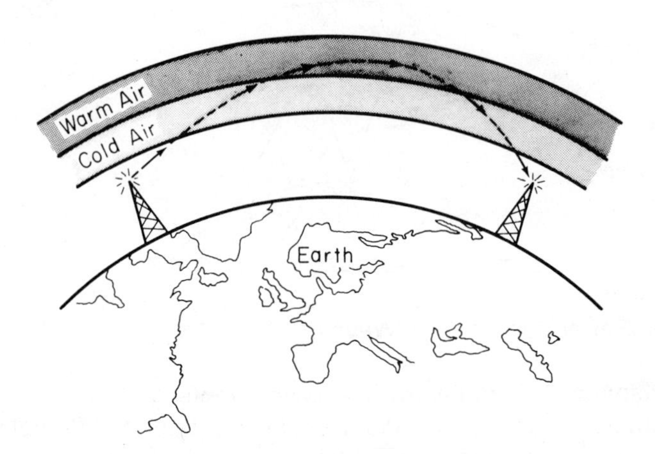


**Fig. 8-36.** An Effect of Temperature Inversion.

A portion of the wavefront reaches the warm air and increases its speed of travel. As it gradually pulls ahead of the slower moving portion, the entire front takes an eliptical path and eventually returns to earth. Transmission from low powered walkie talkies have been received over distances of 3000 mi because of this effect.

### *obstructions*

Electromagnetic waves prefer to travel in straight lines, but objects as well as conditions influence them to take detours.

Good conducting surfaces such as aircraft wings, water towers, and helicopter blades, are noted for causing radio and television interference. A part of the wavefront travels a straight line to the receiving antenna. Another part arrives at the antenna after reflection from some such object. The difference in arrival time throws the signals out of phase. This causes ghosts on television and garbled outputs from radios. This also causes signals to fade in and out.

Electromagnetic waves travel along the surface of the earth and follow the gradual contours fairly well. Sharp obstruction, such as mountains and tall buildings, cause a shadow effect. Antennas in such shadow areas are shielded from the waves.

Another effect similar to reflected signals is caused from sky waves. The constant changing of the ionosphere may cause the refracted sky wave to move about at its point of contact with the earth. This causes fading of signals as the antenna is first in then out of the skip zone. Two sky waves refracted at a slightly different angle will arrive out of phase and cause the same effect.

## CHAPTER 8 REVIEW EXERCISES

1. An open-wire line is constructed for 100 MHz. What is the maximum spacing in in.? (1 m = 39.4 in.)
2. A certain copper coaxial line has the following dimensions:
   (a) The inside diameter of the outer conductor is 0.6 in.
   (b) The outside diameter of the inner conductor is 0.12 in.
   What is the resistance of 10 ft of this line to a frequency of:
   (a) 300 MHz?
   (b) 3000 MHz?
3. Name five types of RF transmission lines?
4. Which two lines suffer the greatest radiation loss?
5. Which two types of wire lines can be laid directly on metal without increasing radiation loss?
6. An infinitely long line has a capacitance of 30 pF and an inductance of 0.1 $\mu$H per section. What is the characteristic impedance?
7. What must be done to a practical line to make it react as an infinitely long line?
8. Label the drawing in Fig. 8-37 to indicate which lines are *E* and which are *H,* and draw arrows on the lines to show their direction.
9. What is the wavelength of a 3000-MHz signal in free space?
10. A transmission line has a propagation constant of 96 percent. How many ft/$\mu$s will RF energy travel in this line?
11. The line in item 10 is cut to two wavelengths of 30 MHz. What is the length of the line in ft?

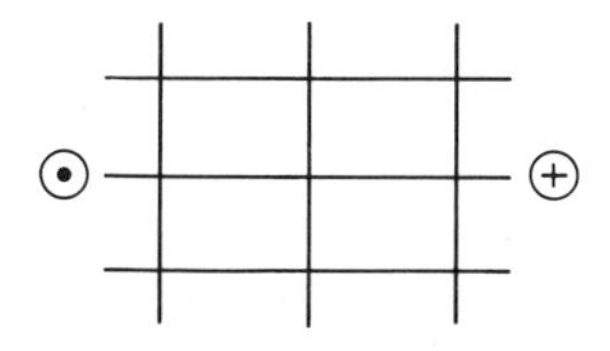

**Fig. 8-37.** End View of a Transmission Line.

12. What constitutes a resonant line?
13. What is a standing wave? How is it formed?
14. A line is resonant to 20 MHz. It is three wavelengths long and terminates in an open.
    (a) How many direct shorts are reflected along the line?
    (b) These shorts are separated by how many in.?
    (c) How many inches separate the open end from the first reflected short?
    (d) What is the total length of the line in in.?
15. In terms of odd and even quarter wavelengths from an open, where are the points of minimum and maximum impedance located?
16. A resonant line is terminated in a short. Describe the values *Z, E,* and *I* at points removed from the short by an:
    (a) Odd number of quarter wavelengths.
    (b) Even number of quarter wavelengths.
17. Draw a diagram to illustrate how a quarter wave short can be connected to eliminate even harmonics from a transmission line.
18. Draw a diagram to illustrate how a quarter wave open may be used to reject a band of frequencies from a transmission line.
19. In item 18, the filter section is 0.985 in. long. What is the center frequency of the rejected band?
20. When a 30-MHz signal is carried on a copper transmission line, how deep will the energy penetrate the line?
21. A dipole with 0.5 $\mu$F of capacitance has 500 V applied. What is the power in the electric field?
22. A yagi antenna consisting of a folded dipole, one director, and one reflector is designed for a frequency of 100 MHz.
    (a) How long is the dipole in in.?
    (b) How long is the reflector and where is it placed in respect to the dipole (in in.)?
    (c) How long is the director and where is it placed in respect to the dipole (in in.)?
23. An antenna is mounted on a 400-ft tower. What is the line of sight distance to the horizon in miles?
24. Differentiate between the reflection of light from a mirror with the refraction of electromagnetic energy from the ionosphere.

# 9

# devices with special applications

The devices discussed here are neither new nor rare. In fact, most of them are rather common and have been in use for a number of years. They find a place in this chapter because of their unique design or special application.

## SOLID-STATE RESISTORS

These resistors are based on the eccentricities of semiconductor material. In the construction of some components, the variable resistivity of a material can be bothersome. This same sensitivity to changes in temperature has been turned to an asset in solid-state resistors. Three of these that are in common use are thermistor, photoresistor, and varistor.

### *thermistor*

The thermistor is a solid-state resistor, and its resistance is dependent upon the temperature. It is based on the semiconductor's large negative temperature coefficient. It is composed of two pieces of wire imbedded in a bead of solid-state material. This is illustrated in Fig. 9-1.

The resistivity of semiconductors is much more sensitive than metals to a change in temperatures. Thermistors are used to detect thermal changes as low as 0.0005 °C. This is a sensitivity about ten times as great as the sensitivity of a thermocouple.

The thermistor's resistance decreases as the temperature increases. In other words, the resistance varies inversely with temperature. A thermistor can be used to compensate for changes in components that have a positive temperature coefficient. In this manner, circuits can be constructed that are completely immune to temperature over a change of 100 °C.

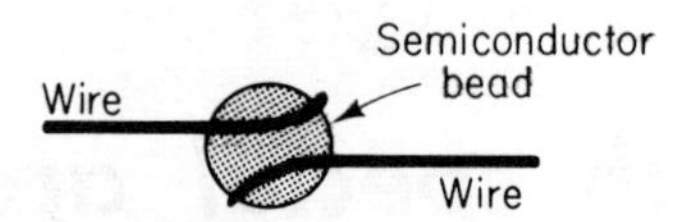


**Fig. 9-1.** Thermistor Construction.

High speed thermistors are designed to store a certain quantity of heat and dissipate this heat rapidly. The heat storing capacity of the body and the rate of dissipation determine its response speed.

Some thermistors are designed with an indirect heating element. This has two applications. It can be used to maintain a constant resistance despite temperature changes. It can also be used in measuring devices. For instance, it could be used in a power meter for radio frequencies. The RF could be applied to the heater coil and the heating effect could be measured. The indirectly heated thermistor is illustrated in Fig. 9-2.

### *photoresistor*

The photoresistor is a light sensitive resistor, and the resistance is inversely proportional to the incidental light. It is generally composed of two metal contacts on a semiconductor body. The semiconductor is a photoconductive device; its conductivity changes as the light intensity varies. A drawing of a photoresistor is shown in Fig. 9-3.

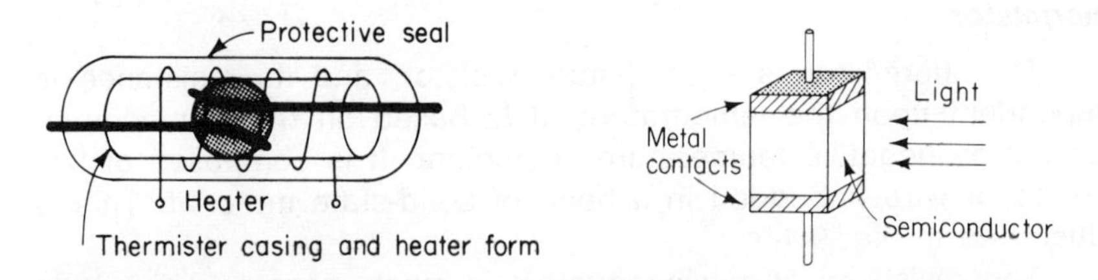


**Fig. 9-2.** Indirectly Heated Thermistor.

**Fig. 9.3.** Photoresistor.

Photoresistors are manufactured in a variety of sizes and shapes. They also have a wide range of characteristics which makes them adaptable to many uses. Wherever light is involved in measuring, sensing, or control, the photoresistor is likely to be used.

### *varistor*

The varistor is constructed from a non-ohmic material. This material is thyrite or silicon carbide. These materials have maximum resistance when there is a steady voltage applied. Within specified operating limits, the resistance drops off rapidly as voltage is either increased or decreased. The semiconductor material is molded into the desired shape with leads similar to the resistor. A voltage–resistance response curve is shown in Fig. 9-4.

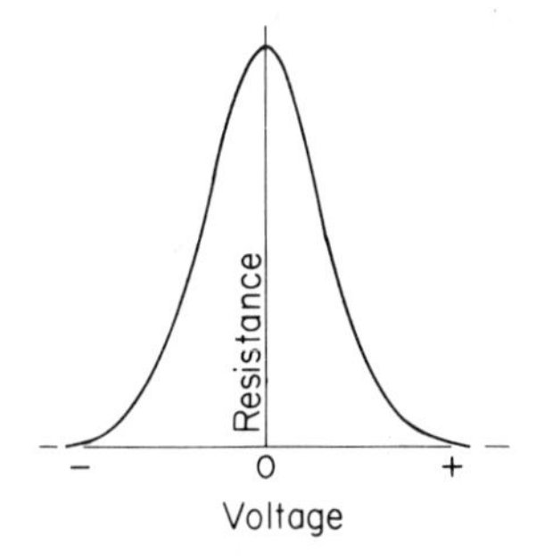


**Fig. 9-4.** Varistor Response Curve.

The first application of a varistor was in lightning arrestor units. It is presently in wide use in power supplies. Its characteristics reactance to voltage changes enables it to protect the equipment from voltage surges. A surge of voltage causes a drop in resistance which effectively eliminates the surge. When the surge has passed, it returns to the normal high resistance.

Some semiconductor diodes also have special characteristics.

## UNIQUE DIODES

These diodes are designed around the characteristics which ordinarily have an adverse effect on semiconductor operations. We will limit this discussion to *p–n* junction diodes, and it will involve four of these. They are: variable capacitance diode, symmetrical Zener diode, tunnel diode, and photodiode.

### *variable capacitance diode*

Even though this device is a semiconductor diode, its application is that of a voltage sensitive capacitor. The purpose of such a capacitor is to automatically control frequency in response to a change in bias. Of course, this is only one of many applications. Any *p–n* junction diode reacts in this manner to some extent, but in the variable capacitance diode the characteristic is enhanced. These diodes are fabricated with a high capacitance and high sensitivity to voltage. Variable capacitance diodes are marketed under such trade names as varicaps, semicaps, and varactors.

The *p*-type material forms one plate of the capacitor, *n*-type material forms the other plate, and the barrier region acts as a dielectric. When reverse bias first blocks conduction of the diode, it exhibits maximum capacitance. From this point, capacitance decreases as reverse bias increases. The limits are set by the breakdown characteristics of the diode.

### *symmetrical zener diode*

The characteristics, and some applications, of the Zener diode were discussed in an earlier chapter. The symmetrical Zener is equivalent to two ordinary Zeners placed back to back. A symbol and characteristic curve of such a diode are shown in Fig. 9-5.

The theory of breakdowns in solids was advanced by a scientist by the name of Zener. It was based on a quantum-mechanical process which explained that carriers can tunnel through a material

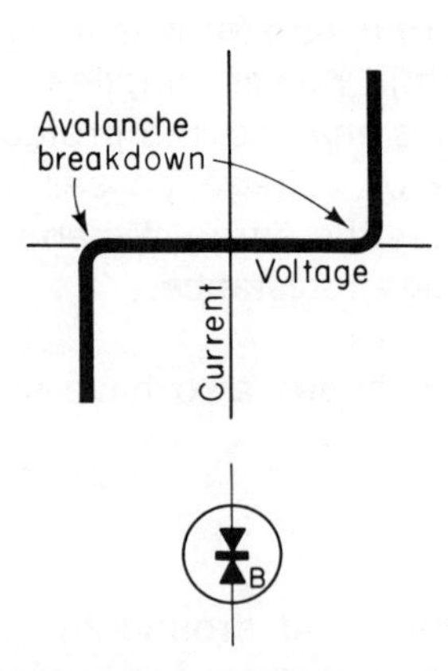


**Fig. 9-5.** Symmetrical Zener Diode (Symbol and Characteristic Curve).

even though they lack sufficient energy to surmount the barrier. The Zener diode bears his name. The Zener effect was further explained and expanded upon at a later date, and in some cases, it is called avalanche breakdown.

The symmetrical Zener has no forward side; voltage in either direction constitutes reverse bias. As a result, it can have avalanche breakdown in either direction.

Another diode of a more recent design also operates on the Zener principle. It is the tunnel diode.

### *tunnel diode*

The tunnel diode was invented in 1958 by Dr. Leo Esaki and is sometimes called the Esaki diode. Even though the electrons effectively tunnel through the barrier region in the fashion of a Zener diode, the characteristics are definitely different. Figure 9-6 shows a circuit symbol for the tunnel diode.

Here we have a definite diode action (at least in one direction), but this action is far from conventional.

The barrier is so narrow, and the barrier potential so high, that the tunneling may be in effect even with zero bias. When voltage is applied (forward direction only) the current rapidly rises to a peak value. If bias continues to increase, the current begins to drop. This is a negative resistance region, and it is linear over a relatively wide change in voltage. As voltage continues to increase, the negative resistance range will pass, and current will start to rise again. Now, we have reached another positive resistance zone, and current will rise linearly with the voltage. These actions are illustrated in Fig. 9-7.

The tunnel diode is both rugged and versatile. It has a high

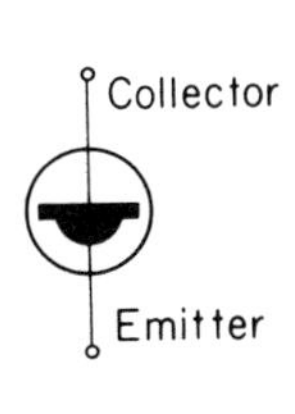


**Fig. 9-6.** Tunnel Diode.

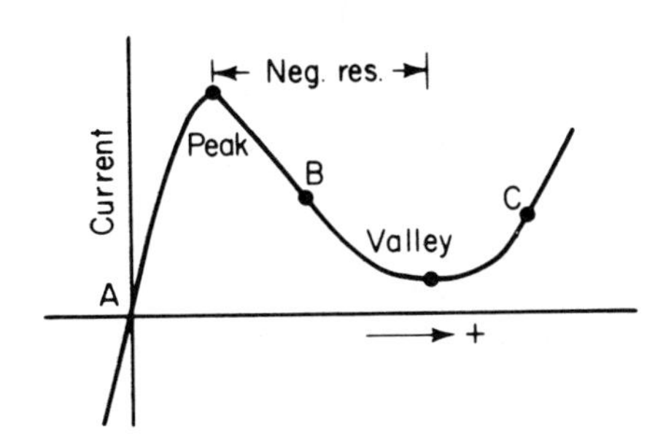


**Fig. 9-7.** Characteristics of the Tunnel Diode.

immunity to external contamination and temperature. This diode may be used as a high speed switch, an amplifier, or an oscillator. It has frequency capabilities in excess of 100 MHz, can handle currents in excess of 5 A, and can switch about 100 times as fast as a transistor.

There are several other schematic symbols in use for the tunnel diode. Some of these are shown in Fig. 9-8.

When a tunnel diode is properly biased, it can be used with a tank circuit to form a free running oscillator. It would need to be biased at the center of the negative resistance portion of its characteristic curve as shown in Fig. 9-9.

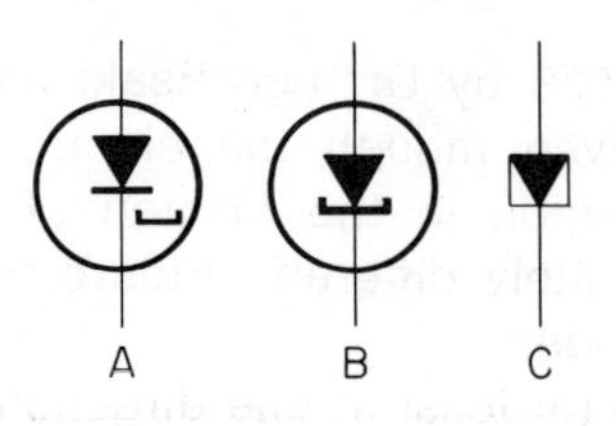


**Fig. 9-8.** Other Symbols for Tunnel Diode.

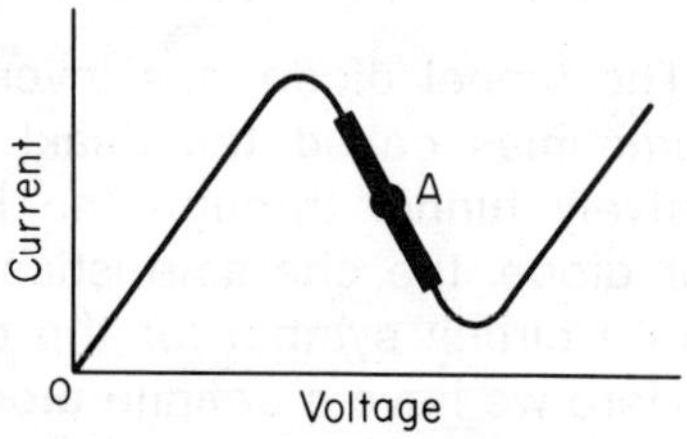

Current
A
0
Voltage

**Fig. 9-9.** Operating Point for Oscillator.

The object of such an arrangement would be the production of sinusoidal waves. By setting the operating point in the center, point *A* on the diagram, a reasonable portion of the linear region is allowed in both directions. The oscillator circuit is illustrated in Fig. 9-10.

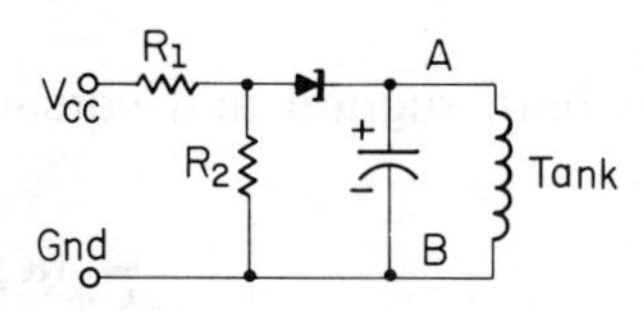


**Fig. 9-10.** Tunnel Diode Oscillator.

The operating point is determined by the voltage divider which is composed of $R_1$ and $R_2$. The positive potential on the collector is the same as the voltage drop across $R_2$. The relative size of these resistors need to be carefully selected.

When power is applied, a surge of current passes from the negative input through the tank capacitor, through the diode, and through $R_1$ to $V_{CC}$. This charges the capacitor in the direction indicated. Thereafter, the conduction of the diode will be controlled by the difference between two values: the voltage across $R_2$ and the voltage across the tank circuit. The fixed bias (voltage $R_2$) sets the operating point, and the tank oscillations swing the operation up and down from this point. This is indicated by the heavy line in Fig. 9-9.

The tank now oscillates at its resonant frequency, and the diode conducts just enough to replace the circuit losses. The diode is a switch which allows an injection of current for a portion of each oscillation. The injection will be equivalent to the energy used. This action sustains oscillations with no unnecessary use of energy.

During the positive alternations (point $A$ positive) forward bias is reduced, and diode current increases. This recharges the tank to peak voltage. On the negative alternations, bias is reduced and diode current decreases. The frequency is determined solely by the size of the components in the tank circuit. Frequencies on the order of 10 THz can be attained.

### *silicon rectifier*

This little diode has come a long way in a short time. Fused junction construction and heat sink connections enable it to replace other types of rectifiers in nearly all applications. Before the silicon rectifier the use of solid-state diodes was impractical in circuits with high inverse voltage. The leakage (inverse current) either became to heavy to tolerate or the diode destroyed itself by avalanche breakdown.

The silicon rectifier is especially designed to withstand very high inverse voltages with no appreciable inverse current. They are featured with ratings such as: peak inverse voltage 15,000 V, leakage current 10 $\mu$A, and temperature $-55$ to $+175$ °C. They can be obtained in single diodes as well as bridge rectifier packages, and they are built for both single- and polyphase power supplies. They come as plug-in models as well as with solder lead wires. Two models are shown in Fig. 9-11.

The plug-in model sometimes features a heat sink bolt as in

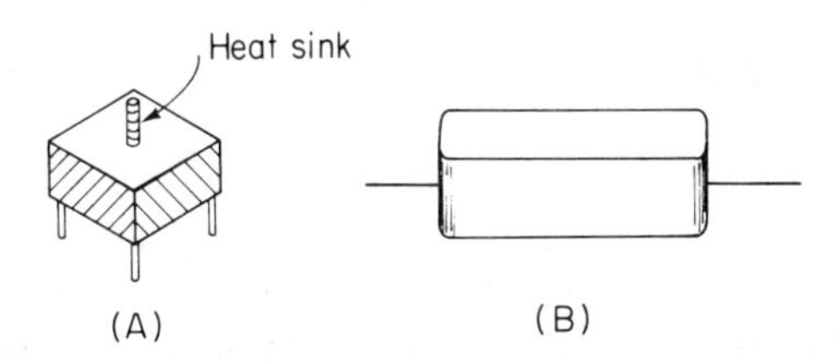


**Fig. 9-11.** Silicon Rectifiers.

Part A. This can be bolted to a metal chassis for more efficient heat dispersal. This model also features solid silver plug in leads.

Another weakness of most rectifiers is their limited ability to handle heavy forward currents. Silicon rectifiers can be obtained with almost any desirable current rating.

The silicon rectifier is a very practical device. It is replacing both electron tubes and metallic rectifiers in modern power supplies.

## *silicon controlled rectifier*

The silicon controlled rectifier (SCR) is often confused with the silicon rectifier (SR) just described. Actually the only relations are the fact that both use silicon and both can perform rectification. The SR is strictly a diode rectifier with the principal use being in power supplies. The SCR is a solid-state, four-layer switch. It belongs to the family of *npnp* or *pnpn* transistors. It may be used as a switch or as an amplifier as well as a rectifier.

Some manufacturers make SCRs that are marketed as silicon controlled switches (SCS), gate turn off switches (GTO), light activated silicon controlled rectifiers (LASCR), and Shockley diodes. The data handbooks more often group them all under the heading of SCR. The structure and a schematic symbol are illustrated in Fig. 9-12.

In practice, the SCR functions as two transistors; one is *npn,* and the other is *pnp.* The equivalent circuit is shown in Fig. 9-13.

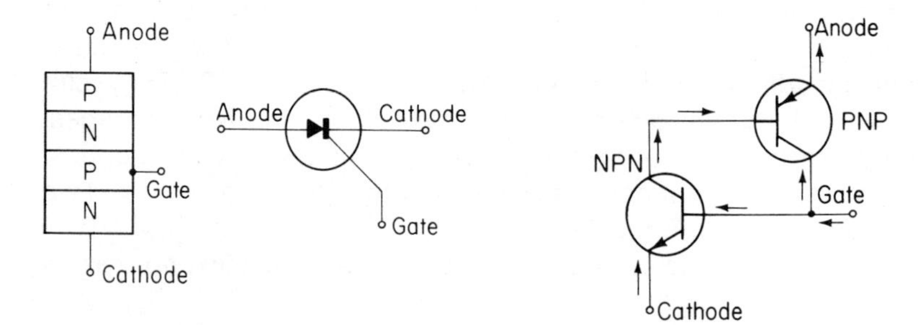


**Fig. 9-12.** SCR Structure and Symbol.

**Fig. 9-13.** Equivalent Circuit of SCR.

The SCR represented here falls roughly into the category of the GTO. With no input at the gate, the anode current is only slightly more than transistor leakage current. This is the off condition of the switch. It can be turned on by applying a pulse of current to the gate. This causes an increase in current. The regenerative feedback loop rapidly drives both transistors to saturation. The switch is now on, and the gate has no control. The current is limited only by the external circuit. Reducing the anode voltage will reduce the current. The switch will automatically cut itself off when the anode current is reduced to a certain critical level. This critical level is commonly referred to as the holding current.

SCRs have many practical applications. The light sensitive ones are used for light dimmers, automatic on–off light control, and in any area requiring light activated switches. Other SCRs are found as amplifiers, time delays, overload protective circuits, light flashers, and many other switching operations.

### *photodiode*

The photodiode has its *p–n* junction exposed to light. The intensity of the light controls the barrier resistance, and in turn, the amount of current.

Figure 9-14 illustrates the structure of the photodiode.

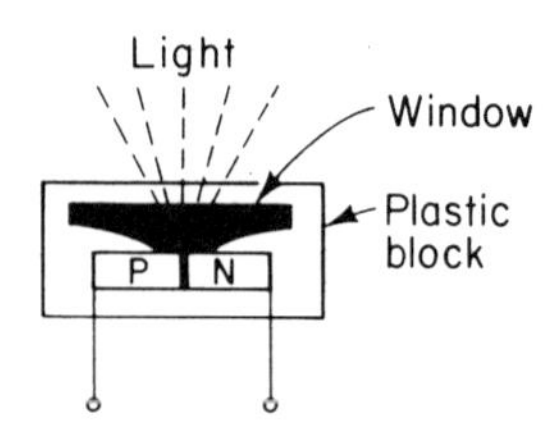


**Fig. 9-14.** Structure of Photodiode.

This device has a window which focuses light on the *p–n* junction. When operated in the dark, the reverse bias causes only a very small current; 1 to 3 $\mu$A. When exposed to light, the same bias produces a much larger current. The photodiode is very sensitive to light. A small change in light intensity produces a very noticeable change in reverse current.

As the technology of electronics advanced over the years several transistors have been set apart by some special characteristic.

## EXTRAORDINARY TRANSISTORS

Some of the factors which limit the operation of most transistors have inspired designers to exploit bothersome characteristics.

These off beat designs have extended the usefulness of transistors for applications in power, high-frequency, switching, and light sensing operations.

### *power transistor*

Most ordinary transistors are severely limited in output power capabilities. These outputs are generally less than 50 mW. The very characteristics of a transistor which make it so attractive to a designer, are the power limiting characteristics. The small size means small elements. The compact, one piece structure means poor heat dissipation.

When a transistor is designed to output more than 50 mW it falls into the power class. Transistors in this class are generally larger and have special methods of dissipating heat. It is estimated that the inside temperature is on the order of 3 to 5 °C higher than the outside temperature for each watt of dissipated power. This means that with an output of 15 W the heat differential is on the order of 45 to 60 °C.

The power transistor must be able to withstand this heat without changing its characteristics. This means larger elements, more volume of semiconductor material, metal casing, and special heat dissipation features. Some are mounted on the metal chassis with metal studs. This provides a heat transfer and more cooling surface. Others accomplish the same thing with detachable cooling fins which can be attached to the metal case. Forced air cooling is often used, and some types use liquid cooling. The liquid may be inside the transistor, or the transistor may be encased in a con-

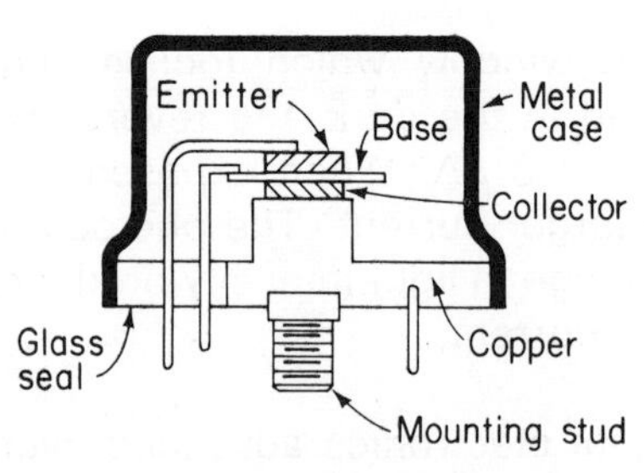


**Fig. 9-15.** Structure of a Power Transistor.

tainer of liquid. Figure 9-15 illustrates the construction of one type of power transistor.

Power transistors have been fabricated that can handle very large collector currents and voltages. These transistors can perform satisfactorily with junction temperatures up to 110 °C. Transistors with power outputs in excess of 150 W have been in use for some time, and there appears to be no reasonable upward limit. All this can still be packaged in a case the size of a dime and a 0.5 in. thick.

### *intrinsic transistor*

One problem in transistor structure, especially at higher frequencies, is the interelement capacitance. Since $X_C$ decreases as frequency increases, there comes a point when even a small capacitance will result in a direct short. The intrinsic transistor is designed to reduce the collector to base capacitance to an absolute minimum.

Since capacitance is inversely proportional to the square of the distance between the plates, moving the elements farther apart might solve the problem. If this is done, care must be exercised not to increase the transit time because high frequencies also require quick transit between elements.

The problem then is to minimize the capacitance without changing the thickness of the base. One means of accomplishing this is to place a layer of pure semiconductor material between the base and the collector. The pure semiconductor is called intrinsic material, and a transistor so constructed is an intrinsic transistor. The structure is illustrated in Fig. 9-16.

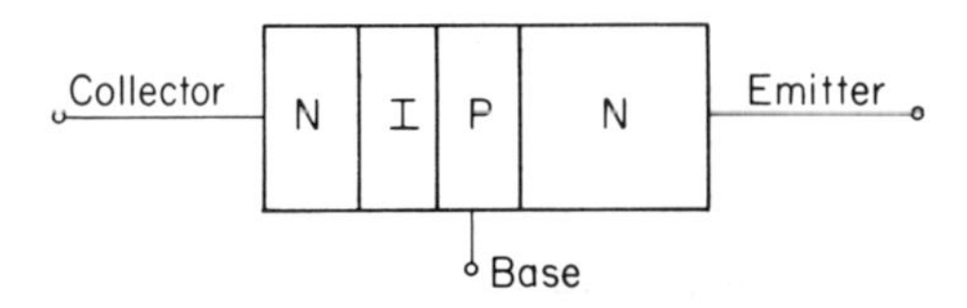


**Fig. 9-16.** Intrinsic Transistor.

This combination is an *npin,* but several others are in use. It could just as well be *pnip, pnin,* or *npip.*

The layer of *I*-type material increases the physical distance between the plates of the capacitor (collector and base). The capacitance is reduced to such an extent that this transistor can handle frequencies up to 100 MHz. The intrinsic material offers no problem except low conductivity. This is compensated for by using a higher collector voltage.

Another transistor solves the same problem in a different manner.

### *tetrode transistor*

Figure 9-17 illustrates a schematic symbol and the structure of the tetrode transistor.

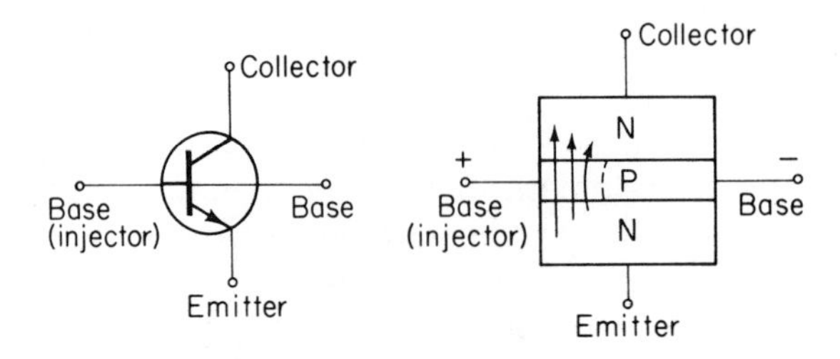


**Fig. 9-17.** Tetrode Transistor.

In this case, the problem of collector to base capacitance has been attacked from the standpoint of plate size. A physical reduction of base size would decrease the carriers and increase the input resistance. So, the physical size was kept the same, and the electric size was reduced.

Two leads were attached to the base of a triode. A potential only is placed on one base lead. This potential will be such that it constitutes cut off reverse bias. Since we have used an *npn* in the drawing, this base lead is negative with respect to the emitter.

The other base lead has forward bias, and also receives the input signal. More than 50 percent of the junction is biased to cut-off. This confines the electron flow to a small area near the injector base lead. Electrically, the contact area has been decreased. So, the size of the capacitor plates has been reduced. The result: smaller plates, less capacitance, higher $X_C$, and much higher frequency handling capability.

Other efforts to solve the collector to base capacitance resulted in the development of the field effect transistor.

### *field effect transistor*

If the collector to base junction could be eliminated, the collector to base capacitance would also be eliminated. "But," you say, "that would also eliminate the transistor." Not necessarily. This is the method of fabricating the field effect transistor.

This transistor is constructed of two pieces of material. One is *n*-type; the other is *p*-type. These materials are arranged in such a manner that one completely surrounds the other as shown in Fig. 9-18.

The doughnut around the single piece of *n*-type material forms one continuous junction. In fact, another name for this transistor

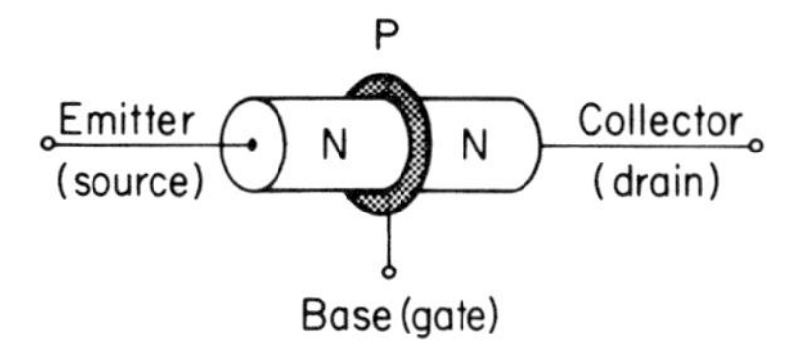


**Fig. 9-18.** Structure of the Field Effect Transistor.

is unipolar. To emphasize its uniqueness of design, other names are sometimes substituted for our standards of emitter, base, and collector. The emitter is called source, the base is called gate, and the collector is called drain.

The action in this transistor is illustrated in Fig. 9-19.

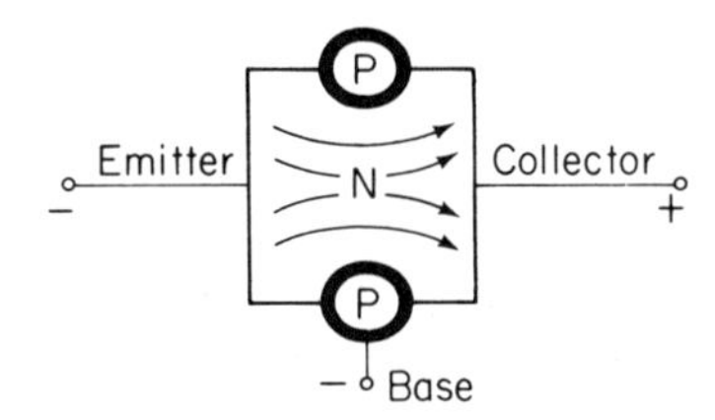


**Fig. 9-19.** Current Through the Field Effect Transistor.

The emitter, which is one end of the *n*-type material, is connected to a negative voltage or to ground. The collector, the other end of the same piece of material, is connected to a positive potential. Electrons pass through the *n*-type material from emitter to collector.

The base is connected to a negative potential (more negative than the emitter). This forms a strong electrostatic field around the center of the *n*-type material. The constriction of this field has a very definite effect on the flow of electrons from emitter to collector.

The input signal is applied to the base. The signal variations raise and lower the negative voltage level. This action causes variations in the field strength, and regulates the movement of electrons between emitter and collector. The collector current will have the same variations as the input signal.

What about the capacitance? Electrons have no path from the base to the collector. The base-collector capacitor has been eliminated. The signal coupling is through a modulated, electrostatic field.

What have we done for the frequency? The field effect transistor has handled frequencies in excess of 500 MHz. This is not recognized as a limit. That frequency can probably be doubled or tripled with improved design.

One of the problems with transistors has been switching speeds, especially where automatic switching action is needed. Let's see how this problem was solved (or at least reduced).

### *four-layer transistor*

The avalanche breakdown of the Zener diode is a fast switching action. The trouble here is getting the switch opened again after it has been closed. Many situations demand a two-way switch. The four-layer transistor is a two-way switch. The construction is illustrated in Fig. 9-20.

This is a three-junction transistor, and it is sometimes called a thyristor. With potentials applied as indicated on the diagram, junctions 1 and 3 are forward biased, but junction 2 is reverse biased. With zero potential, or a small negative, on the base, current from emitter to collector is practically nonexistent. When a positive signal is applied to the base, it causes an almost instantaneous avalanche breakdown of junction 2.

Once the avalanche has been triggered, the signal on the base has no further significance. It has closed the switch, and its job is done. The internal resistance of the transistor is now less than 1 Ω, and it is carrying a high current. The only way to stop this current is to reduce the collector voltage to the point that the avalanche cannot be sustained.

Any dropping device in the collector circuit will reduce the collector voltage. An *RC* network may be used to control the time of the collector voltage drop. Once the collector potential has decreased to a critical level, the avalanche stops, and the resulting negative bias seals off junction 2. This opens the switch, and brings the current to an abrupt stop. Nothing more will happen until the switch is closed again by applying another positive signal to the base.

The signal on the base turns the switch on, the drop in potential on the collector turns it off. So, we have our two way, automatic switch. It can be switched on in about 0.1 μs and off in a slightly longer time. The switching action is somewhat slower when the transistor is designed for heavy currents. This is no problem since heavy current circuits generally require less speed. Some of these transistors are used in switching circuits where load current exceeds 20 A.

Two symbols for the four-layer transistor are illustrated in Fig. 9-21.

Another approach to the problem of fast switching is temperature reduction.

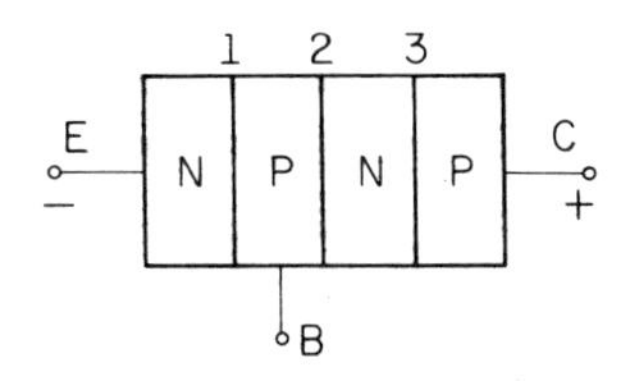


**Fig. 9-20.** Structure of the Four-Layer Transistor.

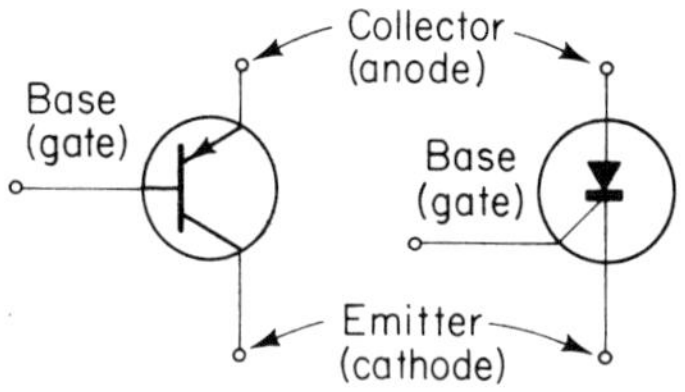

**Fig. 9-21.** Symbols for the Four-Layer Transistor.

## CRYOSARS

This is a coined word. Cryo means frost or icy cold, and it is taken from the Greek, "kryo." The meaning of the other letters are: *s* = switching, *a* = avalanche, and *r* = recombination. The remaining s is simply to make the word plural. Here it applies to almost microscopic semiconductor switches.

The science of cryogenics deals with producing and using very low temperatures. Cryogenic temperatures range from the boiling point of liquid oxygen (—147 °C) to absolute zero (—273 °C or —460 °F). At these temperatures, some strange and wonderful things happen in electronics. Ordinary lead becomes a super conductor with practically zero resistance. Oscillations in an average tank circuit can continue for years.

Another interesting development in this super cold is the cryosar. This is a tiny, semiconductor switch which finds its best temperature to be —269 °C; just 3 °C above absolute zero. (This is the boiling point of liquid helium).

A semiconductor normally needs some thermal agitation to generate carriers. So, at the temperature referred to here, a semiconductor is a very high resistance. We could go so far as to say that the resistance approaches infinity. In the nonconducting state, the cryosar is an open switch with infinite resistance between the two poles.

When it is triggered into conduction, it reaches the avalanche state in a few billionths of a second. The triggering voltage may be as low as 1 V. This is the closed switch, and it is capable of handling high currents. The closed switch has no appreciable energy loss because the resistance is now near zero.

To open the switch again, the activating voltage is removed. This allows almost instant recombination which stops the current with an infinite resistance. A cubic inch of space may house several thousand of these switches.

Another small but useful device is the solar cell.

## SOLAR CELL

One of the problems that has been present since the advent of electronics has been ready sources of power. In aircraft and space capsules, size and weight of the power supply have been the primary sources of concern. Direct conversion of sunlight to electromotive force has always been an interesting prospect. It is estimated that, under favorable conditions, the sunlight delivers about 100 W of power to each square foot of earth surface. That is something on the order of 100 mW/cm$^2$. This is a lot of power, but how can it be harnessed?

One successful device in this endeavor is the solar cell. It is a semiconductor device, circular in shape, about 0.75 in. in diameter, and 0.25 in. thick. Each cell produces about 0.5 V, and they may be connected in series, parallel, and series–parallel combinations. Tiros I, the United States' first weather satellite, carried 9260 solar cells. The Telstar satellite which was launched July 10, 1962, used 3600 solar cells to power its transistorized circuits.

One of the most efficient solar cells is constructed of *p* and *n* silicon. The basic structure is illustrated in Fig. 9-22.

The incoming light striking the *p*-type silicon sets up a forward bias across the *p–n* junction. This difference in potential is a usable dc voltage which can be coupled directly from the metal substrate and collector ring. The cell functions effectively in light that is only $\frac{1}{10{,}000}$ as strong as direct sunlight. Of course, the developed voltage is directly proportional to the light intensity when the light is weak.

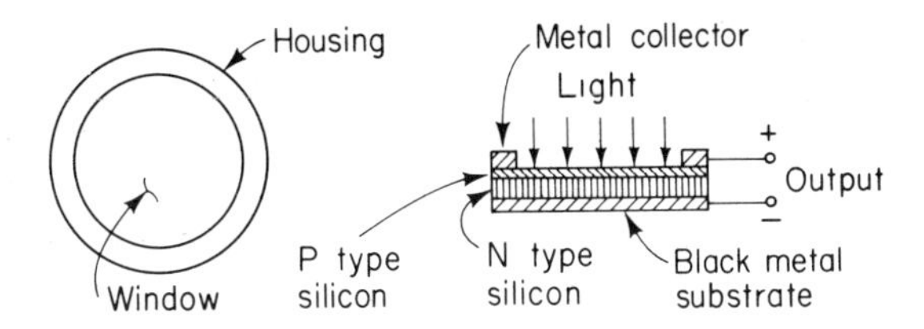


**Fig. 9-22.** Shape and Structure of a Solar Cell.

After a certain critical point is reached in the light intensity, the output holds constant.

One of the most promising aspects of the solar cell is its life expectancy. Barring physical damage, this is several thousand years.

The area of amplification has also produced some equipment worthy of note.

## AMPLIFIERS OF DISTINCTION

One of the driving forces behind miniaturization of electronic components is the desire to operate at an ever higher frequency. This has jumped in multiples of 10; 100 MHz, 1 GHz, 10 GHz, 100 GHz, 1 THz, and 10 THz. This entire range was covered in less than 20 years. Building amplifiers to keep up with the increasing frequency was, in itself, an engineering feat.

Each new frequency jump required new concepts of design and created new problems to be solved. Around the 30 GHz range two problems appeared so substantial that all progress seemed to be blocked. These were: noise and transit time. The maser was developed in the struggle to overcome these problems.

### *maser*

The word maser is an acronym. It was coined by taking the first letters from the key words of "microwave amplification by stimulated emission of radiation." It is based on the principle that all electromagnetic radiation appears in discrete quanta. The first maser was developed at Columbia University in 1957.

This quantum mechanical device eliminates the need for electron flow. Therefore, it bypasses the problems of noise generation and transit time.

First, radiation is stimulated by upsetting the normal energy level of a group of atoms. The natural atom has energy bands consisting of rotating electrons. Each electron exists in a particular orbit which is determined by the energy of that electron. The energy level of the atom is a sum total of the energy levels of all the orbiting electrons. An atom can be forced to a higher energy level by placing it in a strong magnetic field and bombarding it with waves of the proper frequency. This process is known as pumping or population inversion.

The energy level of the group of atoms will stabilize at a point somewhat higher than the normal energy level. An external wave may be used to trigger the atoms to fall back to their customary level. This triggering wave needs to be of a certain frequency and energy level.

The incoming signal releases the atoms from the high energy level. When so released, they spontaneously release energy in order to drop back to their natural state. Thus, radiated emission has been stimulated.

The amplification process is quite simple in concept. The radiated energy from the atoms will be in phase with the triggering wave. The triggering wave is, therefore, amplified. The pumping frequency keeps moving the atoms to a higher energy level, and the triggering frequency keeps releasing them again. Not only has a high amplification been accomplished, but two other important factors have resulted. There has been practically no generation or amplification of noise, and the transit time is a thing of the past. The electrons stay with their atoms and the only movement has been the transfer of energy.

An early model maser used ammonia gas as the radiating material, but the maser quickly advanced to the solid-state stage. The ruby crystal then replaced the gas. Also, the early masers used only two energy levels. This limited the pumping and amplification to two separate timing operations. This meant that amplification was limited to pulses of energy. With the advent of the three energy level maser, it became possible to amplify a continuous wave. Figure 9-23 is a sketch of the functional parts of a solid-state maser.

This drawing is oversimplified but should illustrate the basic principles. The resonators are actually hollow cylinders, but for the frequencies concerned, they function as tuned tank circuits with the associated high amplification. Input and output signals appear to be traveling on ordinary wires, but this is not the case. At these frequencies, waves are propagated through waveguides.

The dc potential provides energy for the coil which in turn, saturates the pump frequency resonator with a strong magnetic field. The pump signal sets up oscillations in the resonator which raises the energy level of the atoms in the ruby rod. The triggering signal is applied to the end of the ruby rod, travels inside the rod, through the pump frequency resonator, and passes into the output resonator. As it travels through the rod, it keeps growing in amplitude as it gains energy from the excited ruby atoms. The output resonator is tuned to the frequency of the triggering signal. It adds a resonant boost to the already greatly amplified signal before passing it on as an amplified output.

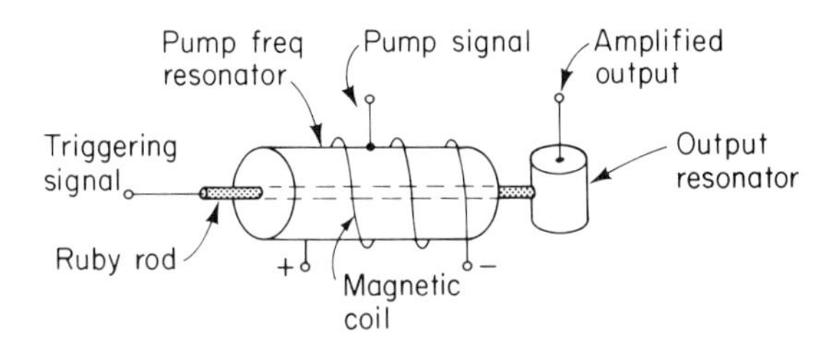


**Fig. 9-23.** Ruby Rod Maser.

Masers solved many problems in microwave amplification but created some of their own. The greatest of these was temperature control. To prevent atoms from dropping back to the lowest energy level after they were primed, a very low temperature was required. The early models were operated at cryogenic temperatures. As the technology advanced, the masers were made to operate at normal room temperature. At this time, they left the laboratory and took their place with other practical electronic devices.

Attempts to impove the maser led to an even greater discovery: the laser.

### *laser*

There is no sharp dividing line between lasers and masers. In fact, the two words are frequently used interchangeably. The principles are the same with the chief difference being the frequencies concerned. The term optical maser has been applied to the laser. It would not be far wrong to think of the laser as simply a high-frequency maser. The acronym laser, was coined from "light amplification by stimulated emission of radiation."

The first laser was developed at the Hughes Aircraft Company. The laser material was the crystal ruby. The ruby is now stimulated by light, amplifies light waves, and has an output of visible light. Many devices have been built over the years which produce visible light. So, what is the difference? Most light sources produce many light frequencies, and these frequencies are constantly changing. This is incoherent light. The laser amplifies a single light frequency (or a very narrow band), and thus, became the first source of coherent light.

What does this have to do with electronics? The only difference between a light wave and a radio wave is a matter of frequency. Both are electromagnetic waves, and both can be made to carry intelligence. The intelligence is impressed on the waves by some type of modulation.

The chart in Fig. 9-24 will help to visualize the relation of the areas in the frequency spectrum. The chart is in terms of terrahertz (THz), and 1 THz is $1 \times 10^{12}$ Hz.

$3 \times 10^{-6}$ THz

| | |
|---|---|
| $3 \times 10^{-4}$ THz | Radio Waves |
| 3 THz | Micro-Waves |
| 300 THz | Infra Red Rays |
| $3 \times 10^{3}$ THz | Visible Light Rays |
| $3 \times 10^{18}$ THz | Ultra Violet, X and Gamma Rays |
| $3 \times 10^{20}$ THz | Cosmic Rays |

**Fig. 9-24.** Frequency Spectrum.

Work in the lower microwave ranges led to the development of the maser. The laser bridged the gap from microwave through visible light rays. Notice that the term changes from wave to ray when the light frequencies are encountered. There appears to be no top limit to frequencies that can be handled by lasers and other similar devices.

Remeber that we are speaking of coherent light, not a jumble of all light frequencies mixed together. Figure 9-25 shows a laser that can separate the light colors.

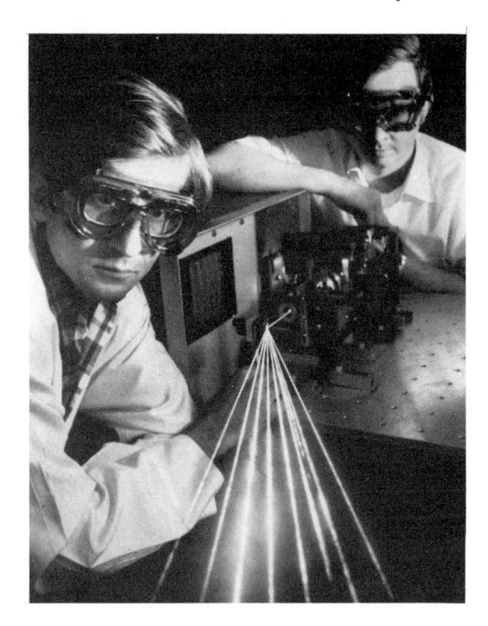

**Fig. 9-25.** Separating Light Colors.

The laser has been tested for its ability to carry many frequencies on a single beam of light. One successful experiment simultaneously impressed all the audio and video from seven television channels onto one laser beam. This included a bandwidth of more than 200 MHz and involved color as well as black and white transmissions. At the receiving end, all seven channels were picked up (on ordinary receivers) with a high degree of fidelity.

Some lasers are very simple and some are extremely sophisticated. Figure 9-26 is an example of a sophisticated version. This photograph shows R.T. Denton of Bell Laboratories as he adjusts one of the two lithium tantalate modulators in the foreground. Each of these modulators is pulse code modulating a single laser beam. The total modulation is 448 million bits per second. The insert shows the lithium tantalate crystal. It is soldered to the top of a copper heat sink to provide thermal stability during operation.

Laser development for many applications is progressing rapidly. Some models are using gas again instead of solid-state crystals. By mixing certain types of gases, one gas can be stimulated by light and made to pump the other by molecular collision. This saves a lot of power. Figure 9-27 shows a large gas laser. This photograph shows C. K. N. Patel adjusting the gas flow into a high powered gas laser at the Bell Laboratories. This laser mixes helium,

**Fig. 9-26.** Intelligent Light. (This Photograph and its Explanation by Courtesy of Bell Laboratories.)

**Fig. 9-27.** High Powered Gas Laser. (Photograph and Description Thereof Courtesy Bell Laboratories.)

carbon dioxide, and nitrogen. It was the first laser to produce a continuous output power in excess of 106 W.

Supercooling is still essential to some high powered lasers, but a great many are now operating at room temperatures. Figure 9-28 is a functional laser with practical applications.

This laser was constructed by a group of scientists at the Bell Laboratories. It is intended to be used by surgeons, much as he would use a scalpel. The hollow, jointed arm can be moved in any direction. Light from the laser source is guided through the arm by prisms mounted inside the corner cubes. It can be used with either pulsed or continuous-wave lasers. It may be used for a variety of cutting jobs apart from its medical applications.

**Fig. 9-28.** The Light Knife. (Photograph and Description Thereof Courtesy Bell Laboratories.)

During 1969, lasers reached an output level of 10 TW in short pulses. Also in 1969, laser beams from the moon were used to measure the distance from there to the earth's receiving station. The accuracy was within $\pm$ 1.5 M. Laser range finders are in use with an accuracy of one part in a million.

Lasers can be very small as well as very large. This is dramatically illustrated in Figs. 9-29 and 9-30.

The same laser is shown in Figs. 9-29 and 9-30. This device is about the size of an average grain of sand. The photographs show the device many times its actual size.

How about transmission distance? A 10-joule (J) laser signal can burn paper and wood at a distance of 1 mi (a joule = 1 W-s). This same beam used for communications can carry intelligence over a distance in excess of 10 million mi and be detected by a 3-in.

lens. This is made possible by the highly concentrated beam which can be radiated with a divergence of less than 0.01 of a degree.

As electronic technology advanced, the need for better measuring devices increased. Many remarkable instruments have been developed to fill this need.

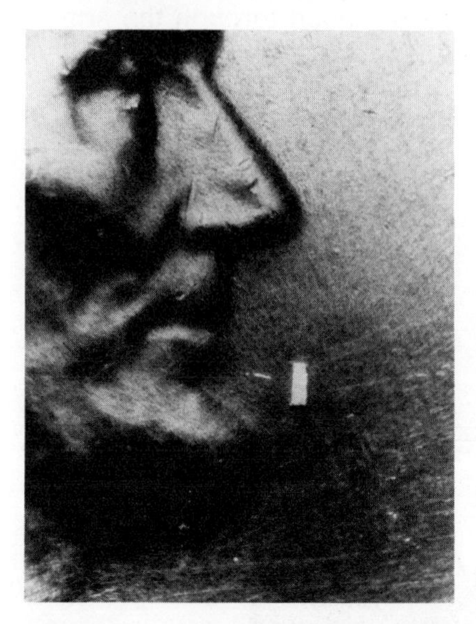

**Fig. 9-29.** A Bright Speck on a Penny. (Photograph and Description Courtesy of Bell Laboratories.)

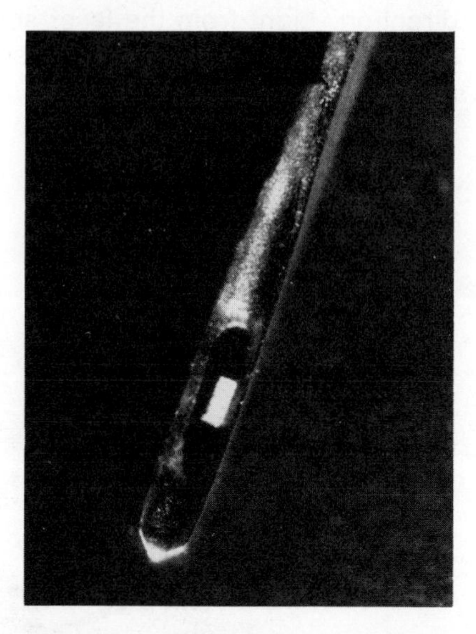

**Fig. 9-30.** Through the Eye of a Needle. (Photograph and Description Courtesy of Bell Laboratories.)

## CHAPTER 9 REVIEW EXERCISES

1. What is the name of a semiconductor resistor with a negative temperature coefficient?
2. List two applications of the device described in item 1.
3. What is the relation between light and resistance where the photoresistor is concerned?
4. Describe the operating characteristics of a varistor.
5. What type of material is used in the varistor?
6. What is the principal function of the variable capacitance diode?
7. What is the difference between a Zener diode and a symetrical Zener diode?

8. Draw a current–voltage characteristic curve for a tunnel diode and label the negative resistance area.
9. List three applications of the tunnel diode.
10. List some advantages of the silicon rectifier over a standard junction diode.
11. Distinguish between the silicon controlled rectifier and the silicon rectifier.
12. Draw the equivalent circuit for the silicon controlled rectifier. Use arrows to indicate current direction.
13. Once an SCR is turned on, how can it be turned off again?
14. What is meant by holding current in an SCR?
15. In a power transistor, what is the inside to outside temperature ratio per watt of power?
16. Interelement capacitance has been a major problem with transistors. State the method of reducing the capacitance in the:
    (a) Intrinsic transistor.
    (b) Tetrode transistor.
    (c) Field Effect transistor.
17. What is the Celsius temperature range for cryogenics?
18. What is a cryosar?
19. What is a solar cell?
20. What is a maser?
21. What does the pumping action accomplish in the maser resonator?
22. How does the maser accomplish amplification?
23. What is the difference between a laser and a maser?

# 10

# microelectronics

There seems to be no clear cut agreement on the precise definition of microelectronics. For the sake of conformity, we'll let it remain indefinite, but under this heading we will cover the electronic trend toward all forms of miniaturization. This move is not new; it is an evolutionary process that started with the invention of the electron tube. Immediately, someone manufactured a smaller, more streamlined model which required less power. But the move has not been entirely voluntary. Necessity has been doing a litle pushing in that direction.

## THE NEED FOR SMALLER COMPONENTS

One of the greatest driving forces behind the move to find higher operating frequencies has been the need to reduce the physical size of the equipment. Let's use the transmitting antenna as an example. The length of the antenna is directly proportional to the wavelength (which is inversely proportional to frequency). The length of the antenna is 0.94 $\lambda/2$. Commercial television frequencies range from 54 MHz to 890 MHz. At the lower end of this frequency band, the antenna must have a length of 8.46 ft. At the top of the band, the antenna length is only 6.1 in. At 10 GHz the length is reduced to slightly more than 1 in., and for 100 GHz it is little more than 0.1 in.

### *decreasing wavelength*

In the high microwave range it is possible to radiate waves from a waveguide without the benefit of an antenna. If antennas were used, they would take on microscopic proportions. At these frequencies, the expression of wavelength in ordinary measurements became awkward because of the extremely small fractions involved. By the same token, frequency expressions became almost meaningless because of the extremely large numbers needed to express the frequencies. For instance 300 THz is $3 \times 10^{14}$ Hz with a wavelength of $1 \times 10^{-3}$ mm.

To avoid these unwieldly figures, a new policy was adopted. Frequency was seldom referred to and wavelength was expressed in angstrom units (Å). An angstrom unit is $1 \times 10^{-7}$ mm. So, a wavelength of $1 \times 10^{4}$ Å is a frequency of 300 THz. More recently the nanometer has replaced the angstrom: 10 Å $=$ 1 nm. With these advances in carrier frequencies showing the way, all circuitry began to grow smaller.

### *miniature electron tubes*

The higher frequencies rapidly pushed the electron tube down to an amazingly small size. The effort of tube manufacturers to keep abreast of the reduced wavelength is characterized by the three ultrahigh-frequency tubes shown in Figure 10-1.

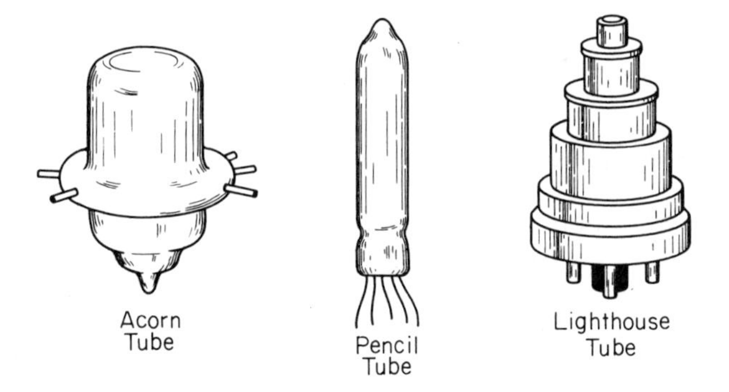


**Fig. 10-1.** Miniature Tubes.

These drawings are close to the actual tube size, and many tubes were made on a much smaller scale. As the tube grew smaller, it grew increasingly more fragile. The solid protective base disappeared and pins were moulded directly into the glass. The pins became problems and had to be reduced in size and shortened. These small tubes marked the beginning of the end of the first generation electronic components. The second generation was being quietly born.

## MEETING THE CHALLENGE

### *transistors*

The second generation circuit components began in the Bell Telephone Laboratories when the transistor was invented. This happened on December 23, 1947 and provided a replacement for the outmoded and battle weary electron tube. This first transistor was also somewhat on the bulky side (see Fig. 2-2).

Its size and crudeness belied its importance. This was a real breakthrough with such far reaching impact that its inventors were awarded the Noble Prize for Physics. This was not an accidental discovery. It came after much work and concentration on a single idea; how to construct a solid-state amplifier. The three men who accomplished this are Dr. William Shockley, Dr. Walter H. Brattain, and Dr. John Bardeen.

The first transistor was a point contact transistor, and Dr. Shockley wasted no time before improving his product. In 1949, he outlined the theory for a junction transistor. A functional model of this was constructed and demonstrated in 1951. The junction transistor proved far superior to the point contact transistor, and the race to improve the product was on again.

### *miniature transistors*

The evolution of the transistor to an ever smaller component progressed at a rapid pace. By 1960, some transistors had become almost microscopic, and the housing of nearly all transistors greatly exaggerated the actual size of the active component. Figure 10-2 dramatizes this trend. This photograph was published by IBM in 1964. It shows 50,000 transistors in a thimble.

### *hybrid circuits*

Meanwhile, the manufacturers began to incorporate many related circuits into a small, rugged, plugable package. This type of circuit arrangement is illustrated in Fig. 10-3.

Part A shows the physical components mounted on a printed

**Fig. 10-2.** Reducing the Transistor. (Photograph Courtesy International Business Machines.)

circuit board. Above each board is a schematic of the actual circuits in the package. Some of the components are not readily discernible because they are chip capacitors and metallic film resistors. The entire board is encapsulated in a tough plastic which adds both strength and protection from contamination.

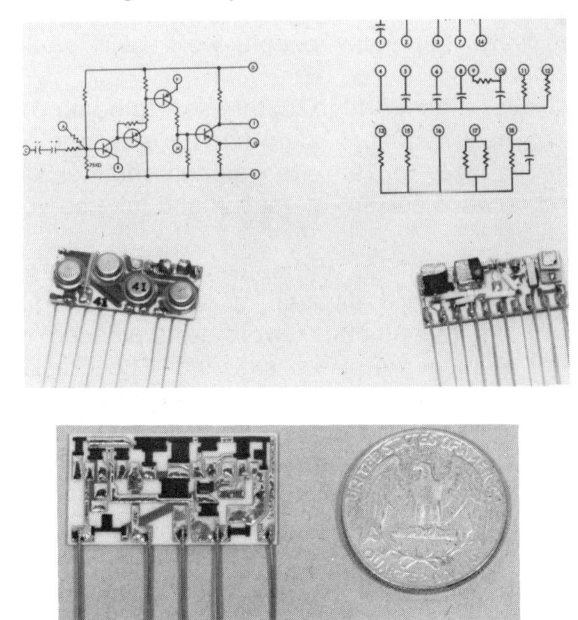

**Fig. 10-3.** Hybrid Circuits. (Photographs Courtesy Sprague Electric Company.)

Part B shows a printed circuit board, and the coin provides some perspective on the actual size.

The hybrid circuit first appeared along with the small transistor radio. In fact, one of these tiny capsules could contain all parts of such a radio with the exception of antenna, controls, power supply and speaker.

It was now possible to construct entire radio receivers with only a few circuits and components external to the encapsulated circuits. Every external wire that was eliminated decreased bulk and interference. Every solder connection that was eliminated removed a spot of potential failure. Instead of a dozen solder joints for each transistor, the whole receiver could be constructed with no more than a dozen separate connections.

### *reducing the power supply*

During this stage of evolution, the silicon rectifier, Zener diode, tunnel diode, and power transistor were doing a great deal for the power supply. The power supply changes were slower in coming because of the unique problems.

How could a solid device be made to handle high current without an intolerable increase in power loss? How could current capabilities be increased without an increase in size? Such questions had no ready answers, but many people were busy seeking the answers.

By the mid sixties, silicon bridge rectifiers were being packaged as hybrid circuits. These rectifiers could handle reasonable current and inverse peak voltage. In 1971, the silicon rectifier could be made to withstand forward current of 1500 A and inverse voltage of 15,000 V.

A mixture of hybrid capsules and conventional circuits marked considerable progress toward smaller more reliable equipment. Both the user and the manufacturer were well pleased with the results, but neither of them were well satisfied. The pressure was on for still smaller and increasingly reliable equipment. The technology kept advancing toward smaller components for the hybrid circuits. Some evidence of this has already been shown in Fig. 10-2. The trend is further illustrated by Fig. 10-4.

The above photographs were published in engineering bulletins and technical papers in 1968 and 1970. As you can see, components were becoming almost microscopic. These small chips were packaged in a manner similar to that previously illustrated and interconnected in numerous ingenious manners. The finished capsules were much the same except for reduced size.

(a) Chip ceramic capacitor mounted on penny

(b) 1200 Chip capacitors in a teaspoon

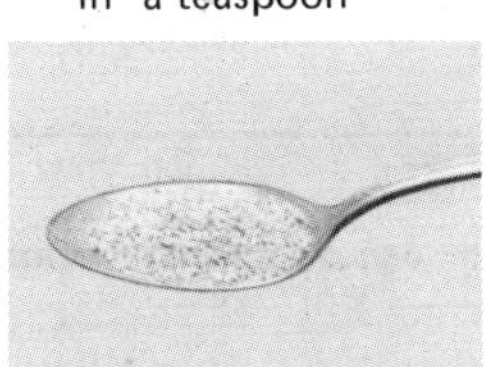

(c) Beam-lead resistor on postage stamp

**Fig. 10-4.** Components for Hybrid Circuits. (Photographs Courtesy Sprague Electric Company.)

Naturally a question arose about the futility of building hundreds of these components on a crystal, breaking them apart, and rearranging them into circuits. Why not build the entire circuit into the crystal? The answer to this question was being sought soon after the first hybrid circuit became a reality. The answer advanced the technology to the next crucial step.

## CIRCUIT INTEGRATION

Circuit integration is a process which places a complete functional circuit on a single chip. The move in this direction started in the early sixties, and the technology advanced as hybrid circuits were being improved. Hybrid circuits had already proven more reliable than the circuits they were connected to. Since most of the failures were in external circuits, it stood to reason that more of the circuit should move into the capsule. The planar transistor emerged about this time and showed the way.

### *circuit construction*

With the planar process, it was possible to place a complete circuit on a single chip. This was the integrated circuit (IC), and it is illustrated in Fig. 10-5.

This is only one example and illustrates a very simple chip. Still, it has resistors, capacitors, diodes, and transistors with the desirable interconnections already completed. The capability of placing any complete circuit on a chip, had now developed. This reduced the need for external connections to four; input, output, dc

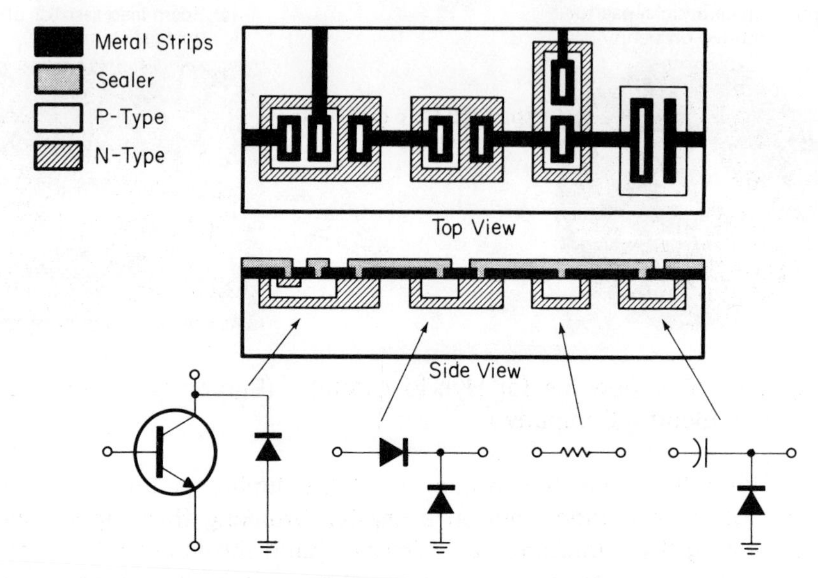


**Fig. 10-5.** Integrated Circuit.

voltage, and ground. This meant fewer connections and larger areas to make these connections.

### *circuit evolution*

The circuit designers had not been standing still. Circuits were reduced in size to match the miniature tubes. This brought the size down considerably, but they were still hampered by long leads, masses of wire, and stray signals. They were reduced again to match the transistor. Large plugable units (2 $ft^3$ in volume) became hand size printed circuit cards. The leads were printed into the card, and components were soldered on.

The printed circuit board was a big step in circuit improvement. There were fewer wires to go bad, fewer solder connections, and less mutual interference.

The next move was to reduce the size of the card, and in each new piece of equipment, they seemed to be smaller than in the last. Weight, size, and power requirements were dropping rapidly, and reliability was still going up.

By this time large, complicated circuits were being reduced to microscopic proportions and fabricated into a single piece of silicon. The silicon wafer was less than 0.25 in. square and only a few thousandths of an inch thick. The tiny circuit was then enclosed in a container with appropriate external leads. One of these circuits is shown in Fig. 10-6.

This is the top of an uncovered metal can. The studs that are connected to the circuit protrude from the bottom of the can as either solderable connections or plugable pins. The diameter of this can is 0.335 in., and as far as the circuit is concerned, most of that space is wasted. This is a resonably complex circuit, yet the circuit components and all interconnections occupy less than $\frac{1}{10}$ of the package area. This means that the circuit occupies an area less than 0.0335 $\times$ 0.0335 in$^2$.

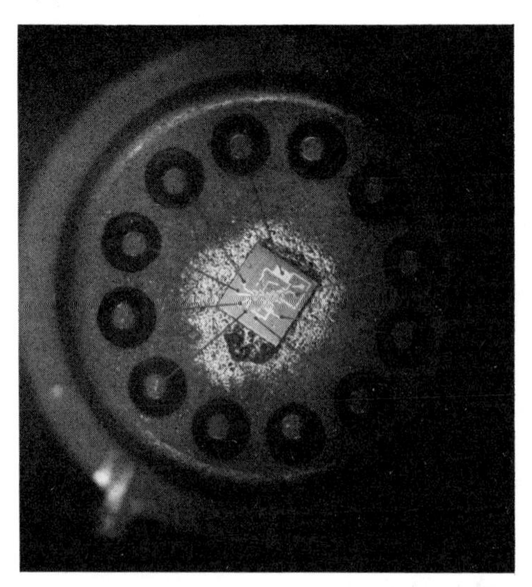

**Fig. 10-6.** Mounting a Small Circuit. (Photograph Courtesy Sprague Electric Company.)

Many obstacles had to be overcome before the state of the art reached this stage. One obstacle was production of microscopic inductors. Standard materials and procedures had been unable to produce a coil. Three developments combined to greatly enhance the advance of IC: Monolithic process, metalized film, and metal oxide semiconductors. These three are hard to separate because they rapidly converged, but we will consider each as a separate technology.

***monolithic process***

First an ingot of silicon is produced. This is a single crystal silicon. The molten element is silicon material at a temperature of 2570 °F. A silicon seed crystal is lowered until it just touches the surface of the melt, and is then slowly raised. The melt clings to the seed crystal, and as it gradually cools, forms into new crystal of the same lattice structure as the seed. These ingots, are from 4 to 6 in. in length and about 1.5 in. in diameter. Silicon wafers are sliced from the ends of the ingots. These wafers are thinner than a page of this book. The thickness is just enough to enable careful handling without causing excessive breakage.

The basic crystal is either *p*- or *n*-type material that was properly doped during the molten state. The wafers that are sliced from this crystal are then placed in a furnace and fired at a temperature of about 2000 °F.

Assuming *p*-type material for the wafers going into the furnace, they will be subjected to gaseous atmosphere of *n*-type silicon while in the furnace. A thin layer of *n*-type material will diffuse into the surface of the *p*-type material. The original *p*-type wafer has now become a substrate.

The surface of the *n*-type layer is now converted to silicon oxide. This thin layer of oxide is a protective film. We now have a three-layer structure ready to receive almost any type of circuit or circuit component. There is still one exception: the inductor.

There follows several repetitions of applying photoresist, photographic screening, exposure, washing, diffusion, etching, etc., to form as many circuit components as may be desired.

The final step in the process is to innnerconnect the components. This is accomplished by etching small openings through the final cover of silicon dioxide. The openings are made at all points to be mutually connected.

The wafer is then placed in a vacuum chamber along with a piece of aluminum. The temperature is raised until the aluminum vaporizes. This vapor deposits a thin sheet of aluminum over the surface of the wafer. There follows another etching process to remove the metal from areas where it is not desired.

Of course, a wafer has ample room for a great many of these microscopic circuits. The technique in mass production is to place as many identical circuits on a wafer as the processor can manage. All of these circuits are formed at the same time. There is little difference in the cost of a wafer with one circuit and a wafer with 100 identical circuits.

The wafer is now scribed with a diamond tipped tool and fractured into tiny chips; each chip containing a complete functional circuit. Each chip is packaged into an appropriate integrated circuit module.

The appearance of these chips is illustrated in Fig. 10-7.

Keep in mind that each chip contains a complete electronic

**Fig. 10-7.** Integrated Circuit Chips. (Photograph Courtesy International Business Machines.)

circuit. Once the technology had advanced to the point that these chips could be produced and packaged, the task was fairly simple. But in microelectronics, simple tasks have a tendency to grow more complex. In this case, the growing complexity was caused from decreasing size.

In 1964, the chip was 0.03 in. square and contained one complete electronic circuit which is equivalent to one stage. In 1966, the chip had changed to $0.1 \times 0.1$ in.$^2$. It contained many circuits, and each circuit was very complex. By 1967, the average chip had enlarged to $0.12 \times 0.12$ in.$^2$, but the quantity and complexity of the circuits increased also. By that time, 254 components to a chip was not unusual.

These chips generally contain several related stages complete with resistors, capacitors, diodes, transistors, and interconnections. The following module was featured in an engineering bulletin by the Sprague Electric Company in 1970. The drawings in Fig. 10-8 and data for the explanation were extracted from that bulletin by special permission.

This sound channel module incorporates a linear monolithic integrated circuit designed for use in television receivers. Description and physical dimensions of the module are given in Part A of Fig. 10-11. Part B is a block diagram of the functional units which the module contains. The circuit details are shown in the schematic of Part C.

The module contains a three stage IF amplifier, a limiter, a bal-

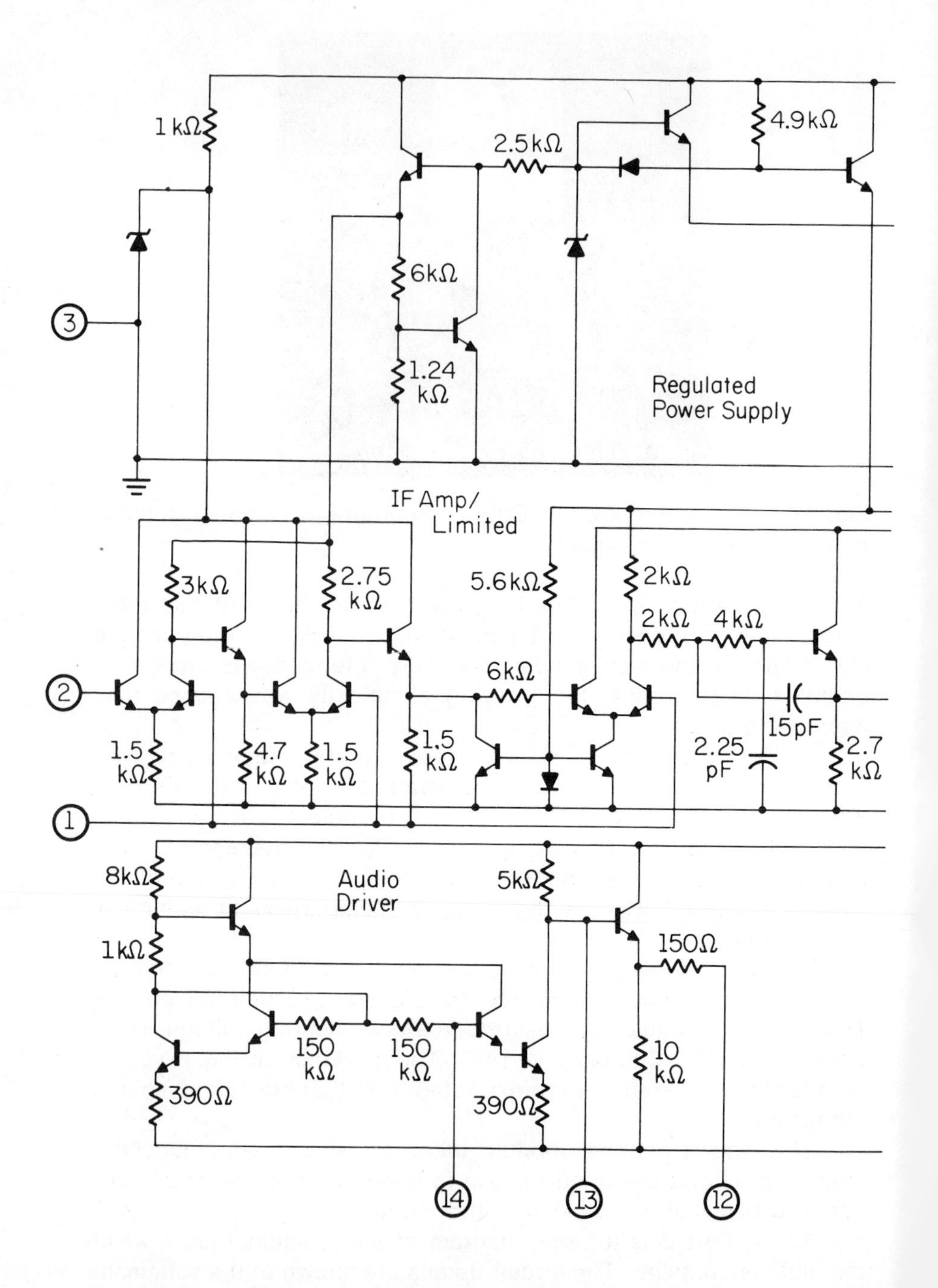


**Fig. 10-8.** An Integrated Circuit Module.

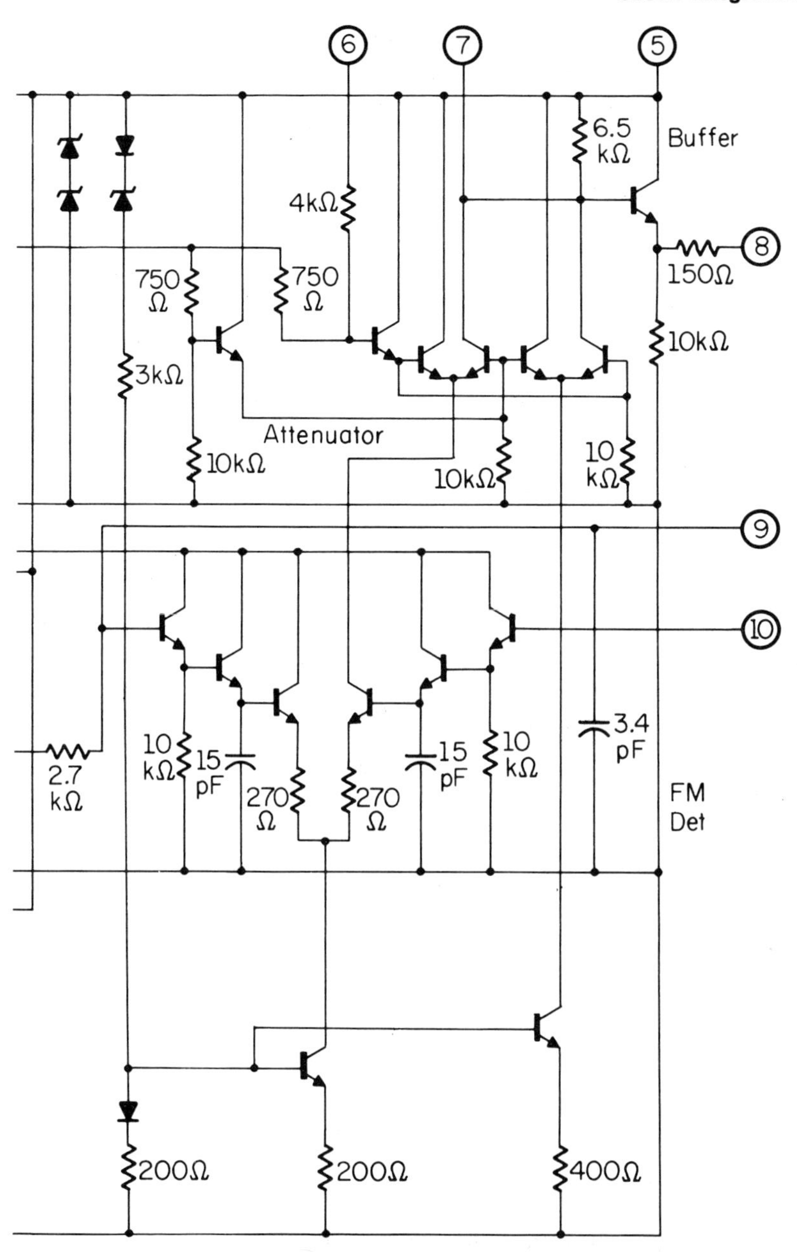


**Fig. 10-8.** Continued.

anced FM detector, an electronic attenuator, a buffer, an audio amplifier-driver, and a Zener diode regulated power supply. The total power requirement is 850 mW with an operating temperature from —40 to +85 °C.

The regulated power supply is innnerconnected and furnishes regulated voltages for all components within the module.

The electronic attenuator performs the function of an automatic volume control. It enables changing volume without causing frequency shifts.

If the receiver has been engineered to accept this module, the remainder of the circuits may be of conventional design.

The module contains 35 *npn*-type transistors, 5 Zener diodes, 4 rectifier diodes, 44 resistors, and 5 capacitors. All of this occupies a volume of less than 0.036 $in.^3$ counting the protective package and the spacing for the seating plane.

As far as related circuits were concerned, the monolithic integrated circuit had come into its own. Figure 10-9 will give you an enlarged view of the inside of an IC module.

This chip is 0.075 $\times$ 0.08 $in.^2$. The leads which connect the circuit to the module pins are about 0.002 in. in diameter.

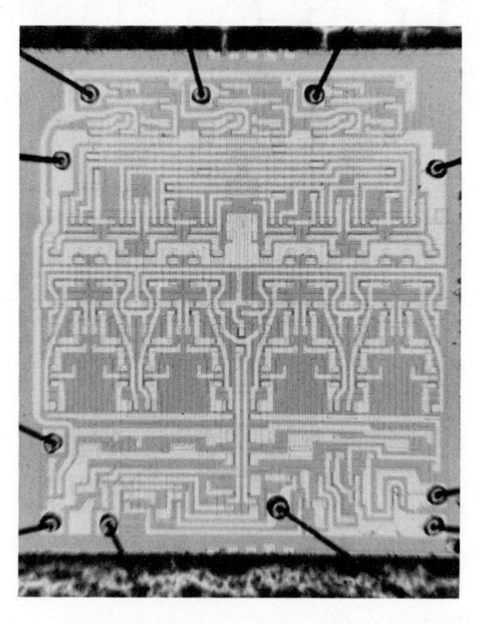

**Fig. 10-9.** Inside an IC Module. (Photograph Courtesy Sprague Electric Company.)

**Fig. 10-10.** A Short History of Circuit Components. (Photograph and Description Courtesy Bell Laboratories.)

Figure 10-10 is a size comparison of a tube, a transistor, and an integrated chip.

Movement is from the large glass tube to the canned transistor to the integrated circuit. The chip on the right contains 22 transistors along with the circuits and other components. It helps to generate the musical dial tones in the touch-tone telephones. The newest integrated circuits have protective coatings which eliminate the need for cans.

The circuit design for a great many circuits had now progressed through the three steps illustrated in Fig. 10-11.

**Fig. 10-11.** Circuit Progression. (Photograph Courtesy Texas Instruments, Inc.)

These are three plugable units which perform an identical counting function. The electron tube model is typical of equipment designed in the early fifties. The huge space in the center of this unit was needed to house the resistors, capacitors, and connecting wires.

The transistorized circuits on the printed circuit board represent typical designs in the late fifties and early sixties.

The IC module in the foreground became popular in the late sixties.

There is a step missing between the transistors and integrated circuits. This was the period in the mid sixties when the new designs were built around the hybrid circuit capsules.

As previously stated, it is difficult to separate the developments in technology during these years of rapid advancement. The later years referred to here under monolithic process actually saw the combination of several other techniques and devices. These included metallic film and metal oxide semiconductors.

### *metallic film*

Many texts and manufacturers refer to thin film and thick film as if they were two vastly different types of film. Actually both names are misnomers. One manufacturer discovered a method of depositing a conducting film on a substrate and called it thin film. Another discovered a different method of doing the same job and called it thick film. The type has no bearing on the thickness of the film. There are many thin films that are thicker than some thick films. Both are films of conducting material covering a specified substrate. Thin film is created by a process of vacuum deposition while thick film is constructed by screen deposition. We will group both into our discussion as metallic film.

With metallic film, it is possible to deposit any desirable pattern of resistors, capacitors, and conductors. The film might be considered as a printed circuit board with built in resistors and capacitors. Many circuits are built by combining the monolithic and metal film technology. Monolithic components can be formed on a flip chip which leaves critical contacts exposed. This chip is turned over with the exposed contact to the surface of the film. When film and chip are properly matched they form the complete circuit.

By 1970, the metal film technology had advanced to a fine art. Resistors and capacitors could be deposited in a wide range of values with extremely close tolerances. It was then possible to construct the inductor on film but this was a new process. Several problems needed to be solved before the solid-state inductor could meet the required state of perfection. The tendency was to avoid the use of inductors when possible, and if an inductor of more than a few millihenries was required, most manufacturers resorted to a discrete inductor.

A spiral inductor could be laid on film but a crossover was necessary to complete the circuit. This was limited to a low inductance. Other possibilities were being examined such as the piezoelectric circuit elements to simulate inductors. This works well in areas where resonant circuits are required. Since frequencies were still climbing there was a possibility that the problem would solve itself.

Depositing transmission lines on film was a well known technique, and a shorted transmission line of less than a half wave length is an inductor. This type of inductor seemed likely to replace many conventional inductors in microcircuits.

### *metal oxide semiconductor*

The metal oxide semiconductor was an outgrowth of the metallic film techniques. It was brought about by combining the processes used in both screen and vacuum deposition. With metal oxide semiconductors (MOS), many new avenues were opened in microelectronics. The process of forming a MOS transistor was similar to established procedures but fewer steps were required. This meant more complex circuits at a lower cost. Surface area of a substrate for forming MOS transistors is only a fraction of that for the standard triode transistor. This opened the door to another drastic reduction in size.

A thin layer of silicon oxide is grown on a substrate of properly doped silicon crystal. The substrate is generally a single crystal silicon wafer of highly resistive *n*-type material, but it may be either *n* or *p*. Steam is used to form a thin layer of silicon oxide on the substrate surface. This film of oxide is approximately 100 nanometers (nm) in thickness.

The wafer is then coated with a photoresist and exposed to ultraviolet light through a high resolution mask. Undesired photoresist is then washed away.

An etching process removes the silicon oxide in the exposed areas. A diffusion furnace is used to diffuse an *n*-type layer over the exposed *p*-type material. A final diffusion spreads a layer of aluminum oxide over the surface. When this oxide is etched from areas where it is not wanted, the transistor contacts are already formed.

The process just described not only produces a transistor, it produces a MOS field effect transistor. What is a MOSFET? It is a device which incorporates most of the advantages of both vacuum tubes and transistors with very few of the disadvantages of either. A cross-sectional view, an equivalent junction, and a schematic symbol are shown in Fig. 10-12.

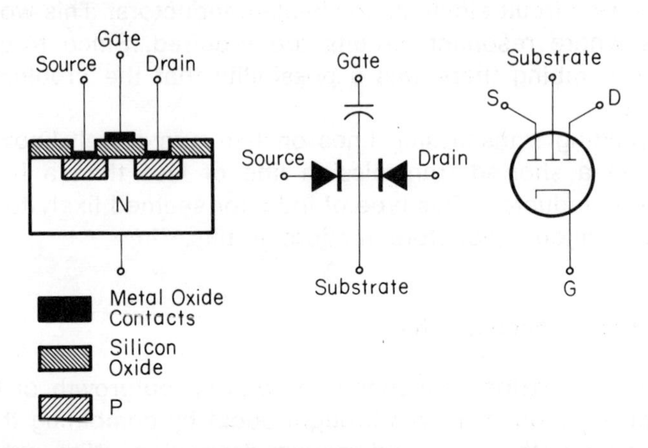


**Fig. 10-12.** MOS Field Effect Transistor.

The source and drain are completely interchangeable as emitter and collector. Electrons may pass in either direction. When the gate has zero bias the forward current is limited to the backward current of one of the diodes. The substrate is connected to a common ground, and it is normally omitted from schematics.

The bias and/or signal at the gate exerts complete control on source to drain current. But this is not a current sensitive device; it is voltage sensitive. In fact, its typical operation very nearly duplicates the characteristics of a pentode electron tube connected as a cathode follower. The important characteristics are:

1. Very high input impedance; about $1 \times 10^{15}\ \Omega$.
2. Low output impednace, a few hundred ohms.
3. Voltage gain less than unity.
4. Sufficient power gain to enable use as a driver.

The process just described forms a *p*-channel MOSFET. The substrate may, of course, be *p* material as well as *n*. In this case, the device would be known as an *n* channel, and the arrow on the symbol would point inward.

The almost infinite input impedance is caused from the method of construction which places a capacitor junction between the gate and the other two elements.

The advent of the MOSFET did not create a new technology so much as it revived an old one. The electron tube oriented engineer picked it up with no problem whatever. The engineers who had worked only with transistors were forced to acquaint themselves with tube characteristics and terminology.

This unique device with its microscopic size seems to have removed most of the obstacles between the present state of the art and large scale integration.

## LARGE SCALE INTEGRATION

Large scale integration (LSI) is a name attached to the next logical step in the process of integration of microcircuits. It encompasses the placement of complete systems on a single chip. In 1969 it was possible to fabricate a chip of $0.2 \times 0.2$ in.$^2$ with as many as 12,000 separate transistors. This was not recognized as the ceiling. Even then speculation was to the effect that this could be increased 10 fold.

### *ultimate barrier*

The limit to the reduction in circuit size appears to be a light barrier. The photographic equipment has been hard pressed to keep pace with microelectronics. Since the accuracy of a line and the minimum thickness of a line on a photographic mask became ever more critical, the photographic equipment grew increasingly complex and sophisticated. The ultimate limit, if any, appears to be the wavelength of the light used to produce the picture.

### *impact of large scale integration*

In the late fifties, a large scale, electron tube computer was delivered to the military. This computer filled a building of 67,000 ft$^2$ of floor space, and little area was wasted. With 50,000 circuits on a chip, the equivalent of this computer can be worn as a cuff link.

As the technology advances, it should cost no more to put a complete television receiver on a chip that it cost in the past to build a transistor. It seems highly unlikely that any person will altogether escape the influence of microelectronics.

The state of the art today is sufficient to furnish each person with a pocket size personal communicator. This device could be both transmitter and receiver complete with both audio and video. By moving to the higher frequencies, each person could be allotted a band of frequencies similar to that used for all public broadcasting today. Direct two-way communications could be maintained between any two people on earth.

## *the challenge*

Before we leave the impression that all discoveries have been made, all problems solved, and that there is nothing more to do, let's stop and think how many years passed before the horse was replaced by the automobile. In addition to building a reliable automobile, many other industries had to grow to support it, roads had to be built, and public acceptance had to be won.

Many years ago an educated man proposed that the U. S. Patent Office be closed on the grounds that everything worthwhile had already been discovered. That was wrong thinking then; it is wrong thinking today; and it will always be wrong thinking.

The field of microelectronics is in its infancy. A great deal of progress has been made, but there has been more talk than progress. Many problems are waiting to be solved. Some of these real problems have not been identified because the engineers involved are too close to see the big pictures. Some problems have been set aside for the sake of economic expediency.

The industry needs fresh minds to face up to the growing challenge. New approaches, new ideas, and new enthusiasm are needed to keep progress from grinding to a halt. A story is told of how the light bulb became frosted on the inside. The experienced engineers knew that it was impossible, and jokingly assigned it as a task to each new engineer. One of these failed to recognize the task as impossible and found a way to do it.

Work in the field of electronics today is one of the most exciting and challenging jobs that can be imagined. Equipment can be found that spans the history from crystal receivers to satellites. The electron tube industry is not dead, yet laser communications is an established fact. No matter what is going on, electronics is involved in some fashion. A career in electronics will require a lot of hard work and hard study, but the person who enjoys a challenge will find it rewarding.

# CHAPTER 10 REVIEW EXERCISES

1. What is the relationship between operating frequency and the size of components such as transmission lines and antennas?
2. A half-wave antenna for 1 MHz is how many feet in length?
3. What type of frequencies can be radiated from a waveguide without an antenna?
4. An angstrom is equivalent to how many millimeters?
5. A wavelength of 1000 Å is equivalent to a frequency of ______.
6. Miniature tubes came about as a result of the tube industry trying to cope with increasing ______.
7. The replacement for the electron tube appeared in 1947 when the ______ was invented.
8. The first transistor was of the ______ ______ type. Its superior was the ______ type which followed soon after.
9. By 1960, some transistors had become almost ______ in size.
10. The first hybrid circuits were composed of a package of several ______ ______ for mounting on a printed circuit board.
11. As circuit and component size decreased, reliability ______.
12. A process which places a complete circuit on a single chip is ______ ______.
13. Name three improvements characteristic of printed circuit cards.
14. A wafer is a cross-sectional ______ from a grown ______ rod.
15. The cost of a wafer is almost independent of the number of ______ it contains.
16. The monolithic integrated technique enabled many related circuits to be placed on a ______ ______.
17. Metallic film is a product of depositing circuits and components on a ______.
18. The only difference between thick film and thin film is the ______ of ______.

19. The vacuum process produces ______ film while the screen process produces ______ film.
20. Which components are better produced on metallic film?
21. A shorted transmission line less than a half wave length is an ______.
22. The metal oxide semiconductor enabled more complex circuits in ______ areas.
23. Forming the MOSFET requires ______ steps than forming a standard triode transistor.
24. The MOSFET combines the advantages of the ______ ______ and the ______.
25. The MOSFET has ______ input impedance, ______ output impedance, ______ power gain, and a voltage gain of less than ______.
26. LSI is a process of placing entire ______ on a single chip.
27. The ultimate barrier to reduction of circuit size appears to lie in the ______ equipment.
28. The field of microelectronics is in its ______.
29. A career in microelectronics should be both challenging and ______.

# appendix

## BIBLIOGRAPHY

Fogiel, M., *Microelectronics,* Research and Education Association, New York, 1968.

Gentry, Gutzwiller, Holonyak, and Von Zastrow, *Semiconductor Controlled Rectifiers,* Prentice-Hall, Inc., Englewood Cliffs, N.J., 1964.

Khambata, A.J., *Introduction to Large Scale Integration,* Wiley Interscience, New York, 1969.

Lathan, D.C., *Transistors and Integrated Circuits,* J. B. Lippincott Co., New York, 1969.

Lenk, J.D., *Handbook of Practical Electronics Tests and Measurements,* Prentice-Hall, Inc., Englewood Cliffs, N.J., 1969.

Levine, S.N., *Principles of Solid-State Microelectronics,* Holt, Rinehart, and Winston, New York, 1963.

Lytel, A., *ABCs of Lasers and Masers,* Howard W. Sams and Co., New York, 1968.

Mandl, M., *Fundamentals of Electronics, 2nd ed.,* Prentice-Hall, Inc., Englewood Cliffs, N.J., 1965.

Marcus, A., *Basic Electronics for Technicians,* Prentice-Hall, Inc., Englewood Cliffs, N.J., 1964.

Mattson, R.H., *Electronics,* John Wiley & Sons, New York, 1966.

Robinson, V., *Electronic Concepts,* Reston Publishing Co., Reston, Virginia, 1972.

Shockley, W., *Electrons and Holes in Semiconductors,* D. Van Nostrand Co., Princeton, N.J., 1950.

Shore, B.H., *Fundamentals of Electronics, 2nd ed.,* McGraw-Hill, Inc., New York, 1970.

Tepper, M., *Basic Radio,* vols. 1–6, John F. Rider Publishing Co., New York, 1968.

# ANSWERS TO REVIEW EXERCISES

## *chapter 1*

(1) Thermionic, photo, secondary, and field. (2) Electrons are boiled from the surface of a metal by application of heat. (3) Surface barrier. (4) It is the potential required to move an electron through the surface barrier of that element. (5)

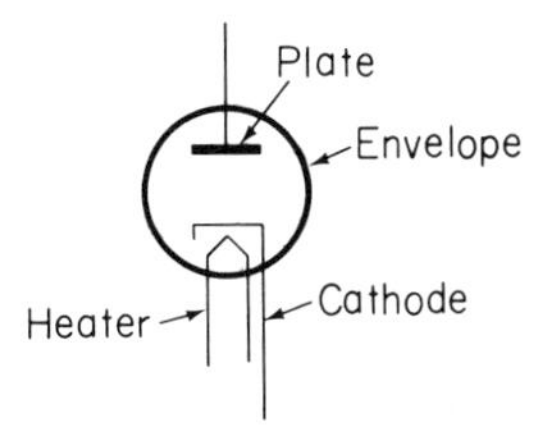


**Fig. A-1.** Parts of a Diode.

(6) When the plate is positive with respect to the cathode. (7)

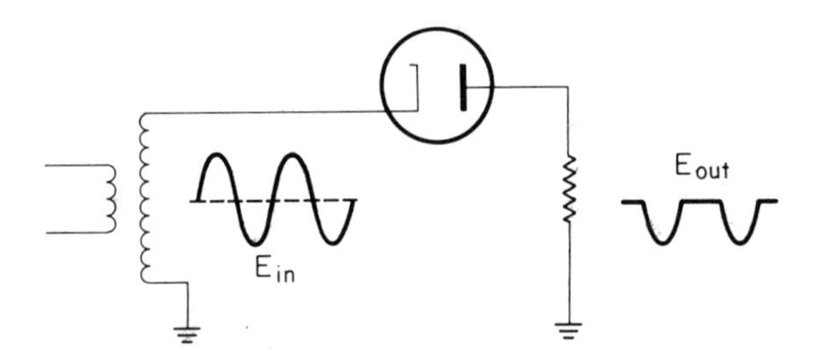


**Fig. A-2.** Negative Half-Wave Rectifier.

(8) 500 Ω. (9) 10 kΩ. (10)

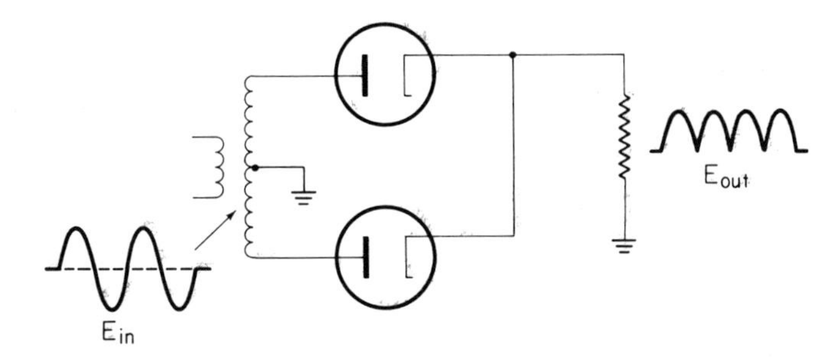


**Fig. A-3.** Positive Full-Wave Rectifier.

(11)

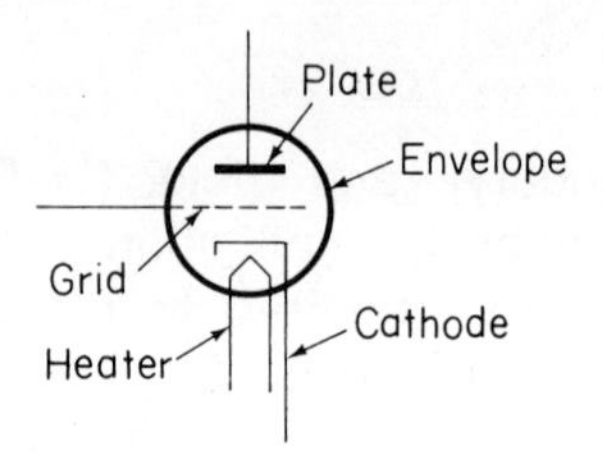


**Fig. A-4.** Parts of a Triode.

(12) The negative potential between the grid and the cathode. (13) ac plate resistance ($r_p$), transconductance ($g_m$), and amplification factor ($\mu$), $r_p = \Delta E_p / \Delta I_p$, $g_m = \Delta I_p / \Delta E_g$, $\mu = \Delta E_p / \Delta E_g$. (14) $\mu = g_m \times r_p$. (15) Fixed, grid leak, and cathode self.

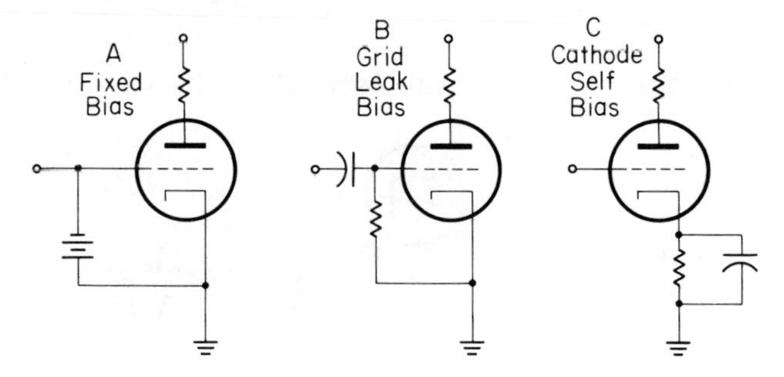


**Fig. A-5.** Illustrating Types of Bias.

(16) A small change in bias causes a large change in plate voltage. (17) It is increasing the amplitude of a signal. The amplification is a ratio of $E_{\text{out}}$ to $E_{\text{in}}$. (18) 19. (19) A line on a family of plate characteristic curves which connects the points of maximum $E_p$ and maximum $I_p$. It passes through all possible values of $E_p$ and $I_p$. (20) 4.3 mA and 150 V. (21) 4.7 mA, 185 V, and 165 V. (22) 3.57 mA, 185 V, and 135 V. (23) —5 V and —7 V. (24) 15. (25) 15. (26) $E_p = 350$ V to $I_p = 10$ mA. (27) 200 V, 4.3 mA, and —6 V. (28)

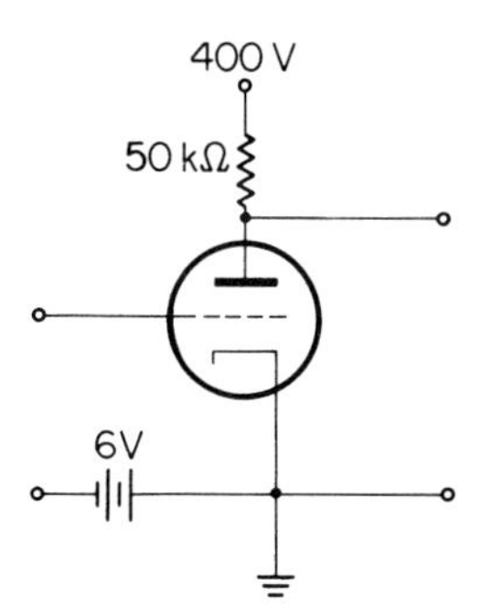


**Fig. A-6.** Amplifier From Load Line.

(29) 175 V, 4.5 mA, 225 V. (30) (a) 4 V. (b) 3.9 mA. (c) 5.2 mA. (d) 140 V. (e) 205 V. (f) 13.75. (31) (a) 12 V. (b) 6 V. (c) 4.242 V. (d) 20 $\mu$s. (e) 50 kHz. (f) 6000 m. (32) (a) 62.5 kHz. (b) 90°.

## *chapter 2*

(1)

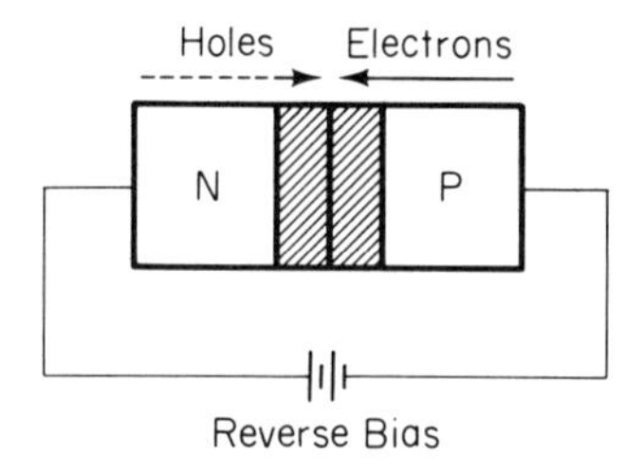


**Fig. A-7.** Results of Reverse Bias.

(2) It contains the same number of electrons as it has protons. (3) When it has eight electrons in the valence shell. (4) (a) Conductor atoms have valence shells of less than four electrons. (b) Insulator atoms have more than four valence electrons; the best insulators have eight. (c) Semiconductor atoms have valence shells of 3, 4, or 5 electrons. (5) (a) It means the ability to combine. (b) Some materials combine valence shells and the atoms share valence electrons with neighboring atoms. (6) This is in reference to the number of electrons in the valence shell of the atoms of the material. Trivalent structures contain three valence electrons; pentavalents contain five. (7) Lithium, silicon, bismuth, sulphur, and neon. (8) Electron-volts. (9) (a) By melting it and while in the molten state

mixing in a small quantity of a pentavalent material. (b) The same process as (a) using a trivalent material. (10) *n* and *p* refer to negative and positive carriers respectively. The current carriers are electrons in *n*-type materials and holes in the *p*-type materials. (11) Growth, diffusion, and alloy. (12) The area near the junction. This area is chemically stable and has been depleted of current carriers. (13)

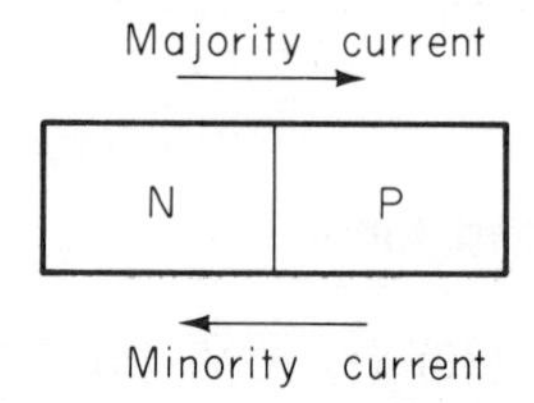


**Fig. A-8.** Indicating Currents Across a Junction.

(14) Minority = reverse and opposite in direction to majority or forward. Majority = forward and is opposite in direction to minority or reverse. (15)

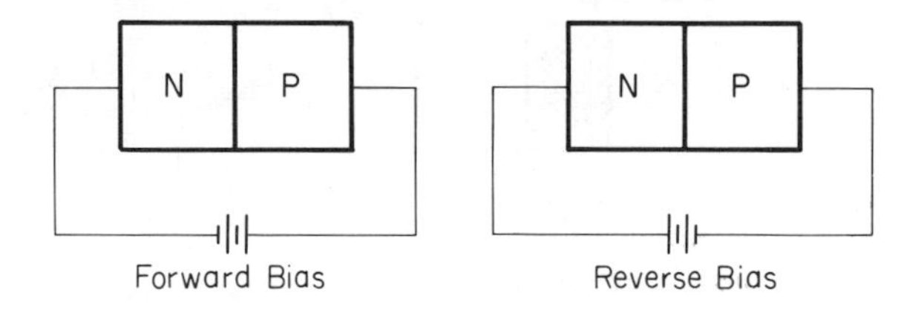


**Fig. A-9.** Biased Diode Junctions.

(16)

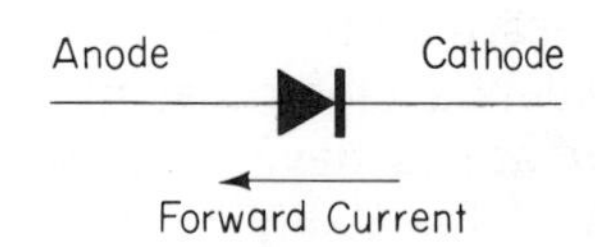


**Fig. A-10.** Schematic Symbol and Current Through a Diode.

(17) 500 Hz. (18) 126.8 V. (19)

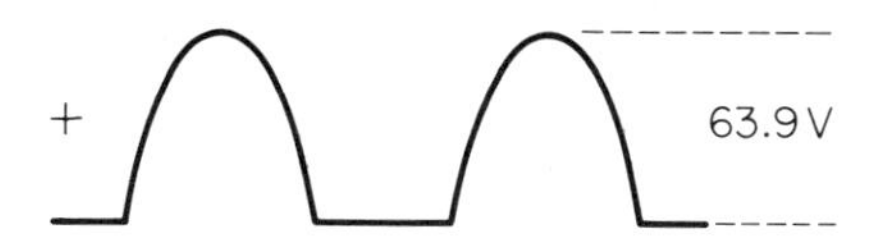


**Fig. A-11.** Positive Rectifier Output.

(20)

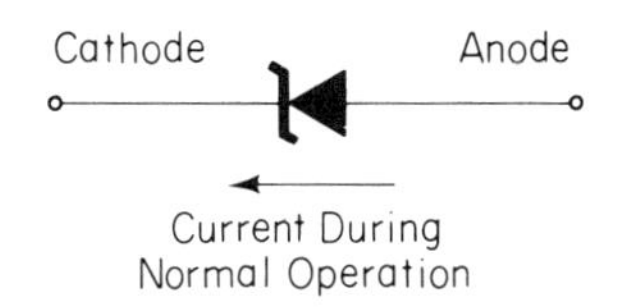


**Fig. A-12.** Zener Symbol and Normal Current.

(21)

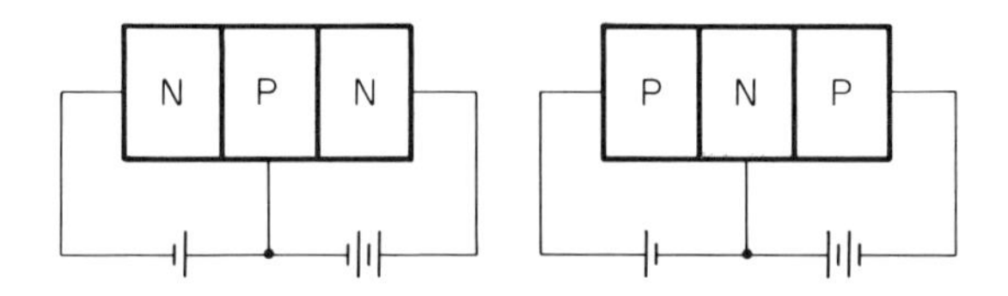


**Fig. A-13.** Properly Biased Triode Junctions.

(22)

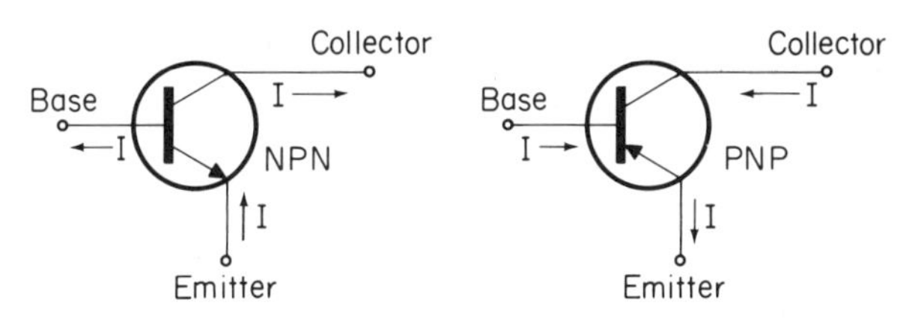


**Fig. A-14.** Symbols and Current.

(23) (a) 98%. (b) 100%. (c) 2%. (24) 490 $\mu$A. (b) 750 mV. (25) (a) $I_e$ increases. (b) $I_c$ increases. (c) $E_c$ swings negative. (26) 50 mW. (27) 1000 °C/W.

## chapter 3

(1)

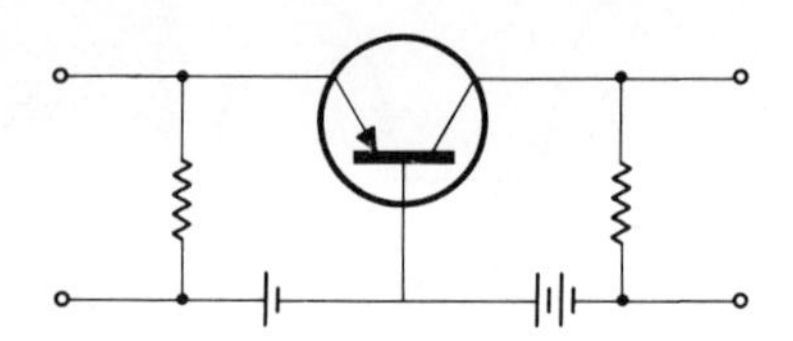

**Fig. A-15.** Properly Biased *pnp* in Common Base Circuit.

(2) Audio, radio, and video (also pulse and power). (3) A, AB, B, and C. (4) It conducts for more than 180° during each input cycle. (5) Input current, output current, input voltage, and output voltage. (6) Hybrid. (7)

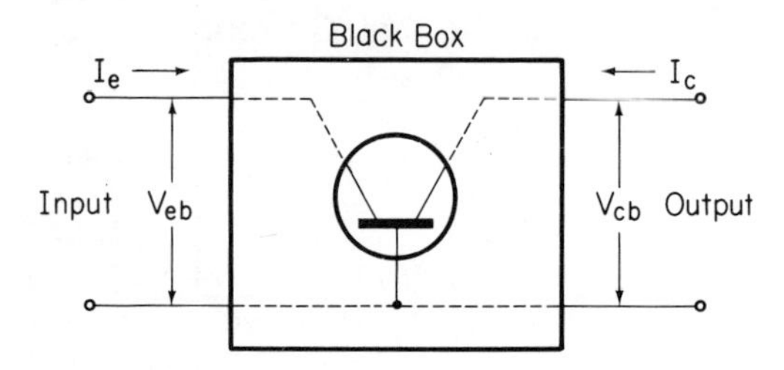


**Fig. A-16.** Black Box Test Circuit.

(8) Using the two voltages as the independent variables. (9) Input resistance, forward current gain, output conductance, and reverse voltage gain. (10) (a) Common emitter. (b) $V_{ce}$. (c) 200 $\mu$A. (d) No. It is much too high. The typical value of $h_{ie}$ for a CE is 500 to 1500 $\Omega$. (11) 40 $\Omega$. (12) 0.9994. (13) (a) $r_{ie}$ = open circuit parameter, input resistance of a CE with output open. (b) $r_{rb}$ = open circuit parameter, reverse transfer resistance of a CB with input open. (c) $r_{oc}$ = open circuit parameter, reverse transfer resistance of a CC with the input open. (14)

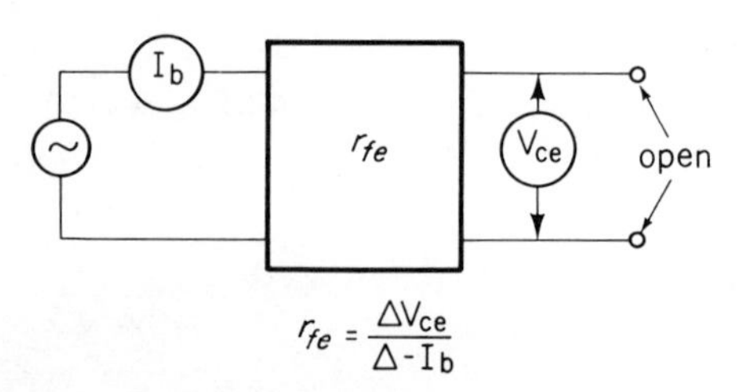


**Fig. A-17.** Test Circuit of $r_{fe}$.

(15)

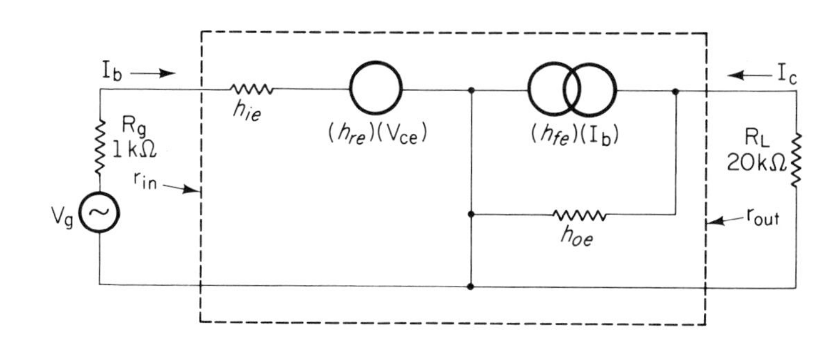


**Fig. A-18.** Equivalent Circuit for Fig. 3-27.

16 (a) $A_i = h_{fe}/(h_{oe})(R_L) + 1$. (b) $A_v = -h_{fe}\ R_L/(h_{ie}\ h_{oe} - h_{fe}\ h_{re})\ R_L + h_{ie}$. (c) $G = (h_{fe})^2\ R_L/(h_{oe}\ R_L + 1)\ [(h_{ie}\ h_{oe} - h_{fe}\ h_{re})\ R_L + h_{ie}]$. (17) (a) 99.8. (b) —416.67. (c) 13888.8. (18) (a) $r_{ie} = 800\ \Omega$. (b) $r_{re} = 24\ \Omega$. (c) $r_{fe} = -2\ \text{M}\Omega$. (d) $r_{oe} = 40\ \text{k}\Omega$.

### *chapter 4*

(1) Maximum output voltage and maximum output current. (2) Input current, output current, and output voltage. (3) Level of supply voltage and resistance of load resistor. (4) The level of input current in the quiescent state. (5) The quiescent values of input current, output current, and output voltage. (6) 6 mA. (7) —24 V. (8) (a) 200 $\mu$A. (b) 2.6 mA. (c) 13 V. (9)

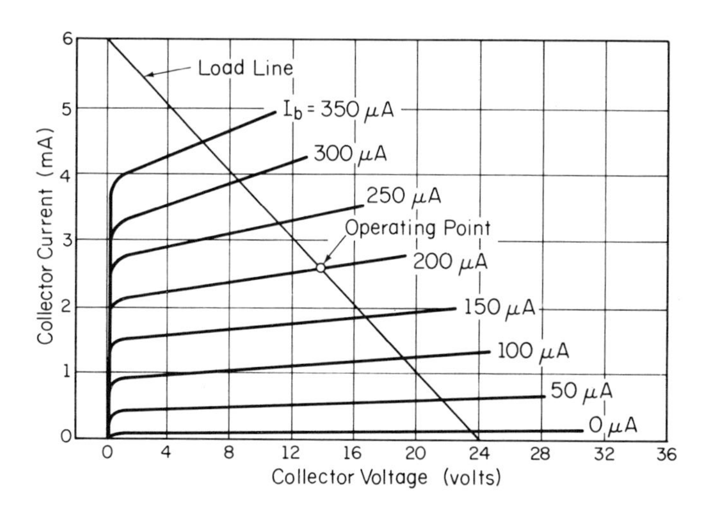


**Fig. A-19.** Load Line for Circuit in Fig. 4-39.

(10) 1.25 mA and 3.25 mA. (11) (a) —15.5 V. (b) —10.5 V. (12) (a) 500. (b) 20. (13) 33.8 mW. (14)

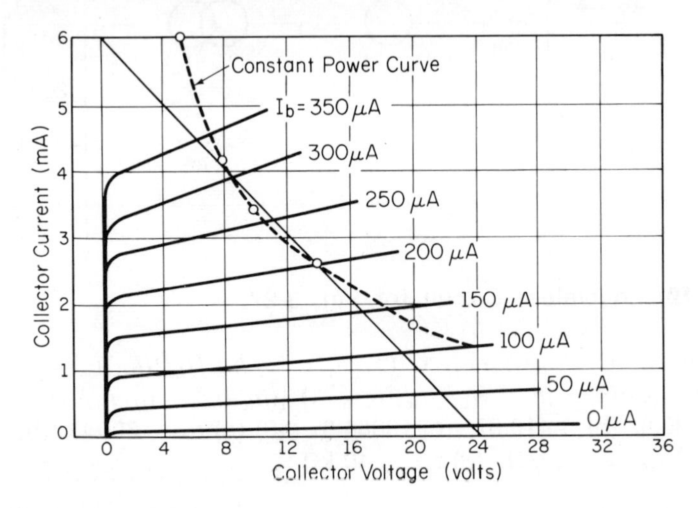


**Fig. A-20.** Constant Power Curve for the Circuit in Fig. 4-39.

(15) 2.5 mA and 4 mA. (16) (1) Increase the resistance of $R_L$. (2) Reduce the level of $V_{CC}$. (17)

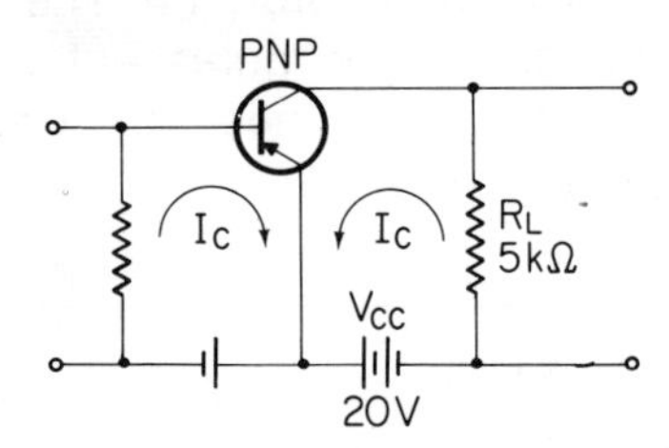


**Fig. A-21.** Circuit for Load Line in Fig. 4-41.

(18) —5 V. (19) —6 V and —14 V. (20) 40.

## *chapter 5*

(1) Batteries and generators. (2) Any situation where the available dc level is either too high or too low. (3) It changes dc to ac and may be used to increase the level of dc. (4) They are 90° out of phase. (5) Each output is 120° out of phase with each of the other two. (6) The step up transformer has more turns on the secondary than it has on the primary. The step down transformer has the great-

est number of turns on the primary. (7) 10:1. (8) (a) 1:2.67. (b) 18.73:1. (c) 9.365:1. (9) (a) 749 mA. (b) 37.46 A. (c) 18.73 A. (10) 30 V. (11) 800. (12)

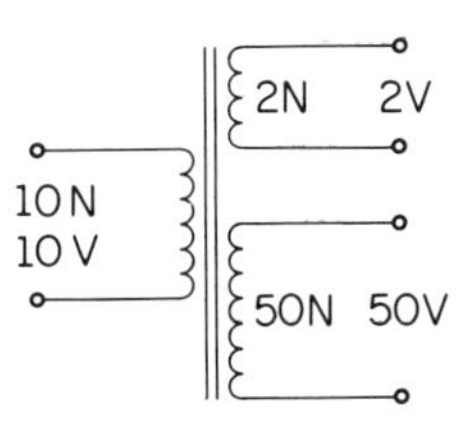


**Fig. A-22.** Relation of Turns and Voltage.

(13) 1:1000. (14) It would need 10 times as many turns on the primary as it has on the secondary. (10:1 step down transformer). (15) (a) 100 V. (b) 2 A. (c) 200 W. (16) 1600 Hz. (17) Mobile (especially airborne). (18)

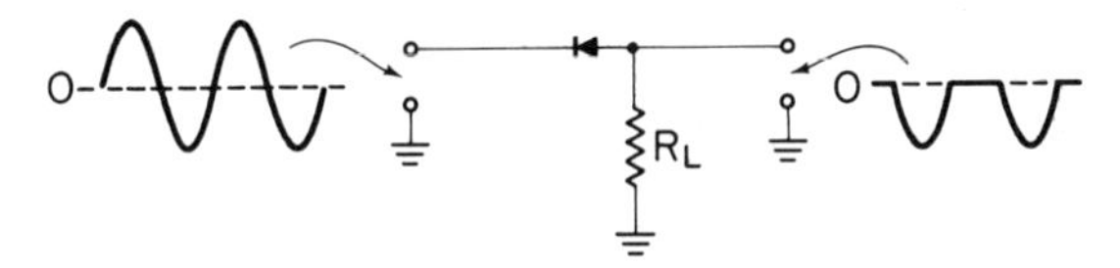


**Fig. A-23.** Half-Wave, Negative Rectifier with Waveshapes.

(19)

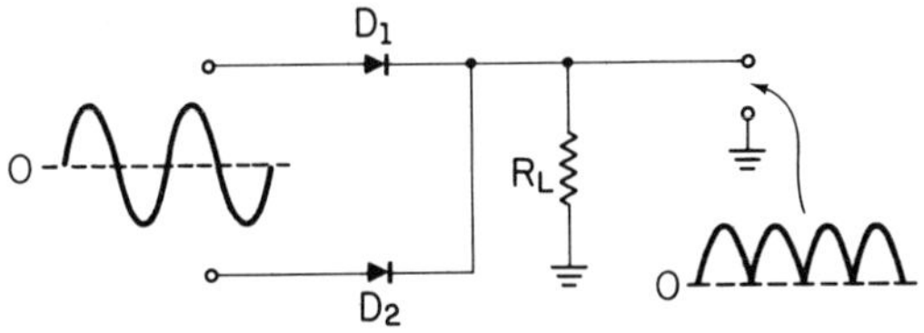

**Fig. A-24.** Full-Wave, Positive Rectifier with Waveshapes.

(20) The bridge rectifier provides more voltage from a given transformer. (21) It changes ac to dc and provides an output in terms of multiples of the peak input voltage. (22) Circuits which require very high voltages and very little current. (23) 5656 V dc. (24) Full-wave with choke input filter. (25) (a) Changing input current. (b) Changing input voltage. (c) Changing load current. (d) Changing load voltage. (26)

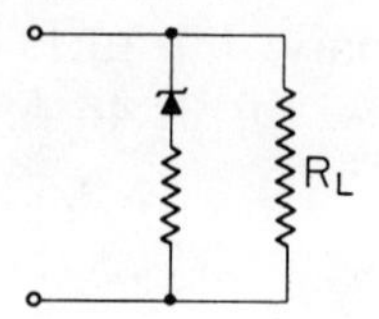


**Fig. A-25.** Simple Zener Shunt.

(27)

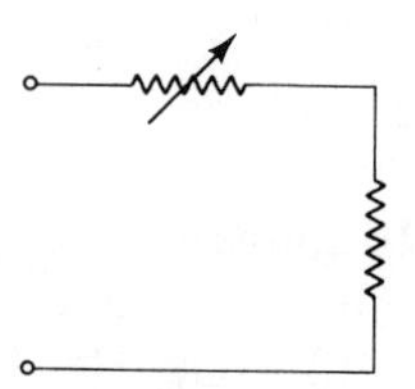

**Fig. A-26.** Variable Resistor in Series with the Load.

(28) To a less positive setting.

***chapter 6***

(1) A limiter prevents a signal from exceeding a specified maximum amplitude while a clamper establishes the starting (or reference) level of a signal. (2) Apply bias to the circuit. (3)

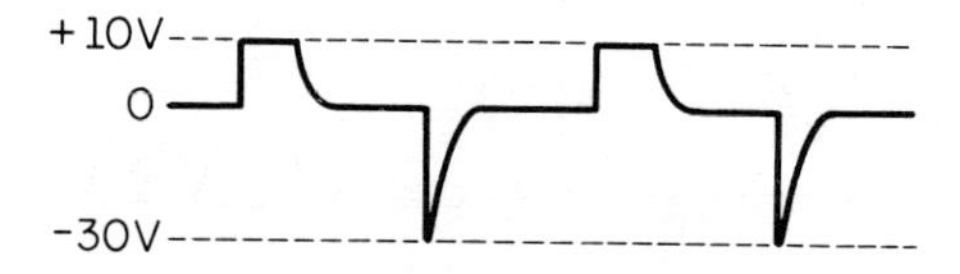


**Fig. A-27.** Output of Fig. 6-46.

(4)

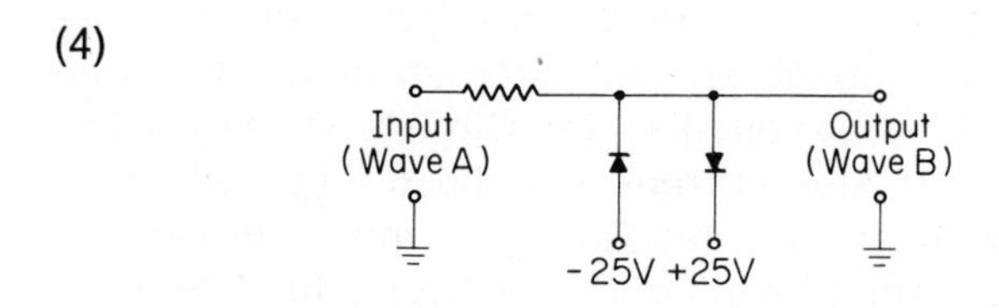


**Fig. A-28.** Biased, Double, Shunt Limiter.

(5) As signals are coupled through capacitors the dc reference is shifted. The clamper restores the proper dc reference level. (6)

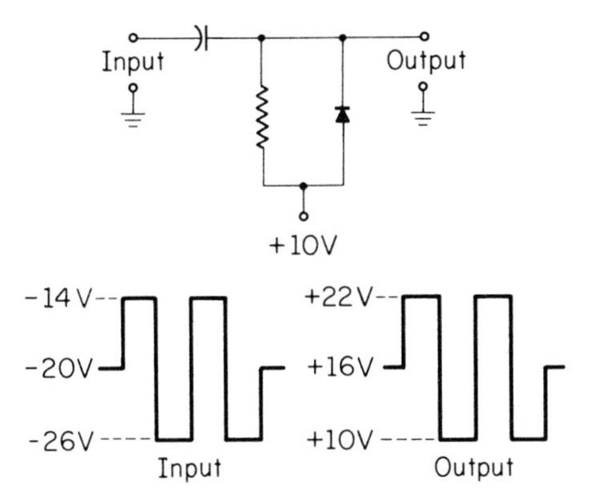


**Fig. A-29.** Clamping to a Bias Level.

(7) About $\frac{1}{50}$. (See graph in Fig. 6-14.) (8) 18.75 s. (9) 375. (10) A gated sweep generator. (11)

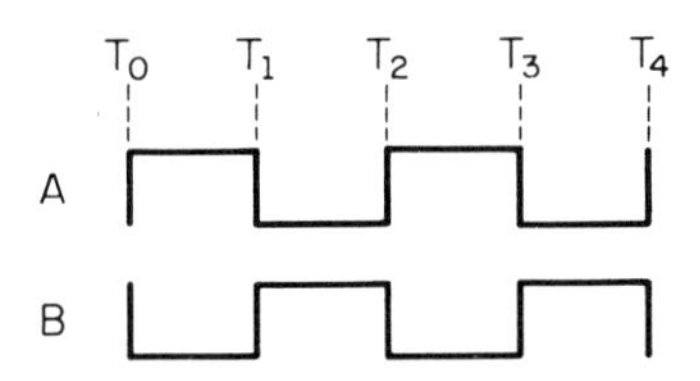


**Fig. A-30.** Outputs of Fig. 6-48.

(12) 6.67 kHz. (13) Monostable multivibrator. (14) 1.5 kΩ. (15) 5 kΩ. (16) 150 ms. (17) Up to seven. (18) 20 μF. (19) Pulses of noise. (20) (a) *RC* time constant of the feedback circuit. (b) Frequency of the synchronizing signal. (21) Continuous wave (CW). (22) The final RF amplifier. (23) It is a process which impresses an intelligence into the amplitude of a carrier wave. (24) It is a process which impresses an intelligence into the frequency of a carrier wave. (25) The RF oscillator. (26) The process of extracting an intelligence from a modulated carrier. (27) The process of mixing two frequencies and extracting the difference. (28) The mixer and the local oscillator. (29) It rectifies and filters the IF in order to extract the audio intelligence. (30) Frequency.

### *chapter 7*

(1) It changes sound waves to electric audio frequency signals which become the modulating waves. (2) The RF power amplifier

amplifies the RF carrier input, but its gain is controlled by the audio modulation signal. (3) They increase the frequency of the continuous RF oscillator signal to the desired transmitter frequency. (4) There are fewer losses in two doubler stages. (5) Center RF carrier, carrier + audio, carrier − audio, and the audio. (6) (a) 1601 kHz. (b) 1599 kHz. (7) 3 kHz. (8) 1598.5 kHz to 1601.5 kHz. (9) (a) 0.833. (b) 83.3%. (10) It constantly samples the center frequency of the carrier and generates a correction signal for the RF oscillator to keep it on frequency. (11) The FM transmitter. (12) It improves the signal to noise ratio of the high-frequency audio signals. (13) (a) Amplitude. (b) Frequency. (14) (a) 51 kHz. (b) 49 kHz. (c) 52 kHz. (d) 48 kHz. (15) When its amplitude is less than one percent of the carrier amplitude. (16) It produces a reference number to the modulation index table. Factors associated with that number in the table will reveal the number of significant side bands and the required band width. (17) 2. (18) (a) 8. (b) 8000 Hz. (19) 1:18. (20) It enables placing more channels in a given frequency spectrum, and a transmitter of a given power can concentrate more power onto the frequencies it transmits. (21) The carrier and one side band. (22) 455 kHz. (23) RF amplifier and local oscillator. (24) The automatic gain control circuit. It produces a dc level directly proportional to average signal strength and uses this dc to control the gain of the RF amplifier, mixer, and IF strip. (25) FM. (26) It removes a distortion that was inserted by the preemphasis network of the transmitter. It does its job by establishing an audio signal amplitude inversely proportional to the frequency. (27) RF amplifier. Mixer. IF amplifiers. Detector. AM. Frequency discriminator. FM. Deemphasis network. FM. Audio amplifier. Speaker.

## *chapter 8*

(1) 1.182. (2) (a) 1.732 Ω. (b) 5.477 Ω. (3) open-wire, stranded wire ribbon, shielded, coaxial, and waveguide. (4) Open-wire and stranded wire ribbon. (5) Shielded and coaxial. (6) 57.5 Ω. (7) Terminate it in its characteristic impedance. (8)

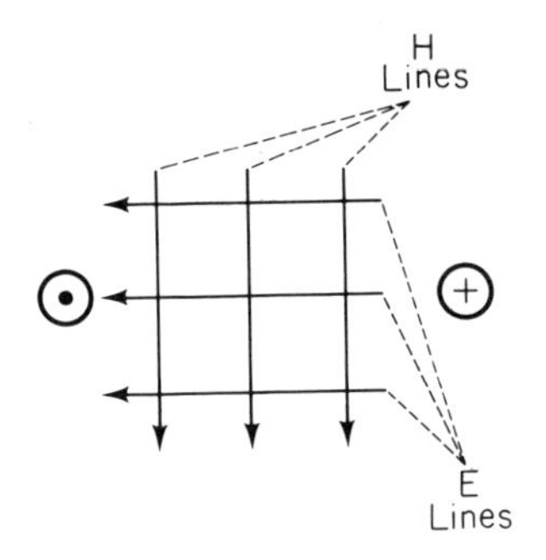


**Fig. A-31.** Line Identify and Direction.

(9) 0.1 m or 3.94 in. (10) 944.64. (11) 63.04. (12) A line terminated in any fashion except in the characteristic impedance. (13) It is a characteristic of a resonant line exhibiting peaks and nodes of potential along the line. It is formed by a combining of the energy waves traveling toward the load and the energy waves being reflected from the load. (14) (a) 6. (b) 295.5. (c) 247.75. (d) 1773. (15) Minimum impedance is located at all odd quarter-wave distances. Maximum impedance is located at all even quarter-wave distance. (16) (a) *Z* and *E* are maximum and *I* is minimum. (b) *Z* and *E* are minimum and *I* is maximum.
(17)

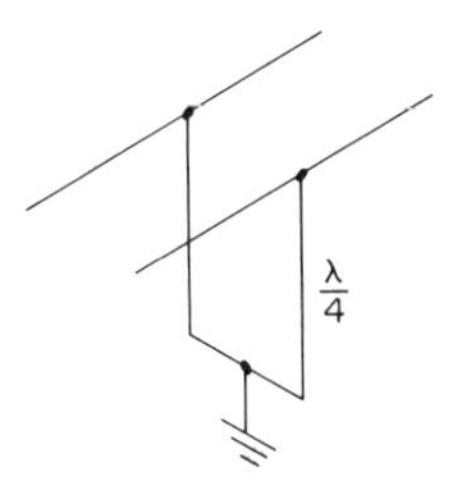


**Fig. A-32.** Even Harmonic Filter.

(18)

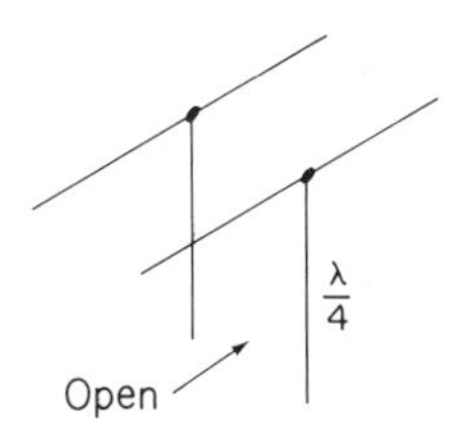


**Fig. A-33.** Band Reject.

(19) 3000 MHz. (20) 0.00121 cm. (21) 62.5 mW. (22) (a) 56.2 in. (b) 59.1 in. long and placed 23.64 in. back of the dipole. (c) 53.4 in. and placed 11.82 in. forward of the dipole. (23) 24.5 mi. (24) Light striking a mirror is reflected at a sharp angle from the surface of the mirror. Electromagnetic energy penetrates the ionosphere for some distance and is gradually curved back toward the earth.

### *chapter 9*

(1) Thermistor. (2) It can be part of a regulator to maintain constant resistance with a temperature change, and it is used in measuring devices. (3) Resistance is inversely proportional to the incident light. (4) It has maximum resistance when a steady voltage is applied. Within limits, the resistance drops for any sudden change in voltage. (5) Non-ohmic materials such as thyrite or silicon carbide. (6) It automatically controls frequency by varying its capacitance with changes of bias. (7) The symetrical Zener is in effect two ordinary Zeners mounted back to back. Voltage in either direction constitutes reverse bias, and it can operate at avalanche breakdown in either direction. (8)

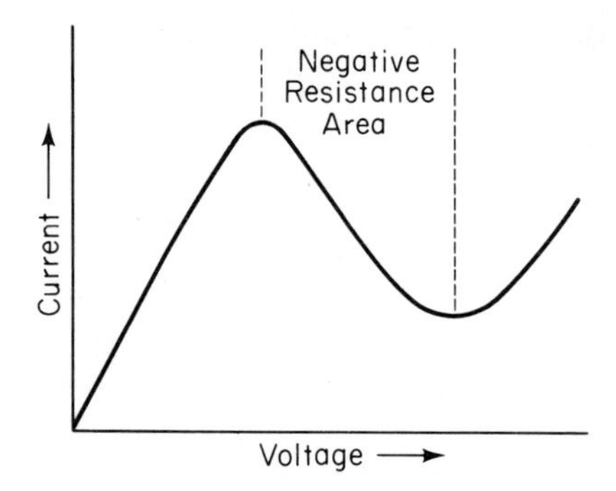


**Fig. A-34.** Negative Resistance Area of the Tunnel Diode.

(9) It is a high-speed switch, an amplifier, or an oscillator. (10) It can withstand very high reverse voltage, has a very high current rating, and has very little leakage current. (11) The silicon rectifier is a diode whose principal use is power supply rectification. The silicon controlled rectifier is a four-layer transistor that may be used as a switch, an amplifier, or a rectifier. (12)

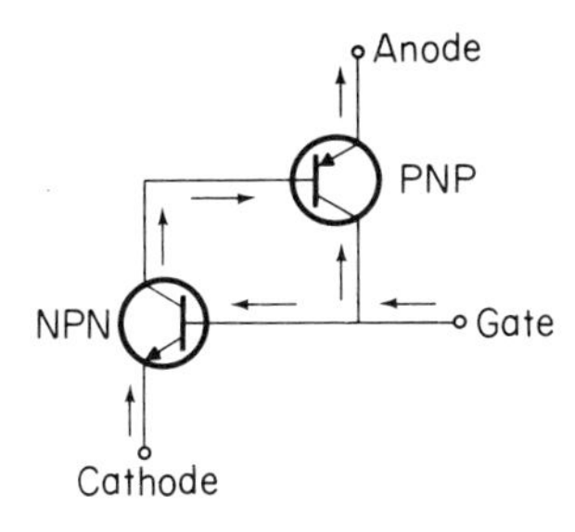


**Fig. A-35.** Function of an SCR.

(13) By reducing the anode voltage. (14) It is the minimum level of anode current that keeps the SCR in a conducting state. (15) 3 to 5:1. (16) (a) The physical distance between the plates was increased. (b) The size of the plates were reduced. (c) The junction was eliminated. (17) —147° to —273°. (18) A very small, very fast electronic switch designed for operation in super cold temperatures. (19) A solid-state device that converts sunlight directly to usable voltage. (20) It is a device that amplifies microwave frequencies by stimulating emission of radiation. (21) It raises the energy level of the atoms. (22) The signal to be amplified passes through the resonator. In passing, it triggers the high energy atoms back to their normal energy state. The released energy is absorbed by the signal. (23) It is a difference in frequency. The laser amplifies light frequencies while the maser amplifies microwave frequencies.

### *chapter 10*

(1) They are inversely proportional; as frequency goes up size comes down. (2) 933.4. (3) High microwave frequencies. (4) $1 \times 10^{-7}$. (5) 3 MHz. (6) Frequencies. (7) Transistor. (8) Point contact; junction. (9) Microscopic. (10) Discreet components. (11) Increased. (12) Circuit integration. (13) Fewer wires, shorter wires, fewer connections, lighter in weight, more reliability, increased ruggedness, and smaller size. (14) Slice; crystal. (15) Circuits. (16) Single chip. (17) Substrate. (18) Method; deposition. (19) Thin; thick. (20) Resistors and capacitors. (21) Inductors. (22) Smaller. (23) Fewer. (24) Electron tube; transistor. (25) High; low; high; unity. (26) Systems. (27) Photographic. (28) Infancy. (29) Rewarding.

# glossary

## A

**accepter element.** A trivalent element which is mixed into pure semiconductor material to form *p*-type material.
**air core.** An inductor with no iron in its magnetic circuit.
**alloy.** A compound formed of two or more metals.
**alpha.** The Greek letter $\alpha$. Used for various mathematical designations including transistor gain of a common base configuration.
**AM.** An abbreviation for amplitude modulation.
**amplification.** The process of increasing the strength of a signal.
**amplification factor.** The maximum gain of a specified electron tube.
**amplifier.** A stage of electronic equipment which increases the strength of a signal.
**amplitude modulation.** A carrier frequency which has intelligence impressed on its amplitude.
**angle of radiation.** The angle between the center of the beam and the earth's surface.
**angstrom unit.** Symbol Å. A measuring unit for extremely small distances. $1Å = 1 \times 10^{-7}$ mm.
**antenna.** A device for coupling electromagnetic energy to or from free space.
**aquadag.** A suspension of graphite in water. Used to form a conductive coating on the inside of cathode ray tubes.
**armature.** The moving part of an electromagnetic circuit.
**array.** A combination of elements on a directional antenna.
**attenuation.** Reduction in the strength of a signal.
**audio frequency.** A frequency in the audiable range between 16 Hz and 20 kHz.

## B

**band.** A group of frequencies between two specified limits.
**bandpass.** Same as band.
**bandwidth.** The number of frequencies in a band. Obtained by sub-

tracting low half-power frequency from high half-power frequency.

**beta.** The Greek letter $\beta$. Designator for gain in a common emitter amplifier. Used as well as other mathematical quantities.

**bias.** A dc voltage used to control either a tube or transistor.

C

**carrier.** (carrier wave). An RF signal used to transmit intelligence.

**carrier frequency.** The frequency of the unmodulated carrier wave.

**cathode.** The emitting element in a solid-state diode or in an electron tube.

**cell.** A single unit of a dc voltage source.

**center frequency.** The frequency of the unmodulated carrier in an FM transmitter.

**coaxial cable.** A transmission line with one conductor running through the center of the other.

**collector.** The transistor electrode used to collect current carriers.

**conduction shell.** An energy level which electrons must possess in order to move freely from atom to atom in a solid.

**control grid.** The control element in an electron tube.

**core.** The center of an inductor.

**coupling.** Means of transferring a signal from one point to another.

**cryogenics.** The science of physical phenomena at very cold temperatures.

**cycle.** A complete set of events in a periodic function.

D

**D'Arsonval movement.** A popular meter movement using a movable coil in a fixed magnetic field.

**demodulation.** The process of extracting the intelligence from a modulated carrier wave.

**detection.** The rectification and filtering portion of demodulation.

**diaphragm.** A flexible sheet which vibrates in response to either sound waves or current variations.

**diffusion.** The joining of materials through a heat process.

**diode.** A solid-state device or electron tube with only two elements.

**directional antenna.** An antenna which radiates or receives energy in one direction better than another.

**discriminator.** A device which produces a voltage amplitude proportional to its input frequency.

**dissipation.** Loss of energy through unused heat.

**distortion.** Unfaithful reproduction of a signal.
**distributive capacitance.** Stray capacitance between parts of a circuit.
**donor element.** A pentavalent element which produces *n*-type material when mixed with a pure material.
**double conversion.** The use of two oscillators and two mixers.
**dynamic characteristics.** Behavior under a specified set of conditions.
**dynamotor.** A combination dc motor and generator. Used to either raise or lower the input voltage.

E

**eddy current.** Circulating currents induced in conducting materials.
**electric field.** A field of force which exists between two points with a difference of potential.
**electrode.** (1) One of the elements in an electron tube. (2) A terminal on a battery.
**electromagnetic energy.** Energy which exists in terms of combined electric and magnetic fields.
**electromagnetic radiation.** Emission of energy through changes in the electric and magnetic fields.
**electromagnetic wave.** A signal composed of combined electric and magnetic fields.
**electron emission.** The ejection of electrons from the surface of a material into the surrounding space.
**electron gun.** Elements in a cathode ray tube designed to emit a narrow beam of electrons.
**electron-volt (eV).** The energy possessed by an electron which has passed through a potential of 1 V.
**element.** In chemistry, one of the basic materials that cannot be chemically subdivided.
**emitter.** The portion of a solid-state device which emits current carriers.
**energy levels.** The levels in an atom where electrons can exist.
**envelope of a wave.** An outline of the modulation signal.
**even harmonic.** Any even multiple of a fundamental frequency.

F

**fade.** A gradual change in signal strength.
**feedback.** Coupling a signal back to a preceding point in the circuit.
**fidelity.** The degree of exactness of reproduction of a signal.

**filament.** The heating element in an electron tube.

**Fleming valve.** A diode tube.

**flip-flop.** A multivibrator requiring a trigger to change from one stable state to another.

**FM.** An abbreviation for frequency modulation.

**frequency.** Periods per unit time.

**frequency modulation.** A method of modulation which impresses intelligence into the frequency of the carrier.

## G

**gain.** The ratio of output to input (voltage, current, or power).

**grid.** The control element in a electron tube.

**ground wave.** Radiated waves which leave the antenna at an angle which brings them in contact with the surface of the earth.

## H

**half-wave antenna.** An antenna that measures $\frac{1}{2}$ wavelength.

**harmonic.** A multiple of a fundamental frequency.

**heat sink.** A device which absorbs and transfers heat.

**hertz.** A unit of measure for frequency. 1 Hz is one period per second.

**hertz antenna.** Any half-wave antenna.

**hole.** The absence of an electron in a crystal valence bond.

**hybrid circuit.** A circuit using two different types of circuit components for the same function. (Tubes and transistors, standard circuits and packaged capsules, standard transistors and integrated components, etc.).

## I

**IF.** An abbreviation for intermediate frequency.

**image.** The effect noticed when one station can be heard in the background of another.

**impedance.** The total opposition of a circuit to ac.

**impurity.** A doping material which changes a pure crystal substance to either *p* or *n*-type material.

**integrated circuit.** A complete circuit on a single semiconductor chip.

**Interference.** Undesired signals or noise which interfere with a selected signal.

**ionosphere.** The upper portion of the earth's atmosphere.

J

**junction.** An area of transistion between two regions in a semiconductor.

L

**lambda.** The Greek letter λ, used to designate wavelength.
**laminated.** Sliced iron cores for transformers.
**laser.** An acronym meaning "light amplification by stimulated emission of radiation."
**LSI.** An abbreviation for large scale integration.

M

**magnetic field.** A force field created by a magnet or an electric current.
**magnetostriction.** Variation in the length of a steel rod due to an alternating magnetic field.
**majority current.** The forward or predominant current in a semiconductor.
**Marconi antenna.** A half-wave antenna composed of a quarter-wave section and a quarter-wave image reflected from ground.
**maser.** An acronym meaning "microwave amplification by stimulated emission of radiation."
**metallic insulator.** A shorted quarter-wave section of transmission line.
**microelectronics.** The technology of constructing circuits and components in extremely small packages.
**microwave.** A wave with a frequnecy above 300 MHz.
**minority current.** Reverse current through a semiconductor.
**mixer.** A stage which mixes received and local oscillator frequencies to obtain an IF.
**modulation.** A process which impresses intelligence on a carrier frequency.
**mu.** The Greek letter $\mu$. Used for amplification factor.

N

**noise.** Interference caused by undesirable random surges of current.

O

**one-shot multivibrator.** A multivibrator having one stable state.
**operating point.** The point on a characteristic curve showing voltage and current levels when no signal is present.
**oxide.** An element combined with oxygen.

P

**parasitic element.** A nonactive antenna element for directing the wave pattern.
**period.** The time required for one complete set of periodic functions. The reciprocal of frequency.
**piezoelectric effect.** The quality of a crystal which causes it to produce a voltage under pressure or to oscillate when voltage is applied across it.
**planar transistor.** A transistor formed by controlling the area of the impurity with an oxide compound.
**polarization.** The orientation of electric lines with respect to the surface of the earth.
**polyethylene.** A tough, flexible plastic used for insulation.
**propagation.** The travel of electromagnetic waves through a medium.

R

**radio frequency.** The band of frequencies from 3 kHz to 10 THz.
**RF.** An abbreviation for radio frequency.
***RC* coupling.** Resistor–capacitor coupling between stages.
**rectification.** The process of converting an ac to a pulsating dc.

S

**SCR.** An abbreviation for silicon controlled rectifier.
**selectivity.** The ability of a receiver to reject undesired signals.
**semiconductor.** A solid material that is neither a good conductor nor a good insulator.
**shielded line.** A transmission line sealed in insulation and surrounded by a metal mesh shield.
**side bands.** The bands of frequencies on either side of the carrier frequency.
**single side band transmission.** A means of suppressing the carrier frequency and one side band while transmitting intelligence on the remaining side band.

**skin effect.** Concentration of current near the surface of a conductor. It becomes more pronounced at high frequencies.
**sky wave.** Electromagnetic wave that is radiated at an elevated angle.
**solar cell.** A cell which converts sunlight directly to voltage.
**space charge.** A collection of electrons between cathode and plate of a tube.
**standing waves.** Stationary waves on a transmission line or an antenna caused from reflections.
**stray capacitance.** See distributive capacitance.
**substrate.** The base on which a microcircuit is formed.
**superheterodyne receiver.** A receiver which beats two frequencies together to extract a different frequency.
**surface barrier.** The barrier on the surface of an element which prevents the surface electrons from escaping.
**synchronization.** Maintaining one operation in step with another.

T

**tolerance.** Permitted variation from a specified standard.

V

**vertical polarized waves.** Waves transmitted with *E* lines perpendicular to the earth's surface.
**video frequency.** Frequencies from 0 to 4 MHz.

W

**wafer.** A thin slice of semiconductor material used as a base for microcircuits.
**wave.** A propagated periodic disturbance.
**waveguide.** A hollow transmission line.
**wavelength.** The distance between any two identical points on a wave.

Y

**Yagi antenna.** A type of antenna array consisting of antenna, directors, and reflectors.

# index